MATHEMATICS

Grade 10
Learner's Book

Busisiwe Goba
Daan van der Lith

CAMBRIDGE
UNIVERSITY PRESS

CAMBRIDGE UNIVERSITY PRESS
Cambridge, New York, Melbourne, Madrid, Cape Town, Singapore, São Paulo

Cambridge University Press
The Water Club, Beach Road, Granger Bay, Cape Town 8005, South Africa

www.cambridge.org
Information on this title: www.cambridge.org/9780521674263

First published 2005
Reprinted 2005, 2006

Printed in South Africa by Paarl Print

ISBN-13 978-0-521-67426-3 paperback
ISBN-10 0-521-67426-3 paperback

Typesetter: A1 Graphics
Illustrators: Robert Hichens, John Hall

Acknowledgements
We are grateful to the following for permission to reproduce photographs:
page 1 Scott Barrow/Imagestate – Photo Access; **page 24** Deirdré Hewitson; **page 32** Henner Frankenfeld – PictureNET Africa; **page 53** Imagestate – Photo Access; **page 54** Karin Duthie/Illustrative Options – africanpictures.net; Roland Metcalfe – iAfrika; **page 55** Cape Town International Convention Centre; **page 84** Deirdré Hewitson; **page 141** C.S.C Naomie/Photonica – Photo Access; **page 142** Deirdré Hewitson; **page 146** Deirdré Hewitson; **page 163** Themba Hadebe – AP Photo (top); NASA (bottom) **page 182** Duif du Toit – Touchline Photos (top); Keith Young – iAfrika (bottom); **page 229** Deirdré Hewitson; **page 247** Eric Miller – africanpictures.net; **page 266** Dave Callow – AP Photo (top); Duif du Toit – Touchline Photos (bottom); **page 233** Michael Agliolo/Imagestate – Photo Access

If you want to know more about this book or any other Cambridge University Press publication, phone us at (021) 412-7800, fax us at (021) 419-8418 or send an e-mail to capctown@cambridgc.org

Contents

MODULE 1: NUMBERS AND NUMBER RELATIONSHIPS 1
Key skills 1

Chapter 1: Numbers and number patterns 2
Key vocabulary 2
Integers, rational numbers and decimals 2
Irrational numbers 5
Decimal places and significant figures 6
Estimating 10
Rounding 11
Indices 11
Number patterns 15
Conjectures and hypotheses 17

Chapter 2: Money matters 23
Key vocabulary 23
Simple interest 23
Rearranging the simple interest formula 26
Compound interest 29
Exponential growth 36
Hire purchase 40
Inflation 44
Exchange rates 46

Language skills 48

Check yourself 49

Review 50

Projects 51

MODULE 2: SHAPE, SPACE AND MEASUREMENT 53
Key skills 53

Chapter 1: Polygons, prisms and cylinders **55**
Key vocabulary 55
Triangles 56
Quadrilaterals 58
The angle sum of a quadrilateral 61
The angle sum of other polygons 62
Area 64
Surface area of prisms and cylinders 68
Volume of prisms and cylinders 71

Chapter 2: Coordinate geometry **75**
Key vocabulary 75
The distance between two points 76
The mid-point of a line segment 80
The gradient of a line segment 84
Gradients of horizontal and vertical lines 87

Chapter 3: Transformation geometry **90**
Key vocabulary 90
Transformations 91
Investigating transformations using coordinates 95

Chapter 4: Trigonometry **100**
Key vocabulary 100
Naming the sides of a triangle 100
The tangent ratio 102
The sine ratio 112
The cosine ratio 115
Sine, cosine or tangent? 118
Degrees, minutes and seconds 120
Other applications of ratio and proportion 122

Language skills 132

Check yourself 133

Review 134

Projects 138

MODULE 3: FUNCTIONS AND ALGEBRA 141
Key skills 141

Chapter 1: Solving and graphing linear equations 143
Key vocabulary 143
Solving linear equations 143
Graphs of linear equations 147
Equations of straight-line graphs 151
Sketching straight-line graphs 153
Solving simultaneous equations using substitution 154
Solving simultaneous equations using elimination 155
Simultaneous equations and problem solving 157
Solving linear inequalities 160

Chapter 2: Solving and graphing non-linear equations 163
Key vocabulary 163
Plotting quadratic graphs 164
Investigating types of parabolas 166
Revision of multiplication and division 169
Revision of factorisation 173
More factorisation 175
Factorising quadratic trinomials 177
Solving quadratic equations algebraically 179
Problem solving and modelling with quadratics 182
Sketching parabolas 185
Plotting hyperbolas 187
Plotting and sketching exponential functions 193
Solving exponential equations 196
Plotting and sketching trigonometric functions 199

Chapter 3: More real-life graphs 205
Key vocabulary 205
Solving simultaneous equations graphically 205
Interpreting graphs 207
Modelling relationships 213

Language skills 215

Check yourself 216

Review 217

Projects 221

MODULE 4: DATA HANDLING 223
Key skills 223

Chapter 1: Conducting a statistical investigation 225
Key vocabulary 225
What are statistics? 225
Stages in a statistical investigation 225
What is a survey? 227

Chapter 2: Organising and displaying data 230
Key vocabulary 230
Frequency distribution tables 230
Bar graphs 233
Histograms 238
Frequency polygons 242
Broken-line graphs 242
Pie-charts 244
Stem-and-leaf plots 248

Chapter 3: Measures of central tendency 251
Key vocabulary 251
Mean, median and mode 251
Measures of dispersion 255
Identifying errors in data 257
Outliers 258
Box plots 260

Chapter 4: Probability and chance 263
Key vocabulary 263
Equally likely events 264
Representing probability 267
Complementary events 272
Independent events 276

Language skills 278

Check yourself 279

Module review 280

Projects 284

Appendix: Rubrics for assessment 286

Symbols and abbreviations

$=$	is equal to
$\neq$	is not equal to
$\approx$	is approximately equal to
$<$	is less than
$\leq$	is less than or equal to
$>$	is greater than
$\geq$	is greater than or equal to
$\sqrt{n}$	the square root of n
LCM	lowest common multiple
HCF	highest common factor
$\hat{}$	angle
Δ	triangle
$\parallel$	is parallel to
$\perp$	is perpendicular to (or just "perpendicular")
π	pi, approximately 3,14
2 : 3	the ratio of 2 to 3
%	per cent (per hundred)
mm	millimetre
cm	centimetre
m	metre
km	kilometre
ha	hectare
g	gram
kg	kilogram
t	tonne
ml	millilitre
ℓ	litre
kl	kilolitre
s	second
min	minute
h	hour
3-D	3-dimensional
r	radius
h	perpendicular height
l	length
b	breadth
A	area
V	volume
I	interest
P	principal
r	annual rate of interest
t	time in years
p.a.	per annum
tan	tangent
sin	sine
cos	cosine
θ	theta (angle)
m	gradient of line $y = mx + b$
b	y-intercept of line $y = mx + b$
$\mathbb{Z}$	the set of integers
$\mathbb{Q}$	the set of rational numbers
$\mathbb{R}$	the set of real numbers
$\mathbb{N}$	the set of natural numbers

MODULE 1

MODULE CHAPTERS

1
Numbers and number patterns

2
Money matters

KEY SKILLS

writing rational numbers as terminating or recurring decimals • identifying between which two integers any irrational number lies • rounding off numbers • investigating number patterns • calculating simple interest • calculating compound interest • calculating exponential growth • understanding hire purchase agreements and calculating costs of buying by hire purchase • understanding inflation and calculating expected costs • understanding exchange rates and converting local currency to foreign currencies and vice versa.

MODULE 1 IN REAL LIFE

an understanding of integers, rational numbers and decimals will help you judge whether your real-life estimates or calculator calculations are reasonable • you will encounter irrational numbers when you measure lengths • index laws are a useful way of simplifying everyday calculations with large numbers • studying patterns and conjectures will improve your ability to think logically • the ability to calculate interest rates and repayments correctly can save you money

NUMBERS AND NUMBER RELATIONSHIPS

In this module, you will work through activities, projects and exercises that involve solving problems and working with numbers in different ways. As you work through the module, you will have opportunities to recognise, describe, represent and work with numbers and their relationships to estimate, calculate and check solutions to problems with confidence (Learning outcome 1).

By the end of this module you will:

- identify rational numbers and convert between terminating or recurring decimals and the form: $\frac{a}{b}$; $a, b \in \mathbb{Z}, b \neq 0$
- simplify expressions using the laws of exponents
- establish between which two integers any simple surd lies
- round irrational and rational numbers off to an appropriate degree of accuracy
- investigate number patterns and make conjectures and generalisations, explain, justify and attempt to prove them
- use simple and compound growth formulas to solve problems (including hire purchase and inflation)
- demonstrate an understanding of the implication of fluctuating foreign exchange rates
- solve non-routine, unseen problems.

Integration

In this module, you will also work with graphs and tables (Learning outcome 2).

CHAPTER 1 NUMBERS AND NUMBER PATTERNS

KEY VOCABULARY

integers • rational numbers • irrational numbers • simplify • decimal • recurring • rounding • fraction • decimal place • significant figure • estimate • approximate • index • exponent • base number • index laws • functional notation • negative index • number patterns • implication • converse • equivalent • hypothesis • conjecture • proof • counter-example

Integers, rational numbers and decimals

You use different kinds of numbers every day in many different situations. It is important to understand and know the different kinds of numbers and how to work with them in your everyday life. This chapter will help you develop your number skills.

Integers

The word **integer** comes from the word **integral**, meaning "whole". The set of integers includes positive numbers, negative numbers and zero. In maths, you use the symbol $\mathbb{Z}$ to name the set of all integers. (The letter $\mathbb{Z}$ comes from the German word *Zahlen*, which means "numbers".)

$\mathbb{Z} = \{\text{integers}\} = \{...; -3; -2; -1; 0; 1; 2; 3; ...\}$

Rational numbers

Sometimes numbers are expressed in relationship to one another. For example, $\frac{2}{3}$ is a number that represents the relationship "two parts out of three" or "two divided by three". The relationship between two or more integers is called a **ratio**. The ancient Greeks developed the system of **positive** rational numbers ("rational" comes from "ratio").

A **rational number** is a number that can be written as a **ratio** or fraction $\frac{a}{b}$, where a and b are integers and $b \neq 0$:

$\mathbb{Q} = \{\text{rational numbers}\}$ $\{\mathbb{Q}$ stands for quotient)

For example, $2\frac{1}{2} = \frac{5}{2}$; $-\frac{1}{3} = \frac{-1}{3}$; $30 \div 24 = \frac{5}{4}$; $4 = \frac{4}{1}$; $-7 = \frac{-7}{1}$

Because every integer a is also a fraction $\frac{a}{1}$, the set $\mathbb{Q}$ of rational numbers contains the set $\mathbb{Z}$ of integers.

Simplifying

You can simplify fractions by dividing the numerator and denominator by their highest common factor (HCF). This is also known as cancelling a fraction to its lowest terms:

$$\frac{24}{30} = \frac{24 \div 6}{30 \div 6} = \frac{4}{5}$$

Decimals

In real life it is useful to express fractions with the same denominator. However, there isn't really a common denominator for all fractions. In Western counting systems, people use **decimals.**

Note: In a decimal or base-10 number system, each digit of a number represents a multiple of a power of 10. For example, $32{,}364 = 3 \times 10^1 + 2 \times 10^0 + 3 \times 10^{-1} + 6 \times 10^{-2} + 4 \times 10^{-3}$.

Decimals are used for measuring length and distance and amounts of money. For example, you would not express a price as $R12\frac{1}{2}$ or $R149\frac{11}{12}$. How would you write these prices?

Terminating decimals

There are two types of decimals: terminating and recurring. A **terminating** decimal is a rational number that can be written as a fraction with a power of 10 as a denominator. This is useful when you want to compare numbers or place them on a number line. Often, the value of the decimal is an approximate value, for example, when you measure distance.

Remember: Measurements never produce an exact answer!

Here are some examples of terminating decimals:

$1\frac{2}{5} = \frac{14}{10} = 1{,}4$

$3\frac{1}{8} = \frac{3\,125}{1\,000} = 3{,}125$

$578\frac{3}{50} = 578 + \frac{6}{100} = 578{,}06$

Recurring decimals

Now look at the fraction $\frac{1}{3}$. What happens when you convert this rational number to a decimal? You cannot write $\frac{1}{3}$ with a denominator that is a power of 10. You can only write it as a **recurring** decimal – a decimal in which some digits are repeated endlessly.

When you divide 1 by 3, the same digit is repeated forever (0,3333...). In other fractions, a few digits are repeated over and over (for example, 13,909090...). Recurring decimals go on forever and are much too long to work with. This is why you always round them off to work with them. You can write a recurring decimal by making a dot over the digit (or digits) that recur (are repeated), to show that they will be repeated endlessly.

Remember: Every recurring decimal can be written as a fraction.

Here are some examples of recurring decimals:

$\frac{2}{3} = 0{,}666666... = 0{,}\dot{6}$

$6\frac{3}{7} = 6{,}428571428571... = 6{,}\dot{4}2857\dot{1}$

$13\frac{10}{11} = 13{,}90909\ ... = 13{,}\dot{9}\dot{0}$

$24\frac{35}{44} = 24{,}79545454545\ ... = 24{,}79\dot{5}\dot{4}$

Investigation

How does your calculator round off a recurring decimal? Work with a partner. Some calculators automatically round off the last digit on the display. Try this on your calculator:

1 3 ÷ 1 1 =
3 ÷ 1 1 =
1 2 ÷ 1 1 =
2 ÷ 1 1 =

a What do you notice? Discuss this with your partner.

b Does your calculator give answers in which the recurring digits keep recurring, or are they rounded off?

c Try some more examples of proper and improper fractions with denominators 7, 9 and 13. What do you notice about the way your calculator works? Does it differ from your partner's? How?

EXAMPLES

Write each recurring decimal as a fraction. Remember to simplify.

1 $0,\dot{5}\dot{1}$ **2** $7,3\dot{4}8\dot{6}$

1 Let $x = 0,\dot{5}\dot{1}$

Then $x = 0,515151$ 1

$\times$ 100: $100x = 51,515151$ 2

Subtract **1** from **2**:

$99x = 51$

$x = \frac{17}{33}$

So $0,\dot{5}\dot{1} = \frac{17}{33}$

Remember: If the cycle length is n, multiply by 10^n and subtract the original equation from the new equation.

2 Let $x = 7,3\dot{4}8\dot{6}$

Then $x = 7,348648$ 1

$\times$ 1 000: $1\,000x = 7\,348,648648$ 2

Subtract **1** from **2**:

$999x = 7\,341,3$

$x = \frac{73\,413}{9\,990}$

$x = \frac{73\,413 \div 27}{9\,990 \div 27}$

So $7,3\dot{4}8\dot{6} = \frac{2\,719}{370}$

Note: Some examples in the exercise show that every terminating decimal can be represented as a recurring decimal with endlessly cycling 9s.
For example,
$1 = 0{,}\dot{9}$
$7 = 6{,}\dot{9}$
$5{,}2 = 5{,}\dot{1}\dot{9}$
$11{,}372 = 11{,}371\dot{9}$

Exercise

1 Express each number as a recurring or terminating decimal:

a $\frac{5}{8}$ **b** $\frac{2}{3}$ **c** $\frac{7}{16}$ **d** $\frac{5}{9}$ **e** $\frac{3}{20}$
f $\frac{7}{12}$ **g** $4\frac{16}{25}$ **h** $5\frac{4}{11}$ **i** $\frac{23}{8}$ **j** $\frac{17}{6}$

2 Express each decimal as a rational number in lowest terms:

a $0{,}15$ **b** $0{,}\dot{7}$ **c** $0{,}108$ **d** $0{,}\dot{1}\dot{8}$ **e** $3{,}12$
f $5{,}\dot{4}\dot{5}$ **g** $1{,}\dot{6}$ **h** $1{,}\dot{2}\dot{1}$ **i** $0{,}2\dot{1}$ **j** $6{,}5\dot{3}$

3 Express each fraction as a recurring or terminating decimal:

a $\frac{11}{6}$ **b** $\frac{13}{12}$ **c** $\frac{7}{15}$ **d** $\frac{16}{37}$ **e** $\frac{2}{27}$
f $\frac{13}{24}$ **g** $3\frac{1}{7}$ **h** $1\frac{3}{14}$ **i** $\frac{27}{13}$ **j** $1\frac{5}{21}$

4 Express each decimal as a rational number. Remember to simplify:

a $0{,}\dot{7}\dot{5}$ **b** $1{,}\dot{0}3\dot{7}$ **c** $4{,}\dot{5}6\dot{7}$ **d** $0{,}\dot{4}35\dot{6}$ **e** $1{,}\dot{9}$
f $2{,}4\dot{9}$ **g** $1{,}5\dot{2}$ **h** $2{,}3\dot{4}\dot{5}$ **i** $7{,}13\dot{8}$ **j** $0{,}11\dot{3}\dot{6}$

5 Write down the recurring decimals for $\frac{1}{7}$, $\frac{2}{7}$, $\frac{3}{7}$, $\frac{4}{7}$, $\frac{5}{7}$ and $\frac{6}{7}$. What is the pattern?

Irrational numbers

There are some numbers that cannot be expressed as the ratio of two integers. These are called **irrational numbers**, for example $\sqrt{2}$, π, $\sqrt{5}$, $\sqrt[3]{2}$, and so on.

Look at the diagram. The Greek mathematician Pythagoras proved that, in a square with each side equal to 1 unit of length, the length of the diagonal is $\sqrt{2}$. Phythagoras' theorem show clearly that fractions do not form a number system that can be used for studying geometry.

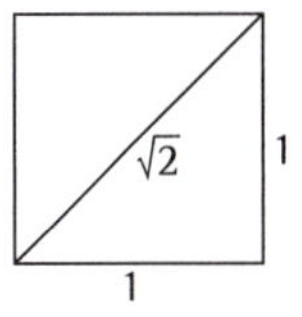

Position of irrational numbers on the number line

The number 2 lies between 1 and 4, and so the irrational number $\sqrt{2}$ lies between $\sqrt{1}$ and $\sqrt{4}$, which means that $\sqrt{2}$ lies between 1 and 2.

This can be expressed as:

$$\sqrt{1} < \sqrt{2} < \sqrt{4}$$

$$\therefore\ 1 < \sqrt{2} < 2$$

and

$\sqrt{4} < \sqrt{5} < \sqrt{9}$

$\therefore\ 2 < \sqrt{5} < 3$

Here are the irrational numbers $\sqrt{2}$, $\sqrt{3}$, $\sqrt{5}$, $\sqrt[3]{2}$ on the number line:

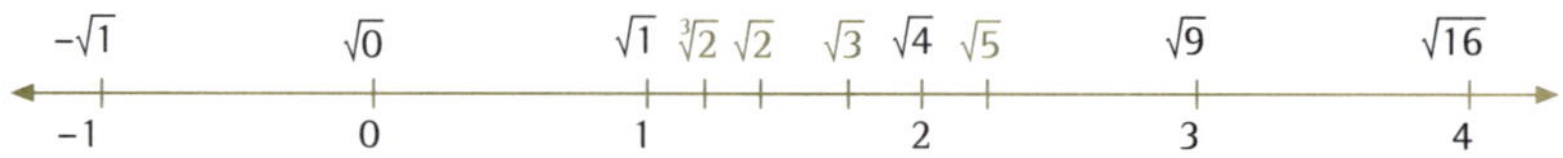

Exercise

1 State which numbers are rational, and express those that are rational as fractions $\frac{p}{q}$ in lowest terms:

a $4\frac{1}{2}$; 5; $-5\frac{3}{4}$; 0; π; $\sqrt{3}$; $\sqrt{4}$; $\sqrt{5}$

b $\sqrt[3]{27}$; $\sqrt[3]{14}$; $\sqrt{\frac{4}{9}}$; -3; $16^{\frac{1}{2}}$; $7^{\frac{1}{2}}$

2 Given that a, b, c and d are integers, with b and d non-zero, simplify the average of $\frac{a}{b}$ and $\frac{c}{d}$ and explain why it is rational.

3 Determine between which two integers these numbers lie:

a $\sqrt{7}$ **b** $\sqrt{20}$ **c** $-\sqrt{8}$

d $\sqrt[3]{9}$ **e** $\sqrt[3]{30}$ **f** $\sqrt[5]{30}$

Decimal places and significant figures

At the beginning of this chapter you learned that the act of measuring can never produce an exact answer. This means that in many calculations that you do in mathematics, your answer does not need to be exact. You may often give an approximate answer, that is, you may round off your answer. When you calculate amounts of money, for example, you always round off answers to 2 decimal places. If you have to divide R52 between 3 people, the exact answer would be R52 ÷ 3 = 17,333... (a recurring decimal). But in real life, you can't split money into smaller amounts than 1 cent, so you would say that each person receives R17,33 and there would be 1 cent left.

Rules for rounding

1 Look at the value of the digit to the right of the specified digit.

2 If the value is 5, 6, 7, 8 or 9, **round up** – add 1 to the specified digit. If, for example, you are rounding to 1 decimal place, 17,589 becomes 17,6. If you are rounding to 2 decimal places, it becomes 17,59.

3 If the value is 0, 1, 2, 3 or 4, **leave the digit unchanged.**

The question will usually state the required degree of accuracy, e.g. to 4 significant figures. The more figures are requested, the more accurate your answer will be.

Significant figures

The first non-zero digit, reading from left to right in a number, is the first **significant figure**, e.g. in 0,00841, 8 is the first significant figure.

EXAMPLES

1 Round 94,738095 to:
 a 2 significant figures **b** 1 significant figure
 c 5 significant figures.

a 95 **b** 90 **c** 94,738

2 Round 0,0064735 to:
 a 4 significant figures **b** 3 significant figures
 c 1 significant figure.

a 0,006474 **b** 0,00647 **c** 0,006

Exercise

1 Write these numbers correct to the number of significant figures indicated in brackets.

a	4,8976	(2)	**b**	0,07874	(3)
c	1 506,892	(5)	**d**	5,523	(1)

2 Calculate and give your answer correct to the number of significant figures indicated in brackets.

a	$4{,}3968 \times 0{,}0007438$	(2)	**b**	$0{,}61135 \div 4{,}1119$	(5)
c	$3{,}4572 \div 0{,}0109$	(3)	**d**	$50\,042 \times 0{,}0067$	(3)

Decimal places

The first digit after the decimal comma is the first (or one) **decimal place**, e.g. in 23,798, 7 is in the first decimal place.

EXAMPLES

1 Round 94,738095 to:

a 2 decimal places **b** 1 decimal place

c 5 decimal places.

a 94,74 **b** 94,7 **c** 94,73810

DISCUSS

In your groups, discuss why or where you would use the number 0,0 in c.

2 Round 0,0064735 to:

a 4 decimal places **b** 3 decimal places

c 1 decimal place.

a 0,0065 **b** 0,006 **c** 0,0

Exercise

1 Use a calculator to find the answers to these problems. Give each answer correct to the number of decimal places indicated in brackets.

a $3,185 \times 0,49$ (2) **b** $0,064 \div 2,536$ (3)

c $0,474 \times 0,0693$ (2) **d** $12,943 \div 6,876$ (4)

e $0,006749 \div 0,000382$ (3) **f** $38,374306 \times 0,007493$ (4)

Exercise

1 Blank CDs cost R4,94 each.

a Estimate, to the nearest rand, the cost of 7 CDs.

b Calculate, to the nearest 10 cents, the cost of 7 CDs.

2 A pile of 10 textbooks of equal thickness is 24,4 cm high.

a Estimate, to the nearest centimetre, the thickness of 1 textbook.

b Calculate, to the nearest millimetre, the thickness of 1 textbook.

3 A man can walk 2,05 metres per second.

a Estimate, to the nearest metre, how far he can walk in 2 minutes.

b Calculate, to the nearest metre, how far he can walk in 2 minutes.

4 From a piece of string 25 metres long, 100 pieces of each 7,825 centimetre long are cut off.

a Estimate the length of the remaining string, to the nearest metre.

b Calculate the length of the remaining piece of string, to the nearest centimetre.

5 The length of the sides of a triangle are 6,75 cm, 7,93 cm and 8,18 cm.

a Estimate the length of the perimeter of the triangle, to the nearest centimetre.

b Calculate the length of the perimeter of the triangle:

i to the nearest centimetre **ii** to the nearest millimetre.

6 A rectangular room is 5 m 75 cm long and 4 m 30 cm wide.

a Estimate the area of the room, to the nearest m^2.

b Calculate the area of the room:

i to the nearest m^2 **ii** to the nearest cm^2.

7 Calculate to 2 decimal places:

a $\pi \times 7{,}5^2$ **b** $\frac{1}{3} \times 6{,}5^3$ **c** $\frac{1}{3} \times \pi \times 6{,}9^2 \times 11$

8 Use a calculator to answer these questions. Give each answer in: **i** fractional form, and **ii** decimal form, correct to 3 decimal places.

a $1 \div 3 + 1$ **b** $2 \div 7 \div 3$

c $1 \div 4 + 2 \div 3$ **d** $3 \div 100 - 23 \div 1\,000$

e $17 + 17 \div 6$ **f** $8 \div 3 + 7 \div 3$

g $2{,}6 \div 15{,}7 \times 3$ **h** $1{,}3 \div 3 - 3 \div 1{,}3$

Estimating

Note: An estimate is a guess of what the answer should be. In the example, the estimate is found by rounding (approximating) each number off to 1 significant figure.

Estimates are important when using a calculator for checking a calculated answer. They allow you to make a quick check for mis-keying, mis-reading or other possible mistakes, by checking to see if an answer is approximately correct.

When estimating an answer before using a calculator it is usually sufficient to round off the answer to 1 significant figure.

EXAMPLE

First estimate the answer to $\frac{3{,}802 + 7{,}45}{3{,}27}$ and then calculate the answer to 1 decimal place.

Note: The symbol ≈ means "approximately equal to".

Round off each number to 1 significant figure:

$\frac{3{,}802 + 7{,}45}{3{,}27} \approx \frac{4 + 7}{3} \approx \frac{11}{3} \approx 4$

Using a calculator:

$\frac{3{,}802 + 7{,}45}{3{,}27} = 3{,}468501529 = 3{,}5$ (to 1 decimal place)

Exercise

1 **i** Estimate each of these to 1 significant figure, without using a calculator.

ii Use a calculator to find each answer to 1 decimal place.

iii Use your answer to **i** to check that each calculator answer is probably correct.

a $1{,}929 \times 95{,}63$ **b** $34{,}77 \div 5{,}34$ **c** $(2{,}099)^2$

d $\frac{3{,}77 \times 9{,}77}{14{,}7}$ **e** $\frac{6{,}25 \times 0{,}87}{2{,}3}$ **f** $\frac{99{,}87 \times 0{,}087}{0{,}93}$

g $(9{,}88)^3$ **h** $0{,}33 \div 0{,}093$ **i** $\sqrt{923}$

j $\sqrt[3]{35{,}937}$

2 Estimate your answer to 1 significant figure and then calculate each to 3 significant figures:

a $3{,}846 \times 3{,}9 - 2{,}35$ **b** $5\,800 \times 0{,}0305$

c $\sqrt{9{,}037138}$ **d** $(3{,}453)^3$

e $(0{,}379 + 5{,}87) \div 2{,}89$ **f** $0{,}833 + (8{,}44 \div 3{,}99)$

g $\sqrt[3]{7{,}9111}$ **h** $\frac{6{,}703 \div 0{,}216}{1{,}954 \div 0{,}0012}$

Rounding

Some calculators simply remove the end digits of a long number. Other calculators round off a number if it has too many digits, e.g. 6200,38181481236 in a calculator display truncates to 6200,381815.

PEER ASSESSMENT

See page 288 (rubric 3).

Investigation

Work with a partner.

To find out whether your calculator rounds off numbers, convert $\frac{2}{3}$ to a decimal:

2 ÷ **3** =

If the display shows 0.666666666 (i.e. the last digit is a 6), then your calculator simply **removes** the end digits of the answer.

If the display shows 0.666666667 (i.e. the last digit is a 7), then your calculator **rounds** up or down.

Discuss with your partner whether there are any disadvantages to rounding. Can you think of examples where rounding can be a problem?

DID YOU KNOW?

Sometimes your decision to round up or down will depend on special circumstances that require you to override the normal rules for rounding. Discuss this in your groups. Can you think of any examples?

Exercise

1 Round off to 2 decimal places:

a	97,847	**b**	6,4396	**c**	297,364	**d**	0,07936
e	2074,968	**f**	5,6984	**g**	0,007798	**h**	194,7439
i	6021,505	**j**	0,8267	**k**	0,05901		

2 **a** Use a calculator to find 3 ÷ 7. Discuss your answer with a partner.

b Write a sentence to explain what you discovered.

Indices

A number, k, may take the form $k = a^m$,
were a is called the **base**,
m is the **exponent** or **index**, and
k is a **number**.

index

base → $2^3 = 8$ ← number

The index laws

Do you remember the index laws you have already dealt with?

Law 1: $a^m \times a^n = a^{m+n}$ — $x^3 \times x^5 = x^8$

Law 2: $a^m \div a^n = a^{m-n}$ — $a^5 \div a^2 = a^3$

Law 3: $(a^m)^n = a^{mn}$ — $(2^3)^5 = 2^{15}$

Law 4: $a^0 = 1,\ a \neq 0$ — $7^0 = 1$

Law 5: $(a \times b)^m = a^m \times b^m = a^m b^m$ — $(p \times q)^3 = p^3 \times q^3 = p^3q^3$

Law 6: $\left(\frac{a}{b}\right)^m = \frac{a^m}{b^m}$ — $\left(\frac{x}{y}\right)^7 = \frac{x^7}{y^7}$

EXAMPLES

1 Simplify, using the first index law:

a $x^7 \times x^4 = x^{7+4}$
$= x^{11}$

b $3a^2b \times 2ab^3 = (3 \times 2)a^{2+1}b^{1+3}$
$= 6a^3b^4$

2 Simplify, using the second index law:

a $p^5 \div p^3 = p^{5-3}$
$= p^2$

b $4x^2y^4 \div 8xy^2 = \frac{4}{8}x^{2-1}y^{4-2}$
$= \frac{1}{2}xy^2$

3 Simplify, using the third, fifth and sixth index laws:

a $(x^4)^2 = x^{4\times2}$
$= x^8$

b $(3x^3)^4 = 3^4x^{3\times4}$
$= 81x^{12}$

c
$$\left(\frac{2x^2}{3y^2z}\right)^3 = \frac{2^3x^{2\times3}}{3^3y^{2\times3}z^{1\times3}} = \frac{8x^6}{27y^6z^3}$$

4 Evaluate, using the fourth index law:

a $4a^0 = 4 \times 1$
$= 4$

b $(4pq)^0 = 1$

c $2p^0 + (3p)^0 = 2 \times 1 + 1$
$= 3$

Exercise

1 Simplify, using the first index law:

a $x^3 \times x^4$ **b** $p^5 \times p^2$

c $t^7 \times t$ **d** $3p^2 \times 2p^3$

e $2s^4 \times 3s^3$ **f** $q^7 \times 2q^7$

g $3a^2b \times 5ab^5$ **h** $2x^2y \times 3xy^2$

i $3v^7w \times 6v^2w$ **j** $2rs^3 \times 3r^4s \times 2r^2s^2$

k $3x^6y^7 \times xy \times 2xy^2$ **l** $3x^4 \times 5xy^2 \times 10y^4$

2 Simplify, using the second index law:

a $a^4 \div a^2$ **b** $\dfrac{c^{10}}{c^6}$ **c** $a^7 \div a^6$

d $3r^5s^2 \div r^3s$ **e** $\dfrac{4p^2q^3}{p^2q}$ **f** $2x^2y^3 \div x$

g $6p^4q^2 \div 3q^2p^2$ **h** $\dfrac{8t^4r}{2t^3}$ **i** $16m^7x^5 \div 8m^2x^4$

j $4x^2y^3 \div 8xy$ **k** $\dfrac{5h^6d^4}{3d^3h^2}$ **l** $16k^3t^7 \div 48kt^5$

3 Simplify, using the third, fifth and sixth index laws:

a $(m^4)^2$ **b** $(t^3)^2$ **c** $(g^4)^2$

d $(2u^2)^2$ **e** $(3r^3)^3$ **f** $(4t^2)^3$

g $\left(\dfrac{a^2}{b^3}\right)^2$ **h** $\left(\dfrac{2t^4}{g^2}\right)^3$ **i** $\left(\dfrac{at^3}{3g^4}\right)^3$

j $\left(\dfrac{2h^2}{3g}\right)^3$ **k** $\left(\dfrac{3a^2b}{2pq^3}\right)^2$ **l** $\left(\dfrac{5p^2q^3}{3r}\right)^4$

4 Evaluate, using the fourth index law:

a $3p^0$ **b** $(4q)^0$ **c** $(2pq^7)^0$

d $10x^0 - 10$ **e** $3a^0 - 2$ **f** $\dfrac{2}{m^0}$

g $\dfrac{u^0}{2} + \dfrac{1}{2}$ **h** $3y^0 - (2t)^0$ **i** $\left(2k - \dfrac{5}{y}\right)^0$

5 Use appropriate index laws to simplify:

a $x^6 \times x^5 \div x^3$ **b** $x^2y \div xy \times xy^2$

c $x^4n^7 \times x^3n^2 \div xn$ **d** $\dfrac{x^2y^3 \times x^2y^4}{x^3y^5}$

e $\dfrac{m^2w \times m^3w^2}{m^4w^3}$ **f** $\dfrac{r^4s^7 \times r^4s^7}{r^4s^7}$

g $\dfrac{16a^8b \times 4ab^7}{32a^7b^6}$ **h** $\dfrac{9x^2y^3 \times 6x^7y^5}{12xy^6}$

i $\dfrac{4x^2y^3 \times 12x^2y^2}{24x^4y}$ **j** $(2p^2q^4)^2 \times pq^3$

k $4(a^2b)^3 \times (3ab)^2$ **l** $(4r^2y)^2 \times r^2y^4 \div 3(ry^2)^3$

m $\dfrac{(7s^2y)^2 \times 3sy^2}{7(sy)^2}$ **n** $\dfrac{4r^2t \times 3(r^2t)^3}{6r^2t}$

o $\dfrac{(2xy)^2 \times 2(x^2y)^3}{8xy^2 \times x^7y^3}$

Negative indices

When working with expressions involving indices, you will often express the solution with indices that are not positive whole numbers.

Consider the second index law: $a^m \div a^n = a^{m-n}$.

$a^0 \div a = a^{0-1} = a^{-1}$ but $a^0 = 1$ so $a^0 \div a = 1 \div a = \frac{1}{a}$ therefore $a^{-1} = \frac{1}{a}$

This leads us to: $a^{-n} = \frac{1}{a^n}$

EXAMPLE

Express these in their simplest form with positive indices:

a $y^{-5} = \dfrac{1}{y^5}$

b $\dfrac{2}{x^{-2}} = 2 \times x^2 = 2x^2$

c $\left(\dfrac{3b^2}{b^4}\right)^4 = \dfrac{3^4b^8}{b^{16}} = 81b^{8-16} = 81b^{-8} = \dfrac{81}{b^8}$

Exercise

1 Express with positive indices:

a x^{-3} **b** $2y^{-9}$ **c** $3t^{-6}$ **d** $\frac{2}{r^{-2}}$

e $\frac{2}{s^{-3}}$ **f** $\frac{4}{m^{-7}}$ **g** $(2t^{-4})^2$ **h** $(3x^{-5})^4$

i $2(x^{-7})^3$ **j** $4a^{-2}b$ **k** $3w^{-4}t^{-2}x^4$ **l** $\frac{1}{3}x^{-2}y^3z$

2 Express in simplest form with positive indices:

a $a^{-4} \times a^3$ **b** $x^2y^3 \times x^{-3}y^{-4}$ **c** $a^{-6}y^4 \times a^3y^{-2}$

d $3t^{-3}w^{-5} \times t^5w^{-2}$ **e** $2a^{-3}b \times 3a^{-2}b^{-3}$ **f** $6a^4b^3 \times 3a^{-6}b$

g $a^3b^4 \div a^2b^7$ **h** $p^2q^3 \div p^7q^2$ **i** $r^3s^7 \div r^4s^7$

j $\frac{a^4b^3}{a^2b^5}$ **k** $\frac{m^3n}{mn^3}$ **l** $\frac{p^2q^2r^4}{pq^2r^5}$

m $\frac{3x^2y}{6xy^2}$ **n** $\frac{4m^3n^4}{7m^2n^7}$ **o** $\frac{12r^4s^6}{9rs}$

3 Simplify and express your answers with positive indices:

a $(a^3b^2)^3 \times (a^2b^4)^{-1}$ **b** $2(x^2y^{-1})^2 \times (3xy^4)^3$

c $(2p^2)^4 \times (3p^2q)^{-2}$ **d** $\frac{2a^2b^2}{a^{-3}} \times \frac{2a^2b^5}{b^4}$

e $\frac{(3rs^2)^4}{r^{-3}s^4} \times \frac{(2r^2s)^2}{s^7}$ **f** $\frac{4(x^{-2}y^4)^2}{x^2y^{-3}} \times \frac{xy^6}{2x^{-2}y}$

g $\left(\frac{a^2b^3}{b^{-2}}\right)^2 \div \left(\frac{ab^4}{a^2}\right)^{-2}$ **h** $\left(\frac{m^4n^{-2}}{r^2}\right)^2 \div \left(\frac{m^{-3}n^2}{r^3}\right)^2$

i $\frac{3(x^2y^{-4})^2}{2(xy^2)^2} \div \frac{(xy)^{-3}}{(3x^{-2}y^4)^2}$

Number patterns

Implication, converse and equivalence

For two statements p and q, "$p \Rightarrow q$" is read as "p implies q".
For example,

- $x - 3 = 4 \Rightarrow x = 7$
- x is divisible by 3 $\Rightarrow$ $2x$ is divisible by 6
- $x = 3 \Rightarrow x^2 = 9$

Note: The word "converse" means "opposite" or "reverse".

In the first two examples the converse (or opposite) statements are also true.
$q \Rightarrow p$ is the converse of $p \Rightarrow q$

The converse of the two examples is:

- $x = 7 \Rightarrow x - 3 = 4$
- $2x$ is divisible by 6 $\Rightarrow$ x is divisible by 3

You can then combine $p \Rightarrow q$ and $q \Rightarrow p$ to write $p \Leftrightarrow q$, which may be read as "p is equivalent to q".

Thus you can write

- $x - 3 = 4 \Leftrightarrow x = 7$
- x is divisible by 3 $\Leftrightarrow$ $2x$ is divisible by 6

However, $x^2 = 9$ does not imply that $x = 3$, i.e. the converse does not hold, because $x^2 = 9 \Leftrightarrow x = 3$ or $x = -3$.

There are many other examples where the implication only holds one way.

For example,

- $x > 3 \Rightarrow x^2 > 9$ but $x^2 > 9$ does not imply that $x > 3$.
- x rational $\Rightarrow x^2$ rational, but x^2 rational does not imply that x is rational.
- x and y even $\Rightarrow x + y$ even, but $x + y$ even does not imply that x and y are even.
- $a = b \Rightarrow a^2 = b^2$ but $a^2 = b^2$ does not imply that $a = b$.
- Quadrilateral *ABCD* is a square $\Rightarrow$ the sides of quadrilateral *ABCD* are of equal length, but the sides of the quadrilateral *ABCD* are of equal length does not imply that quadrilateral *ABCD* is a square.

GROUP ASSESSMENT

See page 288 (rubric 3).

Investigation: Conjectures and counter examples

Work in a group.

Consider numbers of the form $n^2 + n + 11$, where n is a natural number. Look at this table:

n	1	2	3	4	5	6	7	8
$n^2 + n + 11$	13	17	23	31	41	53	67	83

Note: A conjecture is something you assume.

From the table of values, you could make this conjecture:

"numbers of the form $n^2 + n + 11$ are prime numbers".

However, if $n = 11$, it is clear that $n^2 + n + 11$ is divisible by 11. So $n^2 + n + 11$ is not prime. The value of $n^2 + n + 11$ when $n = 11$ provides a **counter example** to your conjecture. The conjecture has been shown to be false.

Consider the conjecture: $(a + b)^2 = a^2 + b^2$ for all a and $b \in \mathbb{R}$.
The values 2 and 3 for a and b respectively provide a counter example.

Consider the conjecture: $x^2 > x$ for all $x \in \mathbb{R}$.
The value $\frac{1}{3}$ for x provides a counter example.

Consider the conjecture: The cube of a natural number is greater than the natural number.
The natural number 1 provides a counter example.

This final conjecture can be "fixed up" by making the conjecture read: "the cube of a natural number greater than 1 is greater than the natural number".

Now try to "fix up" the first two conjectures in your groups. Discuss your answers within your groups.

Conjectures and hypotheses

In previous years, you may have come across proofs in geometry and problem solving. Proofs are an important way to test a hypothesis or conjecture. They are similar to experiments you do in your natural science activities. A proof, like an experiment, follows these steps:

Step 1: State your conjecture or hypothesis.

Step 2: Try to find a counter example that proves the conjecture or hypothesis to be false.

Step 3: If you find a counter example, "fix up" the conjecture to make it accurate.

Step 4: If you cannot find a counter example, try to construct a "chain of reasoning" that enables you to reach the same result from the simplest possible set of assumptions.

EXAMPLES

1 Consider this pattern:

$4^2 - 3^2 + 2^2 - 1^2 = 10$
$5^2 - 4^2 + 3^2 - 2^2 = 14$
$6^2 - 5^2 + 4^2 - 3^2 = 18$

From this you could draw the conjecture that, "if a, b, c, d are consecutive natural numbers with $a < b < c < d$, then $d^2 - c^2 + b^2 - a^2 = d + c + b + a$".

Proof

Let the numbers a, b, c, d be $n - 1$, n, $n + 1$, $n + 2$ respectively.

Then $d + c + b + a = n + 2 + n + 1 + n + n - 1$
$= 4n + 2$

and $d^2 - c^2 + b^2 - a^2 = (n + 2)^2 - (n + 1)^2 + n^2 - (n - 1)^2$
$= n^2 + 4n + 4 - n^2 - 2n - 1 + n^2 - n^2 + 2n - 1$
$= 4n + 2$

The result has been proved. In fact, it has been proved for any four consecutive integers.

It can be seen from the proof that a more general result is true.

If a, b, c, d are consecutive integers, then

$$d^2 - c^2 + b^2 - a^2 = a + b + c + d.$$

The proof has actually added to the understanding of the problem by leading to a generalisation of the original conjecture.

2

$$\begin{array}{r} 321 \\ -\underline{123} \\ 198 \\ +\underline{891} \\ 1\,089 \end{array} \qquad \begin{array}{r} 941 \\ -\underline{149} \\ 792 \\ +\underline{297} \\ 1\,089 \end{array} \qquad \begin{array}{r} 987 \\ -\underline{789} \\ 198 \\ +\underline{891} \\ 1\,089 \end{array} \qquad \begin{array}{r} 980 \\ -\underline{089} \\ 891 \\ +\underline{198} \\ 1\,089 \end{array}$$

Your conjecture here could be:
"Take any 3-digit number whose digits decrease as you read them from left to right. Make another number by reversing the order of the digits. Then subtract the smaller from the larger number. Reverse the order of the digits of the difference and add the number formed to the difference. The result will always be 1 089."

Proof

Let the number be $a \times 10^2 + b \times 10 + c$.
Reverse the digits to get the number $c \times 10^2 + b \times 10 + a$.
It will be assumed $a > c$ without loss of generality.
$a \times 10^2 + b \times 10 + c - c \times 10^2 - b \times 10 - a = (a - c) \times 10^2 + (c - a)$
Now $c - a$ is negative, but $(10 + c - a)$ is positive.
Write $(a - c) \times 10^2 + (c - a) = (a - c - 1) \times 10^2 + 90 + (10 + c - a)$.
The digits are now correctly displayed.
Reverse the order to obtain $(10 + c - a) \times 10^2 + 90 + (a - c - 1)$ and add.
$(10 + c - a + a - c - 1) \times 10^2 + (90 + 90) + (10 + c - a + a - c - 1)$
$= 9 \times 10^2 + 180 + 9$
$= 900 + 180 + 9$
$= 1\,089$

3

x	y
1	7
2	10
3	13
4	16

} 3 } 3 } 3

$y = 3x + 4$

x	y
1	-2
2	-4
3	-6
4	-8

} -2 } -2 } -2

$y = -2x + 0$

From these two tables, you could make the conjecture, "if the difference between any two consecutive y-values is a constant, the relationship between x and y is linear".

Proof

Let the y-values be a; $a + d$; $a + 2d$; ..., where a and d are integers.

x	y
1	a
2	$a + d$
3	$a + 2d$
4	$a + 3d$
...	...

} d } d } d

So the constant difference is d.

From the table it is clear that each y-value can be written as $y = a + (x - 1)d = dx + (a - d)$, which is a linear equation.

GROUP ASSESSMENT

See page 287 (rubric 2).

Investigation: Generalise a proof

Work in a group.

Look at example 2 on page 18 again. Read the example and its proof carefully to make sure you understand it.

As a group, try to generalise the proof for this example. Here are some questions to help you:

1 What happens if you work with a 4-digit number instead of a 3-digit number? Will you still get the same result?

2 Look carefully at the base number. What result do you get if you use a different base?

3 Work out a general proof for the example. Then present your findings to the class.

Exercise

1 Insert $\Rightarrow$ or $\Leftarrow$ to make these statements about integers true:

a p is even ... pq is even

b $p + q$ is odd ... pq is even

c $x = 0$ ... $xy = 0$

d $ab = ac$ ... $b = c$.

2 State the converse of each of these statements and also state whether the converse is true or false.

a If n is odd, then n^2 is odd.

b N is divisible by 3 $\Rightarrow N^2$ is divisible by 9

c $x^2 > 4 \Leftarrow x < -2$

3 The sum of two consecutive odd numbers is divisible by 4. Can you make similar statements about:

a the sum of three consecutive odd numbers

b the sum of four consecutive odd numbers?

Prove your conjectures.

4 Prove that the sum of the squares of five consecutive integers is divisible by 5.

(Take $n - 2$, $n - 1$, n, $n + 1$, $n + 2$ as the integers.)

5 For each of these statements give a counter example that proves that the given statement is false.

a The sum of the squares of two numbers is equal to the square of the sum of these two numbers.

b If a number is even, then it is not divisible by 7.

c $\sqrt{a + b} = \sqrt{a} + \sqrt{b}$ for all $a, b \in R$.

d $\frac{1}{2}\left(a + \frac{b + c}{2}\right) = \frac{1}{2}\left(\frac{a + b}{2} + c\right)$ for all $a, b, c \in \mathbb{R}$.

e The sum of two prime numbers is a prime number.

f $\frac{1}{s + t} = \frac{1}{s} + \frac{1}{t}$ for all $s, t \in \mathbb{R} \setminus \{0\}$.

Note: The symbol \ means "excluding" or "except".

6 Write the converse of each of these statements and state whether each is true or false.

a If $a - b$ is positive, then $a > b$.

b If $x = 0$ and $y = 0$, then $x = y$.

c If $x + y = 0$, then $x = -y$.

d If x is even and y is odd, then xy is even.

e The square of an even number is even.

7 Either prove each of these conjectures for the set of natural numbers or provide a counter example.

a A number $\mathbb{N}$ has an odd number of divisors if and only if it is a perfect square.

b For any odd number, $\mathbb{N}$, there is a number with exactly $\mathbb{N}$ divisors.

c For any number, $\mathbb{N}$, there is a number with exactly $\mathbb{N}$ divisors.

d There are infinitely many numbers with exactly $\mathbb{N}$ divisors ($\mathbb{N} \neq 1$).

8 Prove each of these for $a, b \in \mathbb{Z}$.

a If a and b are even, $a + b$ is even.

b If a and b are odd, $a + b$ is even.

c If a is even and b is odd, ab is even.

d If a and b are odd, ab is odd.

e $a + b$ is even if and only if $a - b$ is even.

f $a + b$ even and $a - b$ even implies that ab is a difference of perfect squares.

9 **a** Show that this relation is linear: $\{(1; -7); (2; -5); (3; -3); \ldots\}$.

b Find the equation of the straight line passing through these points:

x	1	2	3	4
y	1	−2	−5	−8

10 Look at the work you have done in natural science. Give one example of a conjecture or hypothesis you have tested and found true, and one example of a conjecture or hypothesis you have tested and found false.

SELF-ASSESSMENT

As you complete each chapter of each module, copy and complete this rubric in your workbook. Draw up a list of the most important skills and concepts you learned in the chapter you completed. Then list the concepts under the appropriate headings. Ask someone to help you with the concepts you are not sure about. You can list as many concepts as you wish.

Self-assessment rubric

I understand these concepts and skills very well	I understand these skills and concepts	I need more help to understand these skills and concepts

CHAPTER 2 MONEY MATTERS

KEY VOCABULARY

interest • simple interest • principal • interest rate • investment • per annum • home loan • deposit • monthly instalment • balance • total cost • compound interest • micro-lending • exponential growth • hire purchase • inflation • flat interest rate • consumer price index (CPI)

All around you, every day, you see, read and work with number problems involving money. You read advertisements in newspapers and magazines, and hear offers advertised on the radio and on television. People borrow money from banks to buy vehicles and houses, and to finance their businesses. In this chapter, you will learn about simple and compound interest, as well as other aspects of finance, including hire purchase, inflation and exchange rates. You will also learn about exponential growth and decay, and you will relate these concepts to other learning areas such as social science and natural science.

Simple interest

When simple interest is applied to an investment, the value of the investment increases by an agreed percentage at specified regular time intervals. If money is borrowed, the borrower is charged interest as a fee for its use. **Simple interest** is a fixed percentage of the amount invested or borrowed and is **calculated on the original amount.**

Simple interest entails adding a constant amount to the principal at regular intervals. The formula used to calculate simple interest is:

$$\textit{Simple interest, } I = \frac{Prt}{100}$$

where:

P = amount borrowed or invested, called the principal

r = interest rate

t = time in years.

EXAMPLES

1 Themba borrows R7 000 from a bank to build a new wall around his yard and house. The bank charges him 12% simple interest per year. Calculate how much he will pay on his loan over 4 years.

$$\begin{aligned} 12\% \text{ of R7 000} &= \text{R840} \\ \therefore \text{ Interest per year} &= \text{R840} \\ \text{Total interest for 4 years} &= \text{R840} \times 4 \\ &= \text{R3 360} \end{aligned}$$

Note: "Per annum" means "per year".

2 Clara invests R5 000 for 6 years, at 6,25% per annum. Calculate the amount of simple interest she will earn in total.

$$\begin{aligned} I &= \frac{Prt}{100} \\ &= \frac{5\,000 \times 6{,}25 \times 6}{100} \\ &= \text{R1 875,00} \end{aligned}$$

3 Calculate the amount of simple interest earned by investing R1 050 for 9 months at $4\frac{3}{4}$% per annum. Then find the total value of the investment after 9 months.

$$I = \frac{Prt}{100}$$
$$= \frac{1\ 050 \times 4{,}75 \times 0{,}75}{100}$$
$$= \text{R}37{,}41$$

Total value of investment
$= P + I$
$= \text{R}1\ 050 + 37{,}41$
$= \text{R}1\ 087{,}41$

GROUP ASSESSMENT

See page 291 (rubric 7).

Investigation: Micro-lending

The Micro Finance Regulatory Council (MFRC) was established in 1999 to educate and protect South African borrowers and lenders. The MFRC has laid down some rules, for example:

- Lenders must consider whether the borrower will be able to repay a loan and still cover his or her living expenses.
- Lenders must set up a loan agreement with the borrower and clearly explain the terms of the agreement to the borrower, in a language that the borrower understands.
- The borrower has the right to a three-day "cooling-off period". This means that he or she may cancel the loan agreement during this time.
- Lenders are not allowed to keep the borrower's bank card or PIN number as security. Also, they may not give out private information about the borrower, or collect money illegally.

In groups, discuss these questions:

1. Imagine that you want to borrow R10 000. Whose responsibility is it to make sure that you will be able to pay back the money?
2. If a Xhosa-speaking lender is lending money to an Afrikaans-speaking borrower, how would you suggest they make the agreement?
3. Why do you think a "cooling-off period" is important?
4. What advice would you give to someone if a lending institution requires her to hand over her ID or bank card when taking out a loan?

Exercise

1 Calculate the amount of simple interest for each of these:

	Principal	Interest rate	Time
a	R400	5%	4 years
b	R750	8%	5 years
c	R1 000	$7\frac{1}{2}$%	8 years
d	R1 250	$10\frac{1}{4}$%	3 years
e	R2 400	$12\frac{3}{4}$%	15 years
f	R865	15%	$2\frac{1}{2}$ years
g	R599	10%	6 months
h	R85,50	22,5%	9 months
i	R15 000	$33\frac{1}{3}$%	$1\frac{1}{4}$ years

2 Calculate the amount to be repaid for each of these:

	Principal	Interest rate	Time
a	R500	5%	4 years
b	R780	$6\frac{1}{2}$%	3 years
c	R1 200	$7\frac{1}{4}$%	6 months
d	R2 250	$10\frac{3}{4}$%	8 months
e	R2 400	12%	250 days

3 A woman invests a sum of R10 000 in a fixed-term deposit account for 3 years. Calculate:

   a The simple interest earned if the rate of interest is 6,5% per annum.

   b The total value of the investment at the end of the 3 years.

4 Your uncle takes out a personal loan of R20 000 for a period of 5 years, at a simple interest rate of 12% per annum. Find the amount he still owes after 1 year, if he makes payments of R100 every month of that year.

5 Tamara takes out a loan of R1 200 at a simple interest rate of 14,5% per annum. How much does she owe after 3 months?

6 A company invests R1 000 000 in the short-term money market, at 11% simple interest. How much interest does the investment earn in 30 days? Give your answer to the nearest cent.

7 The Usury Act states that unregistered lenders must stick to the following interest limits: on loans less than R10 000, the maximum interest rate is 23% per year. On loans more than R10 000, the maximum limit is 20% per year. Calculate the maximum amount of interest per annum that you should legally pay on these amounts:

a R1 275,00 b R12 500,00 c R9 850,00.

8 A bank offers the following interest rates on term deposits:

	Interest rate (per annum) on term (months)				
Balance	1–<3	3–<6	6–<12	12–<24	24–<36
R20 000–R49 999	2,85%	3,35%	3,85%	4,35%	4,85%
R50 000–R99 999	3,00%	3,50%	4,00%	4,50%	5,00%
R100 000–R199 999	3,40%	3,90%	4,40%	4,90%	5,40%
R200 000 and over	4,00%	4,50%	5,00%	5,50%	6,00%

Use this table to find the interest you would earn on each of these investments. Give your answer to the nearest cent:

a R25 000 for 2 months
b R125 000 for 6 months
c R37 750 for 18 months
d R200 000 for 2 years
e R74 386 for 8 months
f R145 000 for 23 months.

Rearranging the simple interest formula

The formula for simple interest can be rearranged to find any one of the variables when the values of three other variables are known. The formula for simple interest, or the table facility of a graphics calculator, may be used to determine the unknown value. You may want to check whether the bank has calculated the correct amount of interest on your investment at the end of a term. Or you may want to find out how long you need to invest a sum of money to earn a certain amount. In these cases, it is useful to rearrange the formula for simple interest.

EXAMPLES

Remember: 0,6 years ≠ 6 months!

1 Calculate how long you need to invest R5 000 to earn interest of R1 800 at $13\frac{1}{2}$% per annum simple interest.

$$I = \frac{Prt}{100}$$

$$\text{R1 800} = \frac{5\ 000 \times 13{,}5 \times t}{100}$$

$$\therefore \quad t = \frac{1\ 800 \times 100}{5\ 000 \times 13{,}5}$$

$$= 2{,}6 \text{ years}$$

$$= 2 \text{ years } 8 \text{ months}$$

2 Calculate the rate of simple interest if a principal of R8 000 totals R10 900 after 4 years.

$$\textit{Interest}, I = \text{R10 900} - \text{R8 000}$$

$$= \text{R2 900}$$

$$I = \frac{Prt}{100}$$

$$\text{R2 900} = \frac{8\ 000 \times r \times 4}{100}$$

$$r = \frac{2\ 900 \times 100}{8\ 000 \times 4}$$

$$= 9{,}0625\%$$

3 Calculate the principal if an investment totals R14 800 at $8\frac{1}{4}$% after 4 years.

$$\text{R14 800} = P + I$$

Since $I = \frac{Prt}{100}$

then $\text{R14 800} = P + \frac{Prt}{100}$

$$= P + \left(\frac{8{,}25 \times 4 \times P}{100}\right)$$

$$\text{R14 800} = P + 0{,}33P$$

$$= 1{,}33P$$

$$\therefore \quad P = \frac{14\ 800}{1{,}33}$$

$$= \text{R11 127,82}$$

So the formula $I = \frac{Prt}{100}$ can be rewritten as:

$$P = \frac{100I}{rt}$$

$$r = \frac{100I}{Pt}$$

$$t = \frac{100I}{Pr}$$

Exercise

1 Nosimo invests R2 000 in a savings account. The bank pays her simple interest at 7,5% per annum. How long will it take her to earn R975 on her savings?

2 When Sizwe was 12, he put some money in the bank. The amount earns 4,25% simple interest. By the time he was 22, he had earned R514,25 on his money. How much did he originally have?

3 Jackie puts R5 000 in a savings account. After $2\frac{1}{2}$ years, she has a total of R6 500 (including the simple interest she has earned). Calculate the rate of simple interest.

4 Calculate the principal that earns R780 in 100 days at 6,25% per annum simple interest.

5 Calculate the annual rate of simple interest if a principal of R500 amounts to R550 in 8 months.

6 Calculate the time in days for R760 to earn R35 at $4\frac{3}{4}$% simple interest per annum.

7 Calculate the answers to complete the table:

Principal	Rate	Time	Simple interest	Total investment
R600	6%	5 years	**a**	**b**
R880	$6\frac{1}{2}$%	**c**	R171,60	**d**
R1 290	**e**	6 months	R45,15	**f**
g	10%	4 months	R150,00	**h**
R3 600	**i**	200 days	R98,63	**j**
R980	$7\frac{1}{2}$%	**k**	**l**	R1 200,50
m	$7\frac{1}{4}$%	6 months	R52,50	**n**

8 If Geoff invests R30 000 at 10% simple interest per annum until he has R42 000, for how many years will he need to invest the money?

9 Thabo decides to put R5 000 into an investment account, which is paying 5,0% simple interest per annum. If he leaves the money there until it doubles, how long will this take, to the nearest month?

10 A personal loan of R15 000 over a period of 3 years costs R500 per month to repay.

a How much money will be repaid in total?

b What equivalent rate of simple interest is being charged over the 3 years?

11 To buy her first car, Mpho took out a loan for R40 000. She paid it back over a period of 2 years, and this cost her R14 500 in interest. At what simple interest rate was she charged?

12 Over a period of 4 years, an investment earned R913,50 at a simple interest rate of 5,25% per annum. What was the original amount deposited?

13 How long does it take for R2 400 invested at 12% simple interest per annum to earn R360 in interest?

14 Malibongwe belongs to a stokvel of 16 people. The stokvel has a savings account at a bank that pays 1,5% interest per month. From February to November, each person contributes R200,00.

- **a** Work out how much the group has saved by the beginning of December (before interest).
- **b** How much interest did they earn in total?
- **c** How much money does each person get at the beginning of December when they split their savings between them?

Compound interest

In the previous section, you found out that simple interest is calculated on the original amount you borrow. However, in reality, banks and lending institutions charge **compound interest** on loans, that is, the interest charged in each period is added to the principal. This means that the principal increases, and that the next period's interest will then increase too, because the principal is greater.

The opposite is also true. If you invest money, the interest you earn is added to the principal, and the next period's interest will increase too, because the principal has increased.

EXAMPLE

Your brother invests R250 at 10% per annum. The interest is added to his account each year. How much money will he have in his account at the end of the third year?

In the first year:

Interest = R250 × 10% × 1 = R25,00

so at the end of the first year the amount of money in the account is

R250 + R25 = R275

In the second year:

Interest = R275 × 10% × 1 = R27,50

so at the end of the second year the amount of money in the account is

R275 + R27,50 = R302,50

In the third year:

Interest = R302,50 × 10% × 1 = R30,25

so at the end of the third year the amount of money in the account is

R302,50 + R30,25 = R332,75

and so on.

If you continue in this way, you can build up a table of values. Compare the amount of money invested at 10% simple interest with the amount invested at 10% compound interest.

Year (n)	Total R (simple interest)	Total R (compound interest)
1	275,00	275,00
2	300,00	302,50
3	325,00	332,75
4	350,00	366,03
5	375,00	402,63
6	400,00	442,89
7	425,00	487,18
8	450,00	535,90
9	475,00	589,49
10	500,00	648,44

From the table you can see that after the first year, the compound interest is higher, and this advantage to the investor becomes more obvious as time increases. A graph shows the difference between the two investment strategies even more clearly:

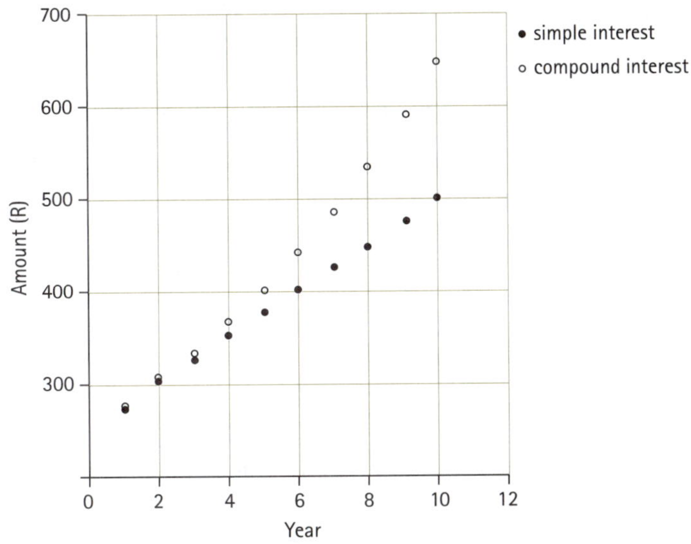

The growth in the investment over time when compound interest applies, is clearly not linear, but rather curved, as shown in the graph.

Calculation of compound interest is very tedious if carried out in the way shown here. But the calculation follows a clear pattern.

In the example, the principal was increased by 10% at the end of each year.

The multiplying factor to increase a quantity by 10% = $(1 + \frac{10}{100}) = 1{,}1$.

Thus, the total amount at the end of:

$$\begin{aligned}
\text{1st year} &= \text{R}250 \times 1{,}1 \\
&= \text{R}275 \\
\text{2nd year} &= \text{R}250 \times 1{,}1 \times 1{,}1 \\
&= \text{R}250 \times (1{,}1)^2 \\
&= \text{R}302{,}50 \\
\text{3rd year} &= \text{R}250 \times 1{,}1 \times 1{,}1 \times 1{,}1 \\
&= \text{R}250 \times (1{,}1)^3 \\
&= \text{R}332{,}75 \\
n\text{-th year} &= \text{R}250 \times (1{,}1)^n
\end{aligned}$$

So the formula for calculating compound interest can be written as:

$$A = P\left(1 + \frac{r}{100}\right)^n$$

where:

A = final amount (R)
P = initial amount (R)
r = rate (%), per time interval
n = number of intervals of time (e.g. years)

That is, if an amount RP is invested at r% compound interest for n years, it will grow to RA.

Investigation: Home loans

Work in a group.

A good example of a loan on which banks charge compound interest over a long period, is a home loan. Find out more about home loans. Then try to answer these questions:

1 How much do different homes and flats cost in your area?
2 Over how many years do banks usually allow people to pay back a home loan?
3 Work out how much interest you would have to pay in total on a house of R330 000 if you took out a loan to pay for a home over the maximum period.
4 What happens if you pay off the house earlier? (This means that you need to pay off more than the specified monthly repayment.) Try to find out how much interest you would **save** if you paid off a house of R330 000 five years earlier than the maximum period.
5 Do you think it is a good idea to buy a house? Or is it better to rent one? Discuss this in your groups and give a reason for your answer.

EXAMPLES

1 Ashraf invests R5 000 at 6% per annum compound interest for 5 years. Calculate the final value of his investment.

$$A = P\left(1 + \frac{r}{100}\right)^n$$
$$= 5\,000\left(1 + \frac{6}{100}\right)^5$$
$$= 5\,000(1{,}06)^5$$
$$= \text{R}6\,691{,}13$$

In practice, the interest is often compounded more frequently than once per year. Common periods are quarterly, monthly or daily.

2 Brenda started running a catering business from home in January 2000. She borrowed R2 000 from the bank to buy baking utensils. The bank charges her 8% interest per year, compounded quarterly (every 3 months, or 4 times a year). How much interest had she paid by January 2003?

If the interest is calculated quarterly, that is, 4 times per year, the rate of interest for a quarter $= \frac{1}{4} \times 8\% = 2\%$

R2 000 will be compounded by 2% each quarter. Therefore the number of time intervals 3 years after taking out the loan $= 4 \times 3 = 12$.

$$A = P\left(1 + \frac{r}{100}\right)^n$$
$$= 2\,000(1{,}02)^{12}$$
$$= \text{R}2\,536{,}48$$
$$\text{Interest} = \text{R}2\,536{,}48 - \text{R}2\,000$$
$$= \text{R}536{,}48$$

3 In March 1999, Joseph borrowed R3 000 at 12% compound interest per year over 5 years, calculated monthly. Work out how much Joseph had paid in total by the time the loan was fully paid off in March 2004.

12% per annum = 1% per month
5 years = 60 months

$$A = P\left(1 + \frac{r}{100}\right)^n$$
$$= 3\,000\left(1 + \frac{1}{100}\right)^{60}$$
$$= 3\,000(1{,}01)^{60}$$
$$= \text{R}5\,450{,}09$$

So Joseph will have paid R5 450,09 in total.

4 Thandeka runs an investment service. She wants to invest some of her clients' money at 11% per annum compound interest. How much money (rounded off to the nearest rand) should she invest if she wants to reach a sum of R100 000 in 5 years' time?

$$A = P\left(1 + \frac{r}{100}\right)^n$$

$$\therefore\ P = \frac{A}{\left(1 + \frac{r}{100}\right)^n}$$

$$= \frac{100\ 000}{(1{,}11)^5}$$

$$= 59\ 345{,}13281$$

Amount to invest = R59 345

PEER ASSESSMENT

See page 291 (rubric 7).

Investigation: Investment options

Work in pairs.

Some investment services offer "ethical investments". These are investments that provide capital to particular kinds of businesses, for example:

- businesses that do not involve gambling or alcohol
- businesses that do not involve meat or animal products
- businesses that adhere to fair trade agreements.

Discuss why you think these investment services offer "ethical investments". What factors influence the type of investments that people choose?

Exercise

1 Lucy invests R3 500 at 5% compound interest per annum for 5 years. Calculate the final amount in the account.

2 Nazeem invests R7 000 at 8% per annum for 4 years. Calculate the total amount of compound interest he earns.

3 James borrows R1 250 at $7\frac{1}{2}$% compound interest per annum for 3 years. Calculate the total amount he needs to repay.

4 Calculate the difference between the simple interest and the compound interest on a loan of R2 000 at 7% over 5 years.

5 Calculate the amount of money that would need to be invested at $6\frac{3}{4}\%$ per annum compound interest to achieve a sum of R36 000 in 4 years.

6 Calculate the interest paid on a loan of R2 750 at 11%, compounded quarterly for 4 years.

7 Calculate the final amount to be paid on a loan of R10 000 at 12% per annum for 5 years, for each of the following conditions:

a simple interest

b compound interest calculated annually

c compound interest calculated quarterly

d compound interest calculated monthly

e compound interest calculated daily.

8 A bank offers the following investment possibilities to customers:

- simple interest at 13% per annum
- compound interest at 12%, calculated annually
- compound interest at 11%, calculated quarterly
- compound interest at 10%, calculated monthly
- compound interest at 9%, calculated daily.

Discuss which option is the best for each of these customers:

a a young couple who wish to invest R5 000 over 5 years to save towards their child's schooling costs

b a micro-lender who invests R65 000 for a period of 1 year.

9 Calculate the amount of money that could be borrowed at $5\frac{1}{2}\%$ per annum compound interest, charged quarterly, over 3 years, so that the total repayments do not exceed R50 000.

10 Joe buys a school uniform for his son on his credit card, with the intention of paying if off over 6 months. The uniform costs R350. If he is charged 18% interest per annum for 6 months, compounded monthly, how much will the uniform cost him in total? (Assume that no payments are made within this time, and give your answer to the nearest cent.)

11 How much money must you invest at 12,5% per annum, compounded monthly, if you know you will need R10 000 in 3 years' time? Give your answer to the nearest R10.

12 Tshidi's parents open an investment account when she is born, in which they deposit R5 000. They intend to withdraw the money on her 18th birthday. If the account pays 8,0% per annum, compounded quarterly, for the first 10 years, and then 6,0% per annum, compounded monthly, for the rest of the term of the investment, how much will she receive? Give your answer to the nearest R10.

Exponential growth

Exponential growth occurs when an initial amount changes regularly by a given percentage.

In natural science, you may have come across bacteria and viruses. After a certain amount of time, each virus splits in two – that is, the number of virus molecules doubles. Look at this diagram:

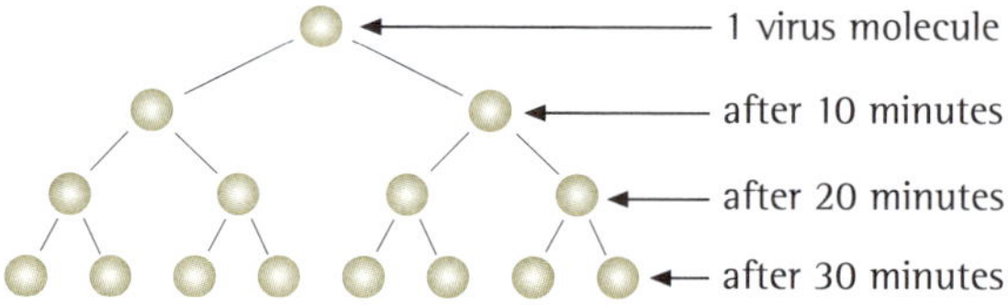

DID YOU KNOW?

Viruses make copies of themselves by taking over host cells – particular cells of the body that the virus is able to attack and use. A virus molecule has a spiky structure that allows it to break the surface of a particular host cell and get inside the cell. Once the virus enters the host cell, it can alter the reproductive enzymes of the cell so that the host cell becomes a kind of reproductive machine for the virus.

We can express what happens in the diagram as a number series:

1; 2; 4; 8; ...

Another way of expressing this number series is by using exponents or indices. These are the powers to which we raise the number with each step:

2^0; 2^1; 2^2; 2^3; and so on.

We call this **exponential growth.**

Some examples you may have seen are:

- Inflation is rising at 5% per year (that is, your money loses 5% of its value after 1 year).
- You invest your money at 17% per annum (that means your money grows by 17% every year).
- Certain algae grow to 3 times their size in a week (so the algae grow by 200% weekly).

These situations can be described by exponential equations, since the original amounts can be multiplied by a common factor over and over again.

EXAMPLES

1 Form exponential equations for these situations:

a Bheki invests his savings at a rate of 14% per annum.

b A city's population is increasing at 12% per year.

Remember: 100% + 14% = 1,14

a Let A = the amount of money at any time

n = the number of years for which the money is invested

A_0 = amount invested

n	0	1	2	3	n
A	A_0	$A_0 \times 1{,}14$	$A_0 \times 1{,}14 \times 1{,}14$	$A_0 \times 1{,}14 \times 1{,}14 \times 1{,}14$	
A (index form)	A_0	$A_0(1{,}14)^1$	$A_0(1{,}14)^2$	$A_0(1{,}14)^3$	$A_0(1{,}14)^n$

$\therefore\ A = A_0\,(1{,}14)^n$, so $A = A_0\,(1 + \frac{14}{100})^n$

Remember: 100% + 12% = 1,12

b Let P = the population at any time

n = the number of years for which the population increases

P_0 = starting population

n	0	1	2	3	n
P	P_0	$P_0 \times 1{,}12$	$P_0 \times 1{,}12 \times 1{,}12$	$P_0 \times 1{,}12 \times 1{,}12 \times 1{,}12$	
P (index form)	P_0	$P_0(1{,}12)^1$	$P_0(1{,}12)^2$	$P_0(1{,}12)^3$	$P_0(1{,}12)^n$

$\therefore\ P = P_0\,(1{,}12)^n$, so $P = P_0\,(1 + \frac{12}{100})^n$

2 House prices are rising at 9% per year. Paul's house is valued at R345 000.

a Determine a rule for the value of Paul's house (V) in n years' time.

b What will the value of his house be

i next year? **ii** in 3 years' time?

c Use trial and error to find when Paul's house will be valued at R650 000 to 1 decimal place.

Remember: Define your variables before writing equations.

Let V = value of Paul's house at any time

V_0 = starting value

n = number of years from now

a $V = V_0(1{,}09)^n$

$\therefore\ V = 345\ 000\ (1{,}09)^n$

b **i** $n = 1$ $\quad V = 345\ 000(1{,}09)^1$
$= 376\ 050$
Next year Paul's house will be valued at R376 050.

ii $n = 3$ $\quad V = 345\ 000(1{,}09)^3$
$= 446\ 785$
In 3 years' time Paul's house will be valued at about R446 800. (It is reasonable to round your answer off to the nearest R100 in this case.)

c

n	5	8	7,5	7,4	7,3
V	530 825	687 434	658 442	652 792	647 191

Paul's house will be valued at R650 000 in about 7,4 years.

GROUP ASSESSMENT

See page 291 (rubric 7).

Investigation: HIV infection rate

Work in groups of about five learners.

Ugandan HIV rate drops

The Ugandan AIDS awareness campaign may not be fashionable, but it seems to be wiping out its AIDS problem, according to a university study.

According to the study, Uganda's HIV infection rate dropped by 50% between 1992 and 2000. In the late 1980s, 21,2% of pregnant women in Uganda were HIV positive. By 2001 this percentage had dropped to 6,2%. Also, in 1989, 18,4% of women aged 15 and over reported that they were having unsafe sex with multiple partners. By 2000, this percentage had dropped to 2,5%.

The Ugandan government promoted abstinence as the best way of preventing HIV infection. Abstinence was promoted through posters, billboards, radio programmes and school awareness campaigns.

But not everyone agrees with this approach. Many people claim that it is unrealistic to expect young people to remain abstinent. In Botswana, Zimbabwe and South Africa, most AIDS awareness campaigns promote the "ABC" approach: abstinence is the most effective choice, followed by being faithful to a single partner and using condoms.

Read the article. Then discuss these questions in your groups:

1 What different methods does the article mention for preventing HIV transmission?
2 What are the advantages and disadvantages of each method?
3 Which method do you think is most popular in South Africa?

4 If your school ran an HIV and AIDS awareness campaign, what methods of prevention do you think you would emphasise as most effective? Why?

5 Why do you think researchers measure the number of pregnant women infected with HIV as an indicator of the spread of the virus?

6 The South African government is trying to convince all South Africans to be tested for HIV/AIDS. Do you think this is a good idea? Why do you say so? Discuss.

Exercise

1 Form exponential equations for the following situations:
   - a Invest an amount of money at 17% per annum.
   - b Inflation is running at 3%.
   - c The population of a town increases by 25% per year.
   - d A cell doubles its size every 30 seconds.

2 Ayesha invests R50 000 in a retirement annuity that has an annual return of 11%.
   - a Determine the rule for the value of her investment (V) after n years.
   - b How much will Ayesha's investment be worth in
     - i 3 years? ii 20 years?
   - c Plot a graph and find approximately when her investment will be worth R100 000.

3 A certain type of bacterium grows according to the equation $N = 3\ 000(2{,}6)^t$, where N is the number of cells present after t hours.
   - a Determine the number of cells present after:
     - i 0 hours ii 3 hours iii 6,7 hours.
   - b If 50 000 bacteria are needed to make a drop of serum, how long will we have to wait to make a drop (to the nearest minute)?

4 The HIV infection rate is the rate at which more people are getting infected with HIV. In a random sample of 1 million people, one person has HIV. He infects two other people, who each infect two other people, and so on.
   - a Draw a diagram to show how quickly the number of infected people grows.

b Sketch a graph to represent the information from your diagram.

c As each person infects two more, the infected population doubles. How many steps would it take on your diagram until the infected population reached 1 million?

d How would the infection rate change if each person infected three others, instead of two.

Hire purchase

One way of purchasing expensive goods such as television sets or lounge suites when you don't have the cash available immediately, is to enter into a hire purchase agreement. This means the purchaser agrees to hire the item from the seller, and to make payments of an agreed amount, at specified time intervals (usually monthly). At the end of the period of the agreement, the purchaser owns the item. If the purchaser stops making payments at any stage of the agreement, the item is returned to the seller, and no money is refunded to the purchaser.

The interest rate charged in these contracts is not always stated explicitly. So it is important to be able to calculate it to find out what you are really paying for an item.

Flat interest rate

The interest paid as a proportion of the original debt is called the **flat rate of interest** in a hire purchase context. It is exactly the same as simple interest.

$$\text{Flat interest rate} = \frac{I}{P} \times 100\ (\%)$$

where

I = total interest paid

P = principal owing after the deposit has been deducted

EXAMPLES

1 A student buys a laptop computer priced at R13 500. She pays a deposit of 20% and interest at 15% per annum on the balance. If she pays off the loan in 12 monthly instalments, then

a how much is each monthly instalment?

b what is the total cost of buying the computer by hire purchase?

a Deposit = 20% of R13 500 = R2 700

Balance = R13 500 − R2 700 = R10 800

Interest on the balance = 15% of R10 800 = R1 620

Thus, total to be repaid = R10 800 + R1 620 = R12 420

Therefore, monthly repayments = $\frac{\text{R12 420}}{12}$ = R1 035

b Total cost of the computer = R2 700 + R12 420 = R15 120

2 A car radio has a market price of R960. A purchaser agrees to a hire purchase contract of R100 deposit and 12 monthly repayments of R100. Calculate:

a the total interest paid **b** the flat rate of interest.

a Total repayments = R100 + 12 × R100 = R1 300

Thus, total interest = R1 300 − R960 = R340

b Flat rate of interest = $\frac{340}{860} \times 100$ = 39,53%

The formula on page 40 gives the flat rate of interest over the period of the contract, and since the contract in this example is for 1 year, the flat rate of interest calculated is also the flat rate of interest per annum. The distinction will become clear in the next example.

3 A hire purchase contract for a sound system requires Josh to pay a deposit of R400 and then make 6 monthly payments of R185. If the price of the sound system is R1 400:

a How much interest will he pay?

b What is the flat rate of interest *per annum* that this represents?

a Total repayments = deposit + repayments

= R400 + 6 × R185

= R1 510

Total interest = total repayments – purchase price
= R1 510 – R1 400
= R110

b Using the formula previously given you find that:

Flat interest rate = $\frac{110}{1\,000} \times 100 = 11\%$

over the period of the contract, which is 6 months, or half a year. To convert this to an **annual** interest rate, you need to multiply by 2, to make up a full year.

Thus:

Flat interest rate per annum = $\frac{110}{1\,000} \times 2 \times 100 = 22\%$

Flat rate of interest formula

In general, to determine the flat rate of interest per annum, you need to take into account the period of the hire purchase agreement. This leads to the following formula:

Flat interest per annum, $r_f = \frac{I}{P} \times \frac{m}{n} \times 100$ (%)

where

I = total interest paid
P = principal owing after the deposit has been deducted
m = number of payments made per year
n = number of payments made in total.

Exercise

1 The cash price of a tennis racquet is R330. To buy it on hire purchase requires a deposit of R30 and 12 equal monthly instalments of R28. Calculate:

a the total cost of buying the racquet by hire purchase

b the extra cost of buying by hire purchase.

2 A bicycle has a marked price of R3 000. It can be bought through hire purchase with a deposit of R600 and 10% interest on the outstanding balance, to be repaid in 10 monthly instalments. Calculate:

a the amount of each monthly instalment

b the total cost of buying the bicycle by hire purchase.

3 A hire purchase agreement offers a car stereo with a marked price of R897, for R87 deposit and R46,80 a month payable over 2 years. Calculate:

a the total hire purchase price

b the amount of interest charged.

4 A fridge is advertised for R5 575 cash or R600 deposit and 24 monthly instalments of R268,75. Calculate the flat rate of interest.

5 An exercise bicycle, which normally costs R750, can be bought through hire purchase with a R200 deposit and R26,40 a month for 30 months. Calculate:

a the amount of interest being charged

b the flat rate of interest.

6 A microwave oven is advertised with a marked price of R576 and the opportunity to buy it on hire purchase, with no deposit and an interest rate of 10% repayable over a year in 4 equal instalments. Calculate:

a the amount of interest

b the total amount to be repaid

c the amount of each instalment.

7 A student buys a tent priced at R200. He pays a deposit of R50 and agrees to repay the balance of R150 plus interest at 14% over the period of 1 year, in 2 half-yearly instalments. Calculate:

a the amount of each instalment

b the total price of the tent.

8 A customer bought a car for R36 010 on hire purchase. A deposit of R4 000 was paid and the loan plus interest is to be repaid over 18 months in 6 quarterly repayments of R6 295,30. Calculate:

a the amount repaid

b the flat interest rate.

9 A student bought a computer costing R22 250 on hire purchase. She traded in an old computer for R2 000 and paid a deposit of R1 500. The balance was paid by monthly instalments of R1 120,50 over 2 years. Calculate:

a the total interest paid

b the flat rate of annual interest.

Inflation

Inflation is sustained increase in the national average price level. Under inflation, the buying power of your money diminishes (gets less) year by year. If inflation is at 10% per year, at the end of the year goods and services will cost 10% more than they did at the beginning of the year.

This table summarises the inflation history of South Africa between 1960 and 2002:

Average inflation rates in South Africa			
1960–64	1,6%	1994	9,0%
1965–69	3,2%	1995	8,7%
1970–74	7,5%	1996	7,4%
1975–79	11,8%	1997	8,5%
1980–84	13,5%	1998	7%
1985–89	15,7%	1999	5,7%
1990	14,3%	2000	5,3%
1991	15,4%	2001	5,8%
1992	13,9%	2002	9,9%
1993	9,7%		

This graph shows the annual inflation rate in South Africa from 1960 to 2002.

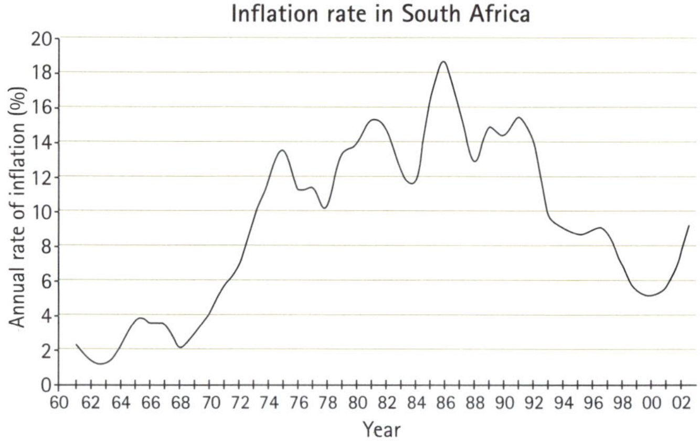

(Based on: *Macro Economics in South Africa*; Van Niekerk, K. M., 1997.)

Inflation and buying power

People will only have more buying power if wages and salaries go up more than inflation. So if someone's wages increase by 8% per year, but inflation is 5% per year, real buying power increases by only 3%.

In manufacturing, employers would perhaps be able to keep the prices of products constant if the productivity of workers increased. However, in practice staff perform at the same level even when their wages are increased. As a result, prices rise to cover wage increases and any other price increases. This results in an upward inflationary spiral.

Inflation can sometimes spiral completely out of control. This happened in Germany after the end of the Second World War in 1945, and in Brazil in the 1990s (it stood at 2 500% in 1994). It is also happening in Zimbabwe, where inflation rose to around 1 100% in 2002, and was still at 450% in May 2004.

EXAMPLE

A movie ticket costs R20. How much will it cost in 3 years' time, if inflation continues to rise at 6% per annum?

The value compounds annually, so use the compound interest formula with $r = 0{,}06$, $n = 3$ and $P = \text{R}20$.

$$\begin{aligned} A &= P(1 + r)^n \\ &= \text{R}20(1 + 0{,}06)^3 \\ &= \text{R}20 \times 1{,}06^3 \\ &= \text{R}23{,}82 \end{aligned}$$

PEER ASSESSMENT

See page 288 (rubric 3).

Investigation: Comparing inflation rates

Work with a partner.

Look at this table comparing inflation rates in different countries in 2002:

Country	Inflation rate (%)	Country	Inflation rate (%)
South Africa	10,2	Italy	2,5
Argentina	25,9	Japan	-0,9
Australia	3	Portugal	3,6
Brazil	9	Spain	3,5
France	1,9	UK	1,6
Germany	1,3	USA	1,6

(Source: Nedcor Economic Unit)

1 How does South Africa's inflation rate compare with the rates of the other countries listed?

2 List the countries in order from the country with the highest inflation rate to the country with the lowest inflation rate. What is South Africa's position on the list?

3 Look in a newspaper to find the latest inflation rates in South Africa. How do they compare with those in 2002?

4 Japan has a negative inflation rate. What does this mean? Is it a good thing or not? Discuss.

5 Does a high inflation rate have any advantages? If so, who benefits most? Discuss.

Exercise

Assume a consistent annual inflation rate of 5% for each of the following:

1 Find the expected cost in 2 years' time of a new car that costs R100 000.

2 How much will a video recorder priced at R950 cost to replace in 4 years' time?

3 An overseas holiday costs R40 000. How much will it cost in 3 years' time?

4 If workers get wage increases to match inflation, what could a worker earning R30 000 p.a. expect to be earning in 2 years' time? Would her financial position be better **in real terms**? Explain.

Exchange rates

The exchange rate is the price of one currency (money unit) in terms of another, for example:

R8,20 = 1 US dollar
R13,11 = 1 British pound
R8,92 = 1 euro
R0,62 = 1 Botswana pula.

Note: "Foreign exchange" means currencies of other countries.

These rates change daily and are determined on the foreign exchange market.

EXAMPLES

1 Thabo is on his way to the US. He buys US dollars from the bank at R8,20 per US dollar. How many dollars does he get for R20 000?

R8,20 = 1 US dollar

So R20 000 = $\frac{20\,000}{8{,}20}$ = 2 439,02 US dollars

2 Prevendra's aunt lives in Botswana. She plans to visit Prevendra in South Africa, so she exchanges 1 000 Botswana pula for rands at R0,62 per Botswana pula. How many rands does she get?

1 pula = R0,62

So 1 000 pula = 1 000 × R0,62 = R620

Investigation: Exchange rates

In 2002, 1 US dollar cost about R13,50. In June 2004, 1 US dollar cost only around R6,50. So 1 US dollar only cost half of what it cost two years earlier.

Try to find some newspaper articles that discuss the effect of such a change in the exchange rate on the export and import of goods between the United States and South Africa.

Then discuss these questions as a class:

1 Who will benefit from this change in interest rate? Think of some examples of people or companies who benefit.

2 Can you think of people and organisations that did not benefit? (Find out what happens if imported products become cheaper than local products.)

Exercise

Use the exchange rates on page 46 to calculate the following.

1 Convert these amounts to rands:

a 23 456 US dollars b 5 294 Botswana pula

c 398 672 British pounds d 59 307 euro.

2 Convert the following amounts to the currency in brackets:

a R23 456 (British pounds)

b R582,39 (euro)

c R654 900 800 (US dollars).

3 Sibusiso exports machines to Botswana. His selling price is R56 782. What amount should he receive in Botswana pula?

4 This table shows some exchange rates from 1997 to 2002. Use the table to help you answer the questions:

Year	US dollars	British pounds	Euro
1997	R4,90	R7,60	R5,20
1998	R5,50	R9,10	R6,10
1999	R6,10	R9,90	R6,50
2000	R7,00	R10,50	R6,40
2001	R7,80	R12,40	R7,80
2002	R8,80	R13,50	R8,80

a A clothing company, Goodness Gracious Apparel, sells locally made and imported clothing in three shops in South Africa. The company also has a small factory where employees make dresses and shirts that are exported to other countries. In 2001, the company had a business meeting to decide whether to discontinue its imported clothing lines. Why do you think they considered discontinuing importing?

b Many export companies started up in 2002. Why do you think this happened?

c Look in the newspaper and find the current exchange rates for the three currencies in the table. Is the rand weaker or stronger now than it was in 2002?

d Draw a bar graph showing the exchange rate between the rand and any five major currencies listed in the newspaper.

Language skills

compound interest	*inflation*	*rational numbers*
conjecture	*integers*	*recurring decimal*
exchange rate	*irrational number*	*simple interest*
flat interest rate	*proof*	

Complete the sentences with words from the list:

1 The positive and negative numbers together with zero are called the … .

2 … can be written as a fraction $\frac{a}{b}$, where a and b are integers and $b \neq 0$.

3 A … is a decimal in which the digits to the right of a certain point cycle endlessly.

4 $\sqrt{3}$ is an ... .

5 A ... requires ... before it can be seen as truth.

6 ... is a fixed percentage of the amount invested or borrowed and is calculated on the original amount.

7 ... calculates the interest for each time period on a sum of money to which the previous amount of interest has been added.

8 The ... is the rate of interest paid as a proportion of the original debt.

9 Sustained increase in the national average price level is called ... .

10 The ... is the price of one currency in terms of another.

Answer these questions after discussing them with a partner:

11 Write as many words as possible derived from:

a ratio **b** estimate.

12 Write a word opposite in meaning to:

a approximate **b** rational **c** positive **d** recurring.

13 Explain how to obtain a fraction equivalent to a given fraction (that is, how to simplify the fraction).

14 For each word, write one sentence showing the mathematical meaning and a second sentence showing an everyday meaning:

a terminate **b** negative **c** rational.

15 Write the opposite of:

a employee **b** inflation **c** decrease.

16 Explain the difference between **simple** and **compound** interest.

17 How should an investor take **inflation** into account when choosing an investment?

18 What is the connection between the **consumer price index** and **inflation**?

SELF-ASSESSMENT

Also see page 293 (rubric 9).

Check yourself

Make a summary of the key facts and formulas in this module under these two headings. It will help if you look at the key skills given at the beginning of this module (see page 1).

Number and number patterns

Money matters

Then complete the Review. The questions in the Review consolidate your knowledge and understanding of the topics in this module.

MODULE 1

REVIEW

1 Express each number as a recurring or terminating decimal:

a $\frac{7}{9}$ b $\frac{7}{8}$

2 Express each decimal as a rational number. Remember to simplify.

a $0,\dot{5}$ b $0,\dot{2}8571\dot{4}$

3 Write down the recurring decimals for $\frac{1}{9}$; $\frac{2}{9}$; $\frac{3}{9}$; $\frac{5}{9}$; $\frac{6}{9}$; $\frac{7}{9}$; $\frac{8}{9}$. What is the pattern?

4 Determine between which two integers these numbers lie:

a $\sqrt{8}$ b $\sqrt[3]{11}$

5 Prove this conjecture: If the sum of the digits of any 3-digit natural number is divisible by 3, the number is divisible by 3.

6 Prove that the sum of any 5 consecutive natural numbers is equal to 5 times the third number.

7 To what amount will R400 increase if invested for 4 years, and simple interest of 3,5% per year is paid?

8 What is the interest rate, per year, if a deposit of R1 500 earns interest of R50 over a period of 6 months?

9 R2 400 is invested at a rate of 4,25% compound interest, paid yearly. What is the value of this investment after 6 years?

10 Calculate the interest earned if R5 000 is invested for 3 years at an interest rate of 4%, given the interest is compounded quarterly.

11 Babalwa buys a car that costs R23 000 under the following arrangement: she pays R5 000 deposit, and then repays R440 per month for the next 5 years.

a How much interest does she pay under this scheme?

b What flat rate of interest does this amount to?

12 How much will a R3 000 holiday cost in 2 years' time, if inflation continues at 4,5% per year?

13 Convert 300 000 British pounds to rands if R13,67 = 1 British pound.

14 Convert R65 432 to US dollars if R9,52 = 1 US dollar.

PROJECTS

ASSESSMENT PORTFOLIO

See page 292 (rubric 8).

1 Consumer price index (CPI)

In this module you learned that inflation is the percentage by which prices increase in a given period. Data analysts work out inflation rates by comparing how much it would cost to buy the same basket of goods at different times. The consumer price index (CPI) is a number used to measure the change in prices. Every month the Central Statistics Service calculates the CPI by measuring the prices of a basket of goods against the base year (the year from which the change is calculated). You could, for example, take January 1995 as your base year. If you then assume that the basket of goods consists of one item and that one item cost R1,00 in January 1995, then the Central Statistics Service would publish the index in 1995 at 100. If today that same item costs R3,50, the index would stand at 350. A group of items would work the same way. If you have 100 items, each would account for 1% of the total index.

a Imagine that you work for the Central Statistics Service, and you are calculating the CPI based on the prices of 15 items. Which goods would you choose to include in your basket? Give reasons for your answers.

b Should you use the same brands every time you recalculate the CPI? Give reasons for your answer.

c Why is the CPI important to consumers?

d If the CPI goes down, will this have an impact on other prices? Discuss this in groups and write a paragraph about the effect of the CPI on the prices of other goods.

ASSESSMENT PORTFOLIO

See page 292 (rubric 8).

2 Comparing simple and compound interest

You will need: a calculator or computer spreadsheet software, graph paper.

1 Use your calculator and the simple interest formula on R100 at 10% p.a. flat. Then use the constant function to add on this same amount of interest year by year for 10 years. Plot the points on a graph as shown.

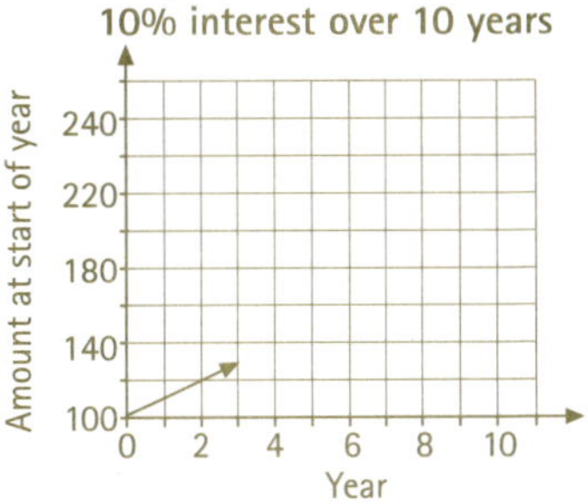

PROJECTS

2 Use the compound interest formula and a calculator or spreadsheet to find the compounded amount to which R100 grows year by year for 10 years when interest is compounded:

a annually **b** half-yearly **c** quarterly.

Plot the curves for **a** to **c** on your graph from question 1, using various colours to highlight your presentation.

3 Compile a report of your findings, using the graph and final figures. Comment on the relative advantages of the different interest arrangements.

SELF-ASSESSMENT

Copy the rubric into your workbook and put a tick (✓) in the column that describes how you feel about the work in this module.

Self-assessment rubric

How well do you understand these concepts from Module 1?	I understand this concept very well	I understand this concept	I need more help to understand this concept
Terminating and recurring decimals			
Irrational numbers			
Calculating percentages			
Estimating and rounding			
Simple interest			
Compound interest			
Positive exponents			
Hire purchase agreements			
Inflation			
Currency conversion			

MODULE 2

MODULE CHAPTERS

1 Polygons, prisms and cylinders

2 Coordinate geometry

3 Transformation geometry

4 Trigonometry

KEY SKILLS

investigating properties of triangles • investigating properties of quadrilaterals • finding surface areas of prisms and cylinders • investigating the effect of multiplying a dimension by k on the surface area of prisms and cylinders • finding volumes of prisms and cylinders • investigating the effect of multiplying a dimension by k on the volume of prisms and cylinders • finding the distance between two points • calculating the gradient of a line segment • investigating translations and reflections • defining and using the trigonometric functions • solving real-life problems in two dimensions • using scale drawings to solve real-life problems • interpreting maps • making models and drawing building plans • researching the history of geometry and trigonometry

SHAPE, SPACE AND MEASUREMENT

In this module, you will work through investigations, projects and exercises that will help you develop your ability to describe, represent, analyse and explain properties of shapes in two-dimensional and three-dimensional space with justification (Learning outcome 3).

By the end of this module you will:

- understand and determine the effect on the volume and surface area of right prisms and cylinders, of multiplying any dimension by a constant factor k.
- – through investigation, produce conjectures and generalisations related to triangles, quadrilaterals, and other polygons, and attempt to validate, justify, explain or prove them, using any logical method (Euclidian, coordinate and/or transformation)
  - disprove false conjectures by producing counter-examples
  - investigate alternative definitions of various polygons (including the isosceles, equilateral and right-angled triangle, the kite, parallelogram, rectangle, rhombus and square)
- represent geometric figures on a Cartesian coordinate system, and derive and apply, for any two points $(x_1; y_1)$ and $(x_2; y_2)$, a formula for calculating:
  - the distance between the two points
  - the gradient of the line segment joining the points
  - the coordinates of the mid-point of the line segment joining the points
- investigate, generalise and apply the effect of these transformations of the point $(x; y)$
  - a translation of p units horizontally and q units vertically
  - a reflection in the x-axis, the y-axis or the line $y = x$

MODULE 2 IN REAL LIFE

• properties of triangles and quadrilaterals are useful when designing buildings, textiles and tiling patterns • surface areas of prisms and cylinders are used to calculate cost for designs of containers and packaging materials • the volume of cylinders is an important aspect of various technical applications, e.g. fluid mechanics • in coordinate geometry you link algebraic and geometric concepts • trigonometry is used to calculate distances in scale drawings and plans, but also to planets in space • the sine function is applied in swinging pendulums, voltage changes in an alternating current circuit, and waves (water, light and sound waves)

- understand that the similarity of triangles is fundamental to the trigonometric functions sin θ, cos θ and tan θ, and be able to define and use functions
- solve problems in two dimensions by using the trigonometric functions (sin θ, cos θ and tan θ) in right-angled triangles and by constructing and interpreting geometric and trigonometric models (including scale drawings, maps and building plans)
- demonstrate an appreciation of the contributions to the history of the development and use of geometry and trigonometry by various cultures through a project.

Integration

In this module, you will also work with graphs and relationships between variables (Learning outcome 2).

Architects and designers apply their knowledge of shapes and measurement when designing buildings.

CHAPTER 1 POLYGONS, PRISMS AND CYLINDERS

KEY VOCABULARY

polygon • prism • cylinder • geometry • two-dimensional • three-dimensional • design • architecture • scalene • isosceles • equilateral • right-angled • quadrilateral • regular • rhombus • parallelogram • rectangle • square • trapezium • kite • pentagon • bisect • collinear • perpendicular • diagonal • parallel • line of symmetry • variable • exterior angle • interior angle • perimeter • area • volume • cubic units

In this chapter you will solve many different geometrical problems involving two- and three-dimensional shapes. People have used geometry since ancient times to mark out land, and to design items ranging from buildings and ships to furniture and simple functional items you use every day.

This photograph and model show the newly built Cape Town International Convention Centre. Can you identify the different geometric shapes that the architects used in their design? Why do you think they chose these shapes?

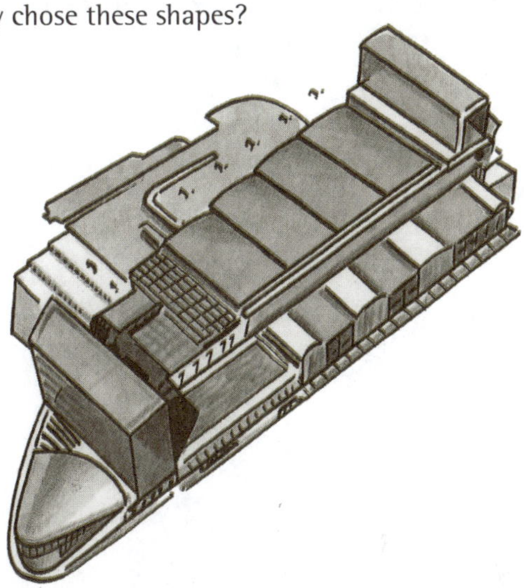

Model of the Cape Town International Convention Centre.

Top view of the Cape Town International Convention Centre.

Triangles

A **triangle** is a plane figure with three straight sides.

PEER ASSESSMENT

See page 290 (rubric 5).

Investigation: Triangles

Work in groups of four learners.

You will need: a ruler, a protractor.

1 Construct each of these triangles, using any necessary instruments (each member of your group can construct one triangle):

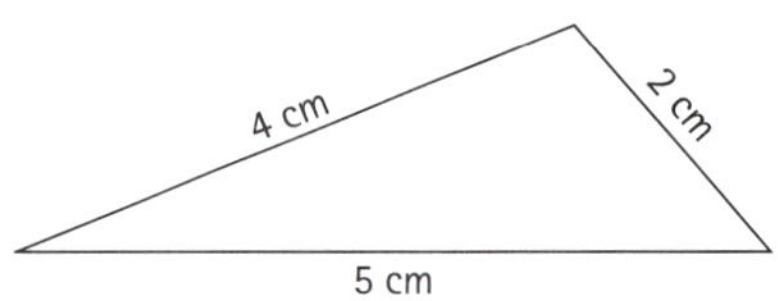

Scalene triangle

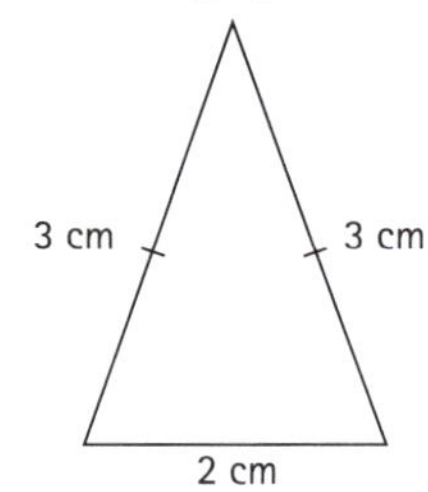

Isosceles triangle

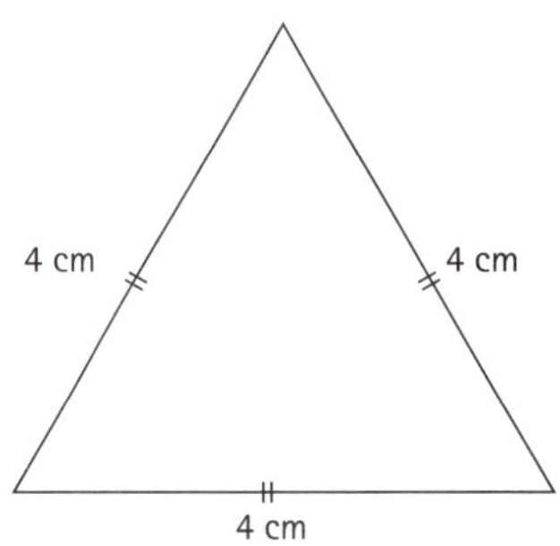

Equilateral triangle

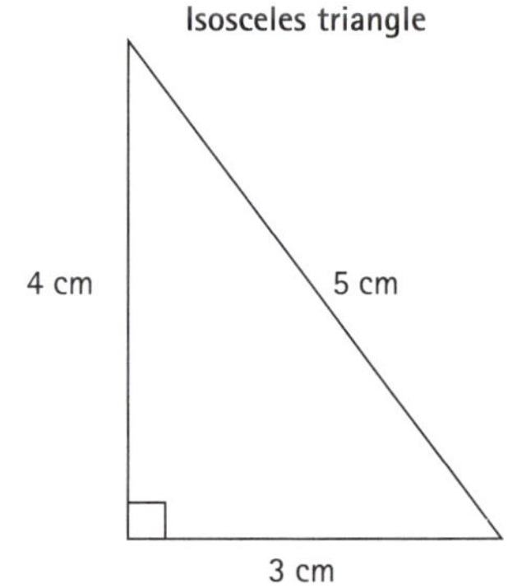

Right-angled triangle

Copy this table into your book. Then complete the table by measuring the necessary angles.

	Scalene	Isosceles	Equilateral	Right-angled
Are angles all equal?				
Are only two angles equal?				
Are all sides equal?				
Are only two sides equal?				
What are the sizes of the angles?				
What is the sum of the angles?				
How many lines of symmetry?				

Use the table to help you answer these questions:

2 Which shape has a right angle?

3 What is the sum of the other two angles in the right-angled triangle?

4 Is it possible to have a right-angled triangle that is isosceles?

5 Is it possible to have a right-angled triangle that is equilateral?

6 Copy and complete this flow chart showing the relationships between the different types of triangles. Place each triangle name in the appropriate box.

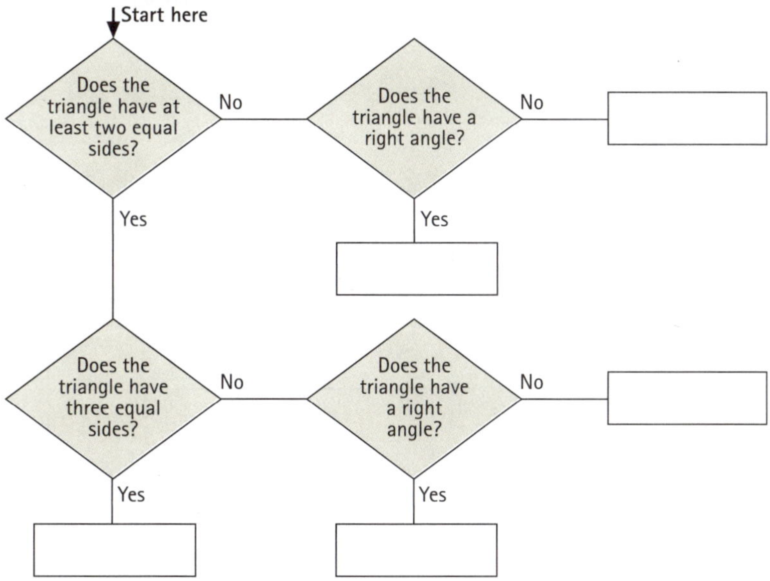

7 Match each definition with the correct name:

Definition	Name
a a triangle with exactly two equal sides	**i** isosceles triangle
b a triangle with exactly three equal sides	**ii** scalene triangle
c a triangle with three sides of different lengths	**iii** right-angled triangle
d a triangle with one right angle	**iv** equilateral triangle

8 As a class, discuss the definitions of triangles. An alternative definition for an equilateral triangle could be: "an equilateral triangle is a triangle with all angles **equal to 60°**". Discuss and write down alternative definitions for the other triangles.

Quadrilaterals

A **quadrilateral** is a plane figure with four straight sides.

PEER ASSESSMENT

See page 290 (rubric 5).

Investigation: Quadrilaterals

Work in groups of four learners.

You will need: a ruler, a protractor.

1 Construct each of these quadrilaterals using any necessary instruments (each learner constructs one or two quadrilaterals):

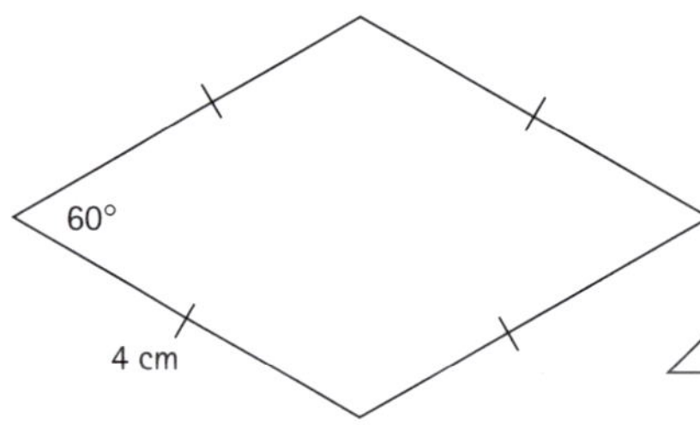

Rhombus

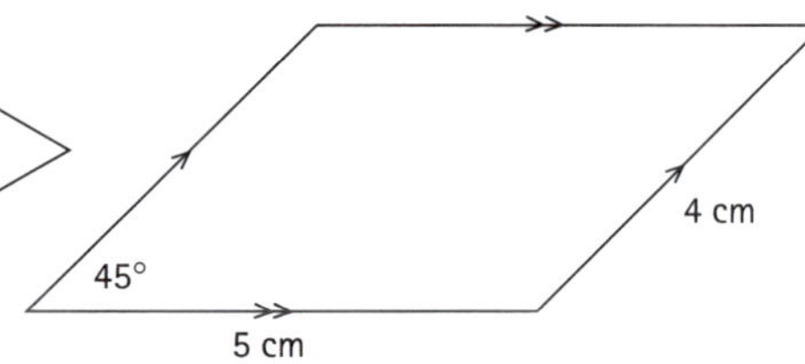

Parallelogram

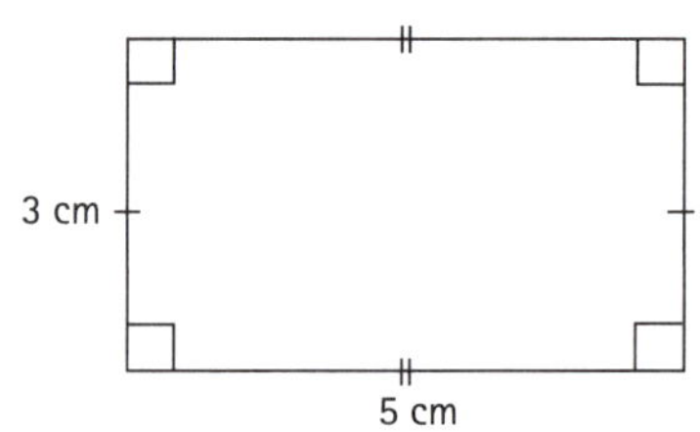

Rectangle

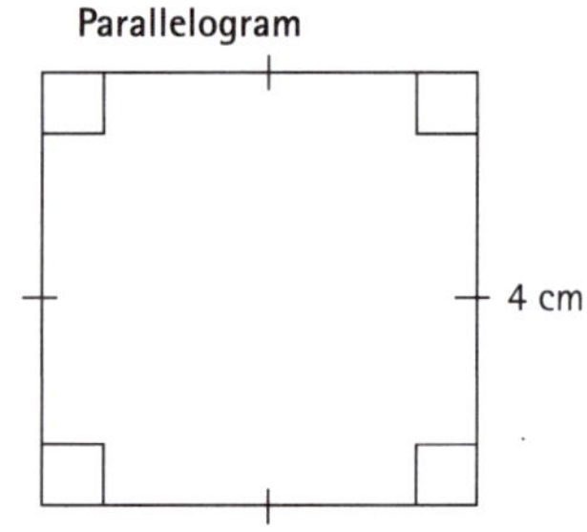

Square

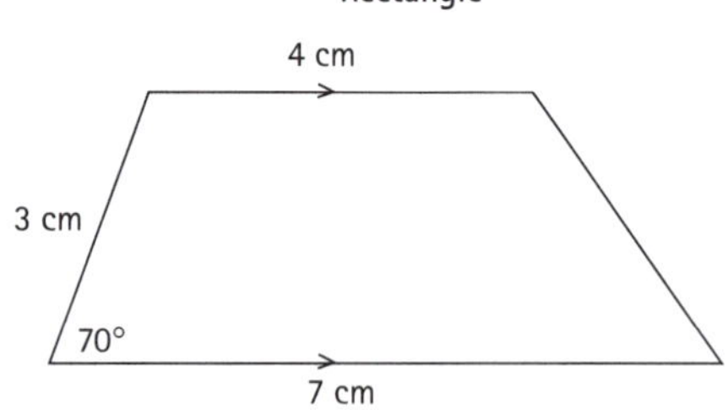

Trapezium

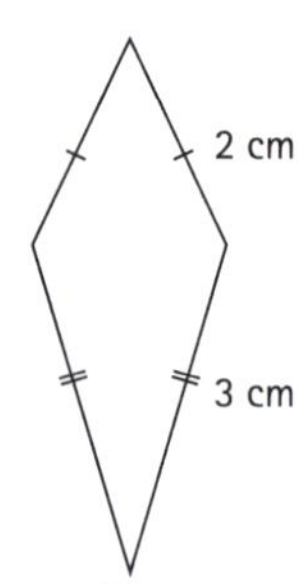

Kite

Copy this table into your workbook and then complete it by answering questions 2 to 5.

	Rhombus	Parallelogram	Rectangle	Square	Trapezium	Kite
Do diagonals bisect angles?						
Do diagonals bisect each other?						
Are diagonals perpendicular?						
Are diagonals equal?						
Are opposite sides equal?						
Are opposite angles equal?						
How many lines of symmetry?						

Note: "Bisect" means to cut exactly in half.

2 Draw in the diagonals of each shape, and use a protractor to determine whether the diagonals bisect the angles they pass through.

3 Use the ruler to determine whether the diagonals of each shape bisect each other.

4 Use a protractor or set-square to determine whether the diagonals cross at right angles.

5 Measure the diagonals, opposite sides and opposite angles to complete the table.

Use the table to help you answer these questions:

6 **a** Which shapes have equal diagonals that also bisect each other?

b What other features do these shapes have in common?

7 The two diagonals of a shape intersect at right angles. What could the shape be? If the two diagonals are not equal, which shape can it not be?

8 Copy and complete the flow chart on page 60, which shows the relationships between different quadrilaterals. Place each quadrilateral name in the appropriate box.

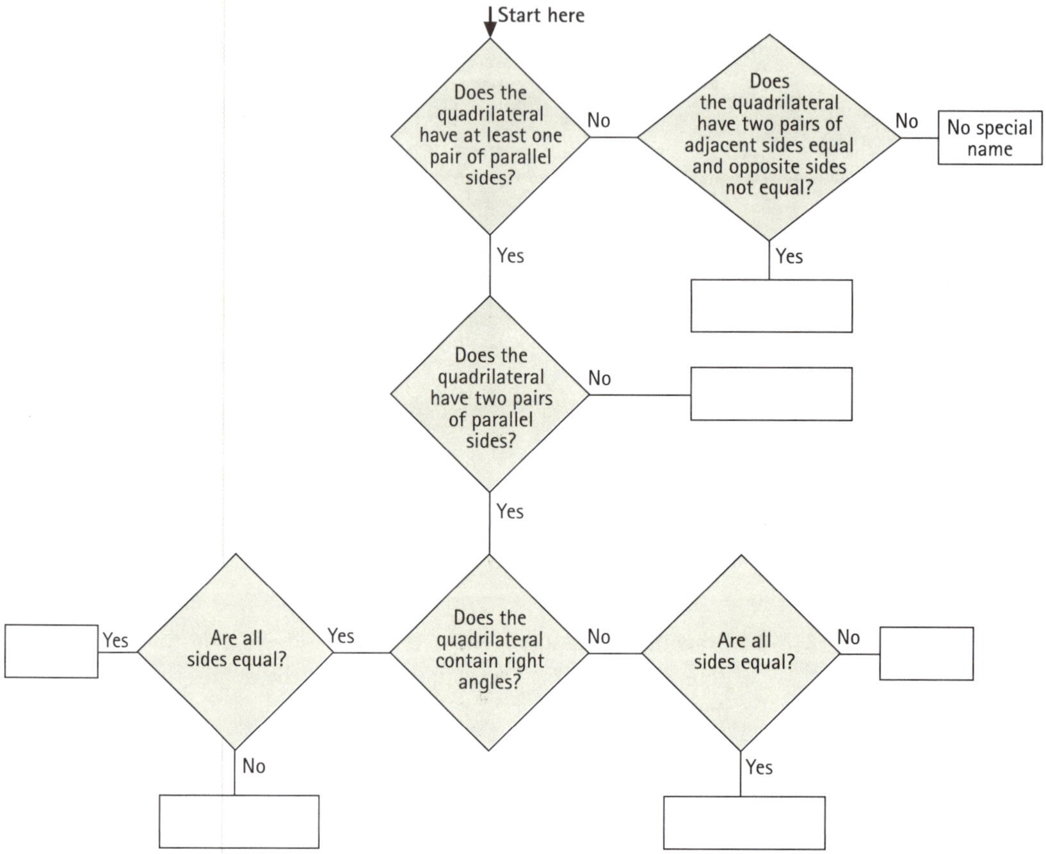

9 Match each definition with the correct name:

	Definition		Name
a	a quadrilateral with both pairs of opposite sides parallel	**i**	trapezium
b	a parallelogram with all sides equal and one interior angle equal to 90° (therefore all interior angles are 90°)	**ii**	rectangle
c	a quadrilateral with exactly one pair of opposite sides parallel	**iii**	square
d	a parallelogram with one interior angle equal to 90° (therefore all interior angles are 90°)	**iv**	kite
e	a parallelogram with one pair of adjacent sides equal in length (therefore all sides are equal)	**v**	parallelogram
f	a quadrilateral with two pairs of adjacent sides equal and opposite sides not equal in length	**vi**	rhombus

10 As a class, discuss the definitions of quadrilaterals. Alternative definitions for a square could be: "a square is a rectangle with a pair of adjacent sides equal in length", or "a square is a rhombus with one angle equal to 90°". Discuss and write down alternative definitions for the other quadrilaterals.

The angle sum of a quadrilateral

If you split a quadrilateral into two triangles as shown, you can calculate the angle sum:

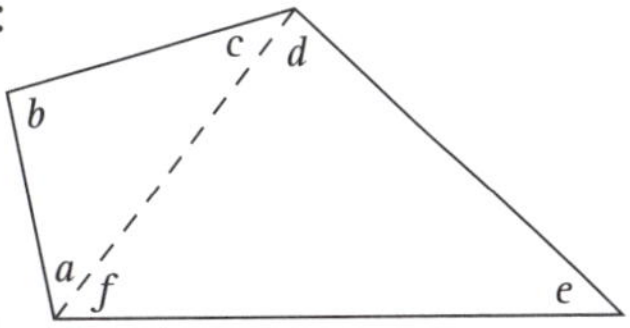

Angle sum = $a + b + c + d + e + f$
but $a + b + c = 180°$
and $d + e + f = 180°$

So the angle sum = 180° + 180° = 360°

∴ Angle sum of a quadrilateral = 360°

EXAMPLES

Find the value of each variable, giving reasons:

1 $x + 120° + 90° + 52° = 360°$
(angle sum of quadrilateral)

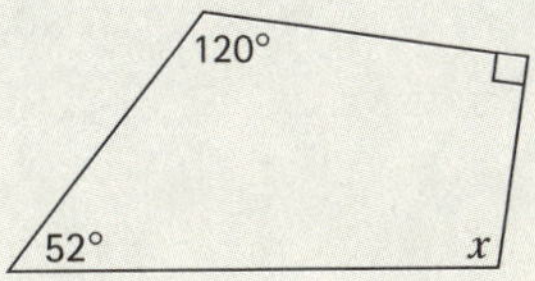

$x + 262° = 360°$
$x = 98°$

2 $3a + 2a + 86° + 109° = 360°$
(angle sum of quadrilateral)

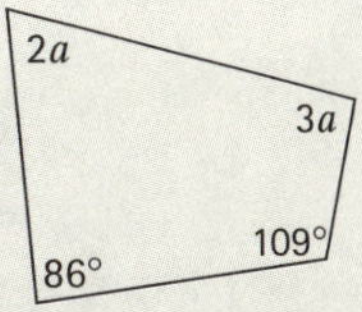

$5a + 195° = 360°$
$5a = 165°$
$a = 33°$

Exercise

Find the value of each variable, giving reasons:

1

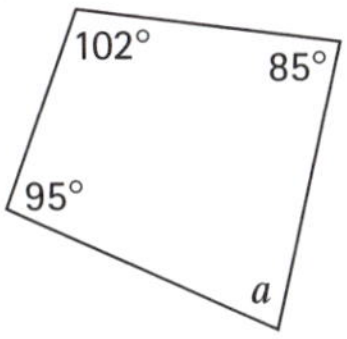

2

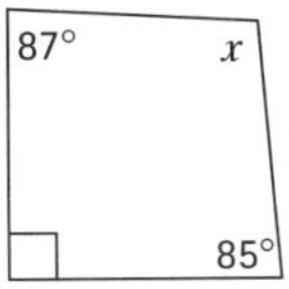

3

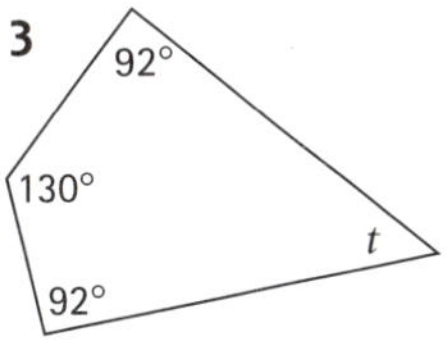

4

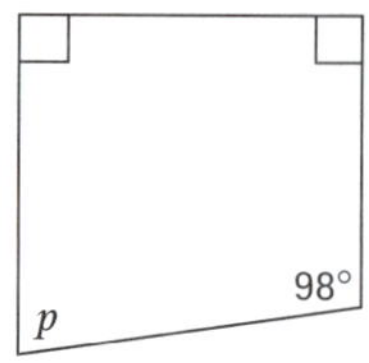

5

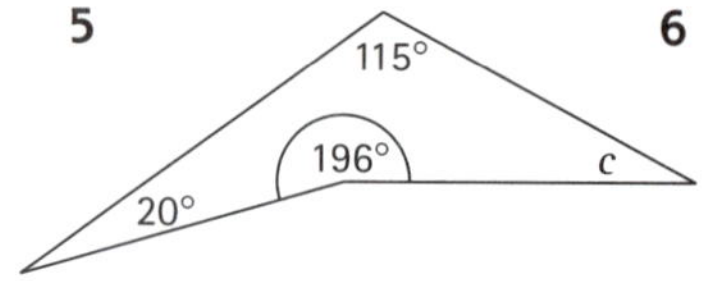

6

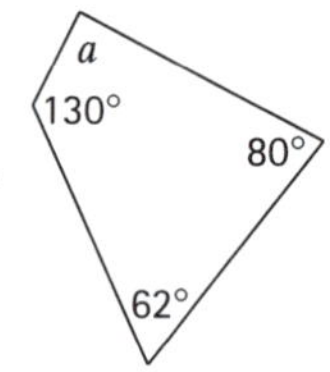

Challenge:

7

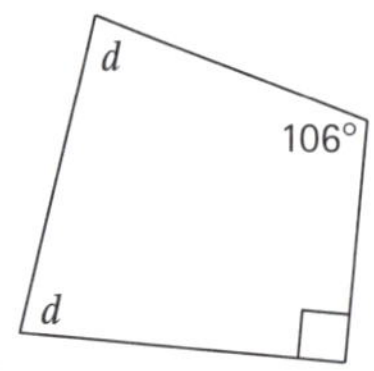

8

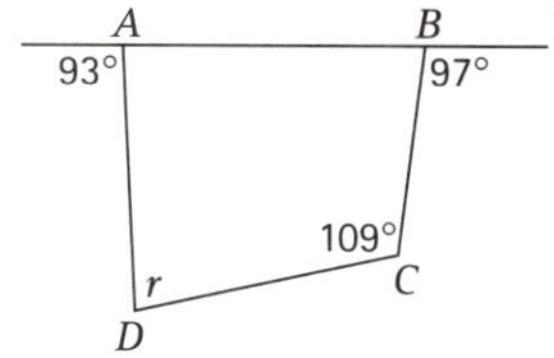

9

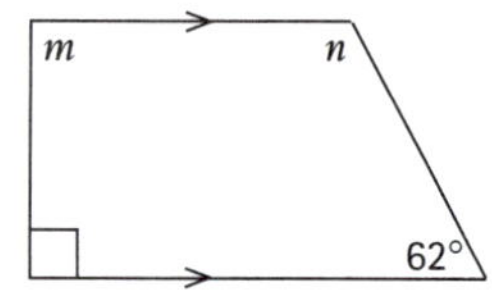

The angle sum of other polygons

PEER ASSESSMENT

See page 290 (rubric 5).

Investigation: The angle sum of polygons

Work with a partner.

To find the angle sum of a quadrilateral, you divided the figure into triangles.

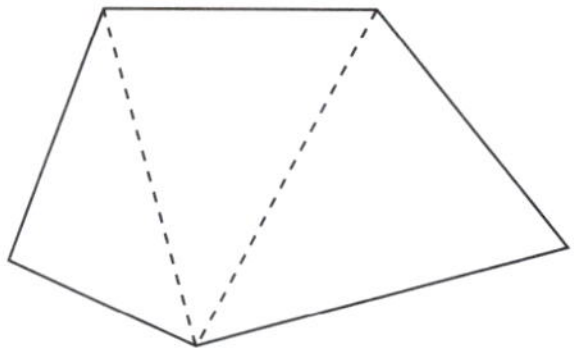

Remember: A vertex is a corner.

Investigate the angle sums of other polygons using the same method. (Make sure that all the triangles meet at one **vertex**.) Record your results in a table:

Polygon	Number of sides	Number of triangles	Angle sum
Triangle	3	1	180°
Quadrilateral	4	2	360°
Pentagon	5		

1 What pattern do you see in the results? Discuss this with your partner.

2 Use the pattern to predict the angle sum for a polygon with:

a 10 sides **b** 25 sides.

3 Try to write an expression for the angle sum of a polygon with n sides.

4 "A **regular** polygon has all sides and all interior angles equal." Use this fact to calculate the size of **each** interior angle of:

   a a regular hexagon

   b a regular octagon

   c a regular polygon with 14 sides.

5 Write an expression for the size of **each** interior angle of a polygon with n sides.

6 Swap your results with another pair and check each other's answers.

Investigation: Computer application

If you have access to geometrical software such as CAD, experiment to see how regular polygons can be drawn.

1 Write instructions using a repeat command. (How many times will you need to repeat?)

2 Work out the angle through which you need to turn between one side and the next. (This will be equal to the **exterior angle** of the polygon.)

3 What is the relationship between the number of sides and the turning angle?

4 How could this method be used to draw a circle? How could the circle be made larger? How could it be made smaller?

The formula for the angle sum of a polygon

The investigation on page 62 should have shown you that:

the angle sum of a polygon with n sides is equal to $(n - 2) \times 180°$.

Use this formula to check your results in the table on page 62 for the investigation.

Exercise

Find the angle sum of each polygon:

1

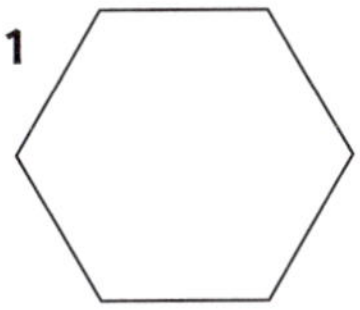

2

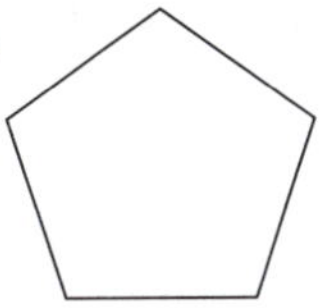

3

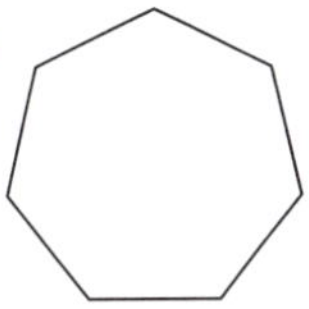

4 a nonagon (9 sides) **5** a dodecagon (12 sides).

Find the value of the variable in each of these regular polygons:

6

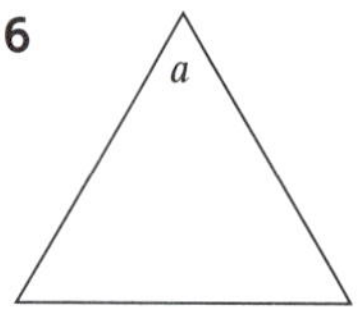

7

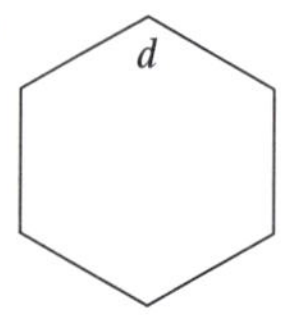

8

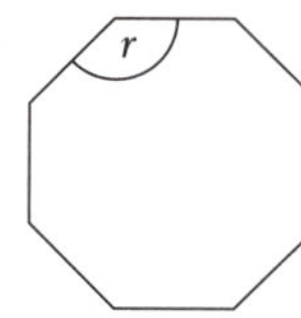

9 What is the size of each **exterior** angle of a regular octagon?

10 What is the **sum** of its exterior angles?

11 Explain why this must always be the case for any regular polygon.

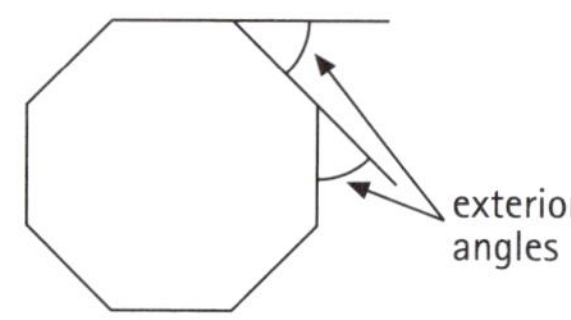

12 What is the size of each exterior angle of a regular polygon with 15 sides?

13 What is the size of each of the interior angles of the same polygon?

14 How many sides would a regular polygon have if the exterior angle was 20°?

Area

Area is the amount of space that a flat shape takes up. The area of a two-dimensional shape can be defined as the number of square units contained within its boundaries. For example, the diagram below is a sketch of a small room. If you want to buy tiles to cover the floor of the room, you need to calculate the area to make sure you have enough tiles to cover the floor space. To calculate the area of a rectangle, multiply length by breadth:

$$\text{Area of a rectangle} = \text{length} \times \text{breadth}$$
$$\text{area of this room} = 2 \text{ m} \times 3 \text{ m}$$
$$= 6 \text{ m}^2$$

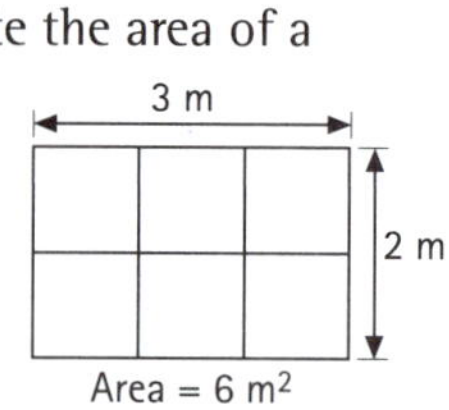

For rectangles and other basic shapes, you will consider the area formulas that you came across in previous years to help you work out the number of square units contained within the shapes.

PEER ASSESSMENT

See page 291 (rubric 7).

Investigation: Why calculating area is important

Work in groups.

Discuss why people in these jobs will find it important to know how to calculate area accurately:

- interior decorator
- painter
- mosaic artist
- gardener
- tailor.

Common area formulas

Square:

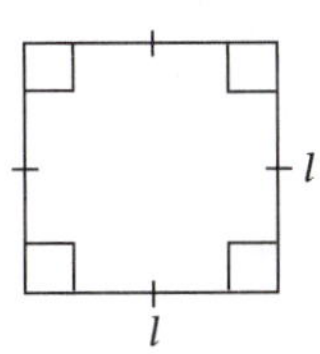

Area = l^2

Rectangle:

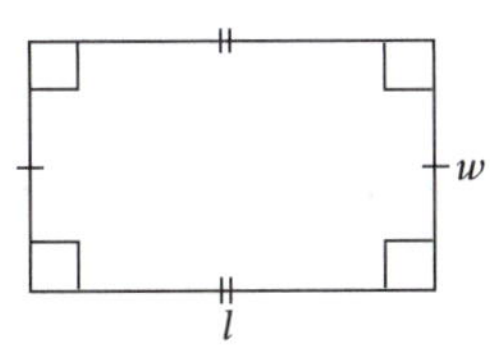

Area = $w \times l = wl$

Triangle:

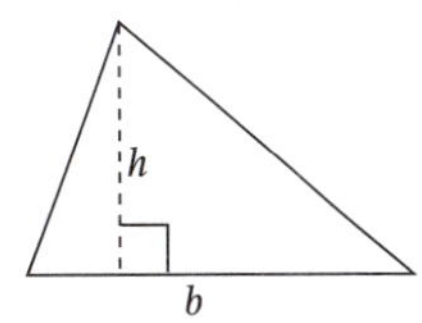

Area = $\frac{1}{2}bh$

Rhombus:

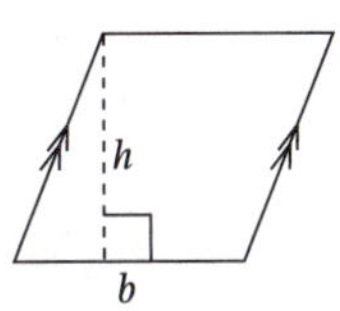

Area = bh

Parallelogram:

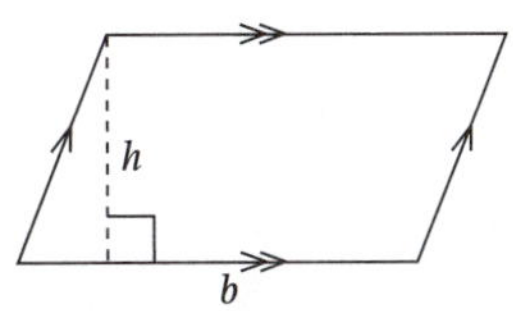

Area = bh

Trapezium:

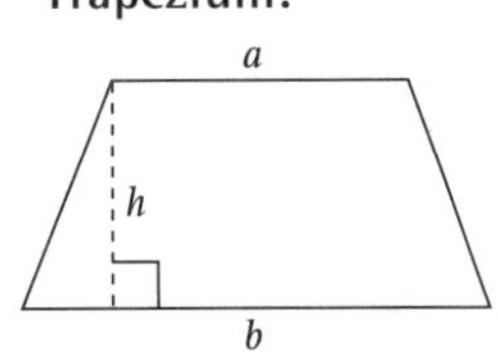

Area = $\frac{1}{2}(a + b)h$

You should also know that the area of a circle is πr^2:

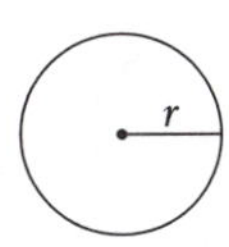

Area = πr^2

EXAMPLES

1 Find the area of these basic shapes to 2 decimal places:

a

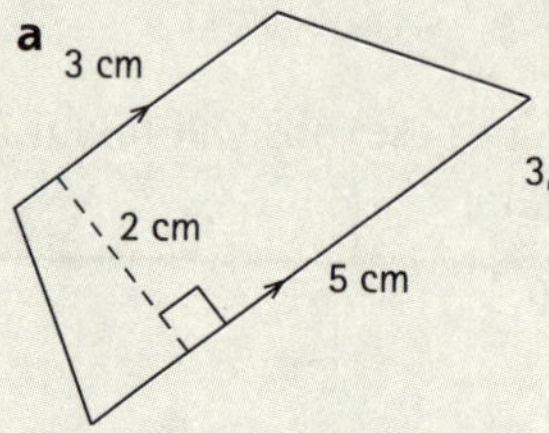

b

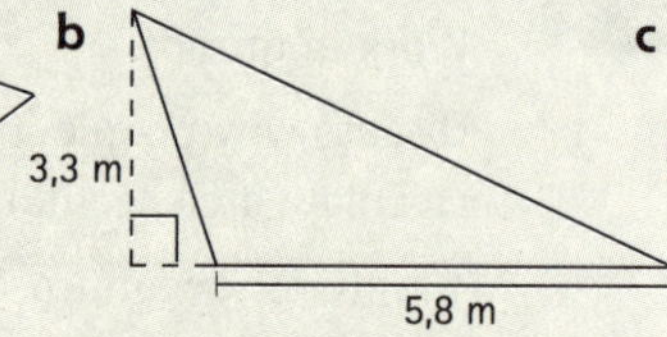

c

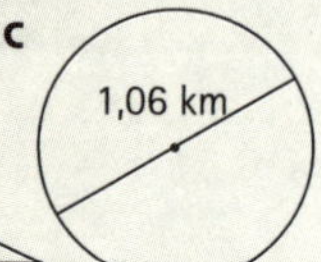

a $A = \frac{1}{2}(a + b)h$
$= \frac{1}{2}(3 + 5)2$
$= 8 \text{ cm}^2$

b $A = \frac{1}{2}bh$
$= \frac{1}{2}(5,8)(3,3)$
$= 9,75 \text{ m}^2$

c $A = \pi r^2$
$= \frac{1}{2}(0,53)^2$
$= 0,88 \text{ km}^2$

2 Find the value of the variable for these basic shapes:

a

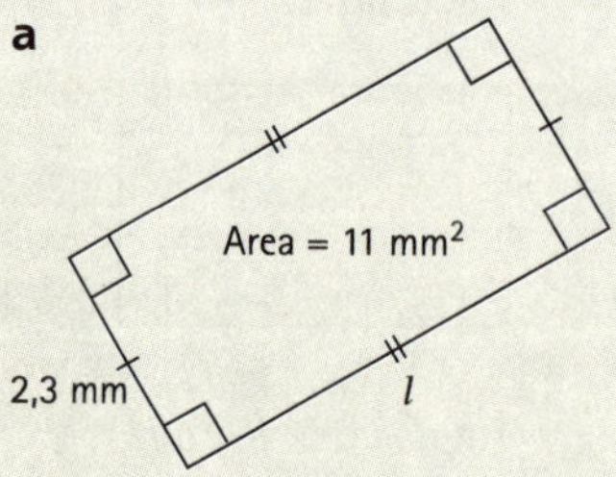

b

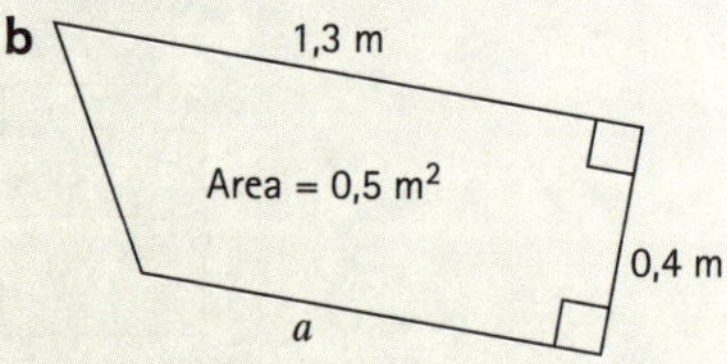

a $A = lw$
$11 = l \times 2,3$
$\therefore l = \frac{11}{2,3}$
$\approx 4,78 \text{ mm}$

b $A = \frac{1}{2}(a + b)h$
$0,5 = \frac{1}{2}(a + 1,3) \times 0,4$
$0,5 = 0,2(a + 1,3)$
$2,5 = a + 1,3$
$\therefore a = 1,2 \text{ m}$

Exercise

1 Find the area of these basic shapes. Round off your answers to 2 decimal places.

a

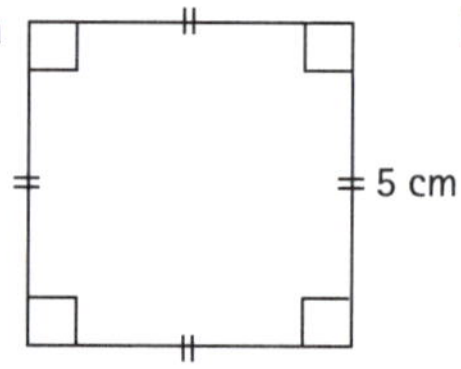

b

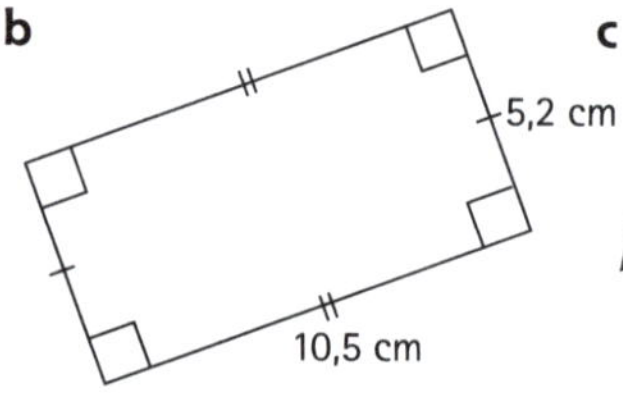

c

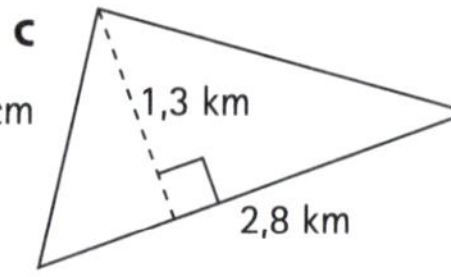

d

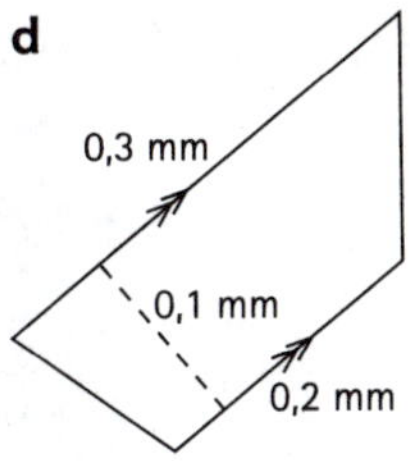

e

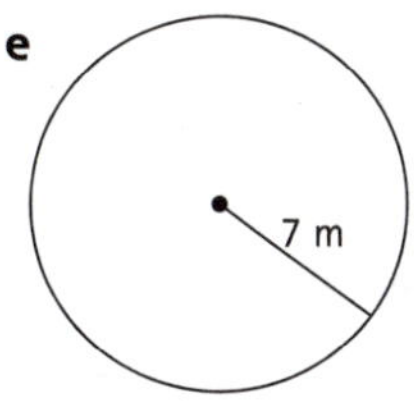

f

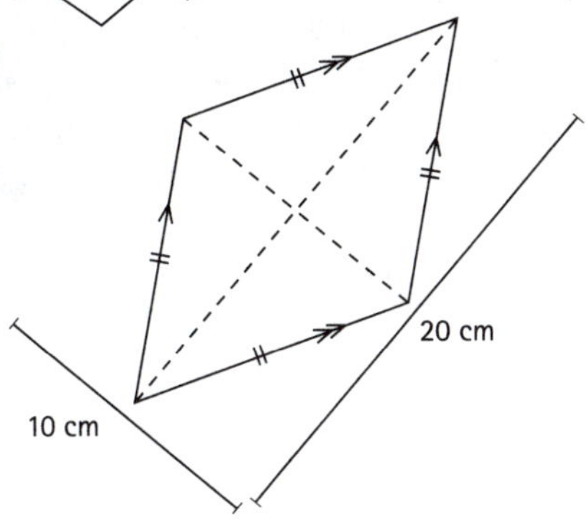

g

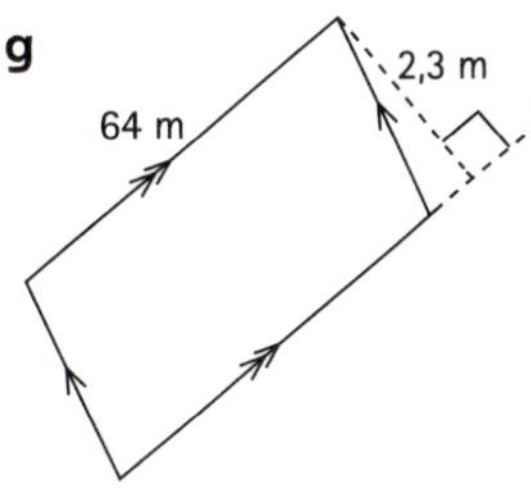

h

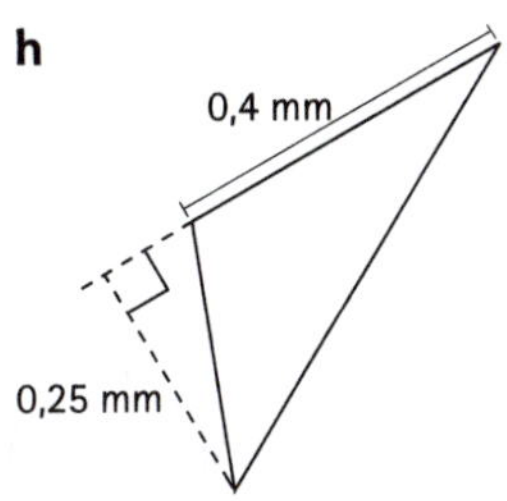

i

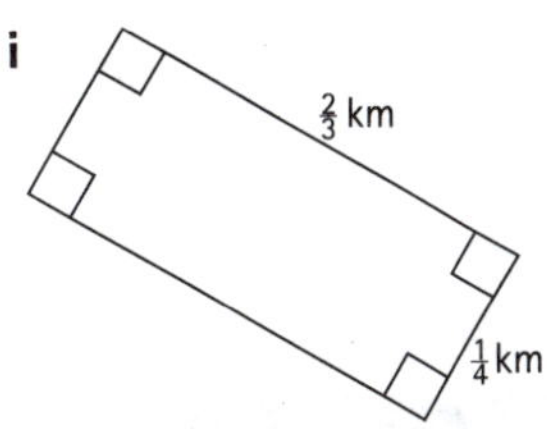

2 Vuyo works for a community development programme that helps people in townships to make the most of their land and to grow sustainable vegetable gardens. Vuyo is asked to work with a piece of land that consists of a semi-circular area with diameter of 6,5 m and a triangular area with length 8,2 m, as shown in the sketch.

a Find the total area of the piece of land.

b If Vuyo needs to lay down 25 kg of soil and 10 kg of compost per square metre of land, work out how much soil and compost he needs altogether for the piece of land.

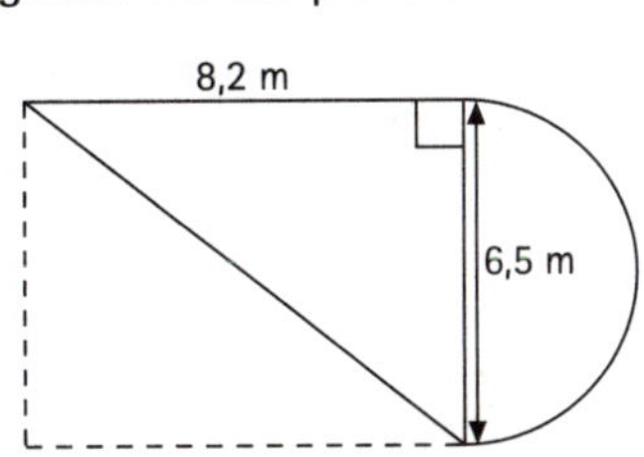

Surface area of prisms and cylinders

The total surface area of a three-dimensional object, e.g. a rainwater tank, can be found by finding the sum of the areas of each of the shapes that make up the surface of the object.

While considering the general case for some special objects, formulas can be developed and remembered for later use.

The **cylinder** is an example of this.

Diagram

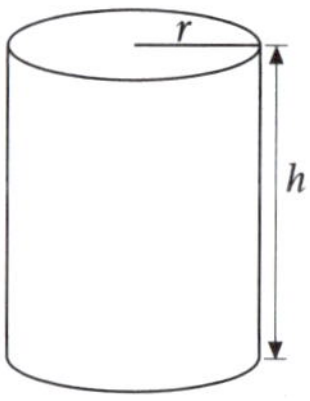

Net

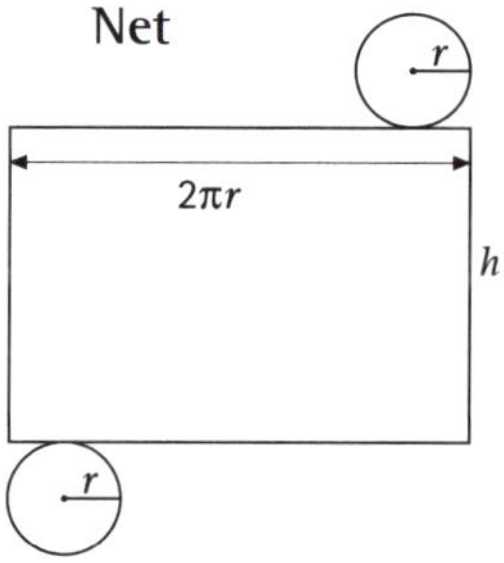

Formula

Total surface area

= 2 circles + 1 rectangle

$= 2\pi r^2 + 2\pi rh$

$= 2\pi r\,(r + h)$

EXAMPLES

Find the total surface area of these objects by first drawing a net of the object. Round off to 2 decimal places each time.

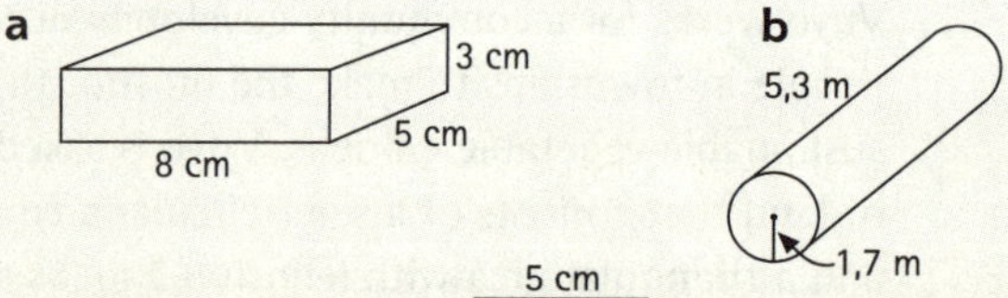

a

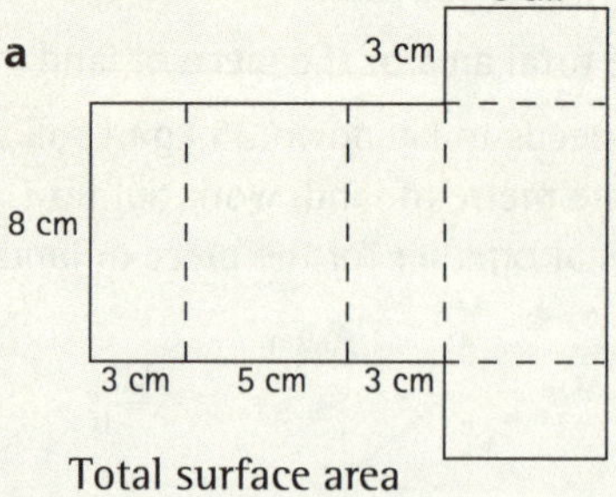

Total surface area

$= 2 \times (8 \times 3) + 2 \times (5 \times 3) + 2 \times (8 \times 5)$

$= 158 \text{ cm}^2$

b

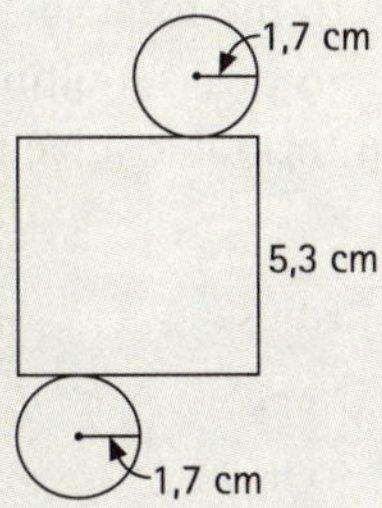

Total surface area
= 2 circles + 1 rectangle
$= 2\pi r^2 + 2\pi r \times h$
$= 2\pi(1,7)^2 + 2\pi(1,7) \times 5,3$
$\approx 74,77\ m^2$

PEER ASSESSMENT

See page 290 (rubric 5).

Investigation: Predicting surface area

Work in groups.

1 Predict what will happen to the total surface area of a rectangular prism if one of its dimensions is doubled (multiplied by 2). Check your prediction by completing this table:

		Length	Breadth	Height	Area
a	Original prism	3 cm	2 cm	5 cm	
b	Only length doubled	6 cm	2 cm	5 cm	
c	Only breadth doubled	3 cm	4 cm	5 cm	
d	Only height doubled	3 cm	2 cm	10 cm	

2 Predict what will happen to the total surface area of a rectangular prism if one of its dimensions is multiplied by 5. Check your prediction by completing this table:

		Length	Breadth	Height	Area
a	Original prism	4 cm	3 cm	5 cm	
b	Only length times 5	20 cm	3 cm	5 cm	
c	Only breadth times 5	4 cm	15 cm	5 cm	
d	Only height times 5	4 cm	3 cm	25 cm	

3 Generalise the results of questions 1 and 2 by completing this table:

		Length	Breadth	Height	Area
a	Original prism	l	b	h	
b	Only length times k	kl	b	h	
c	Only breadth times k	l	kb	h	
d	Only height times k	l	b	kh	

4 Predict what will happen to the total surface area of a closed cylinder if one of its dimensions is multiplied by 3. Check your prediction by completing this table (give the answers as multiples of π):

		Radius	Height	Area
a	Original cylinder	2 cm	5 cm	
b	Only radius times 3	6 cm	5 cm	
c	Only height times 3	2 cm	15 cm	

5 Predict what will happen to the total surface area of a closed cylinder if one of its dimensions is multiplied by k. Check your prediction by completing this table (give the answers as multiples of π):

		Radius	Height	Area
a	Original cylinder	r	h	
b	Only radius times k	kr	h	
c	Only height times k	r	kh	

6 Discuss your findings in your group. Write a short summary of your findings and present it to the class.

Exercise

1 Find the total surface area of these objects by first drawing a net. Round off your answers to 2 decimal places.

a

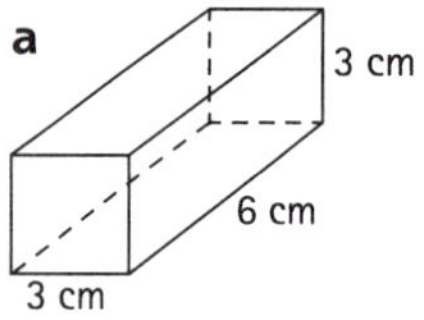

b

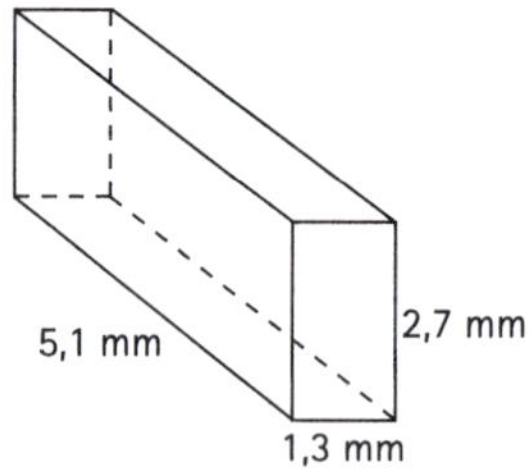

c

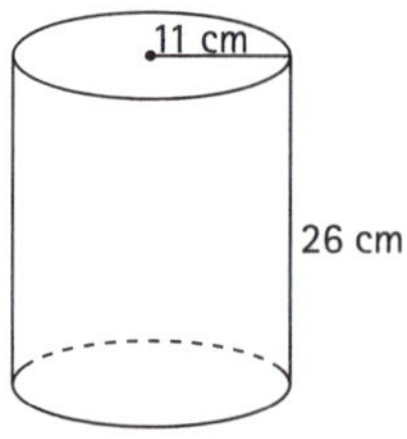

d

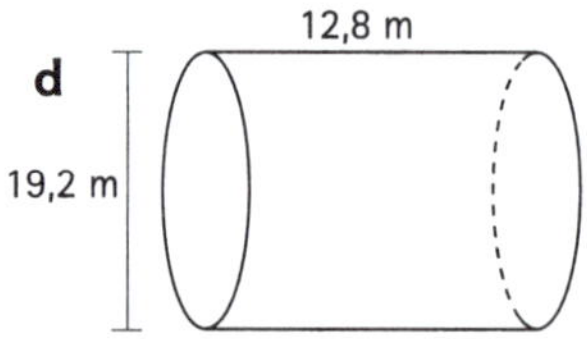

2 Find the total surface area of these objects by first drawing a net. Round off your answers to 2 decimal places.

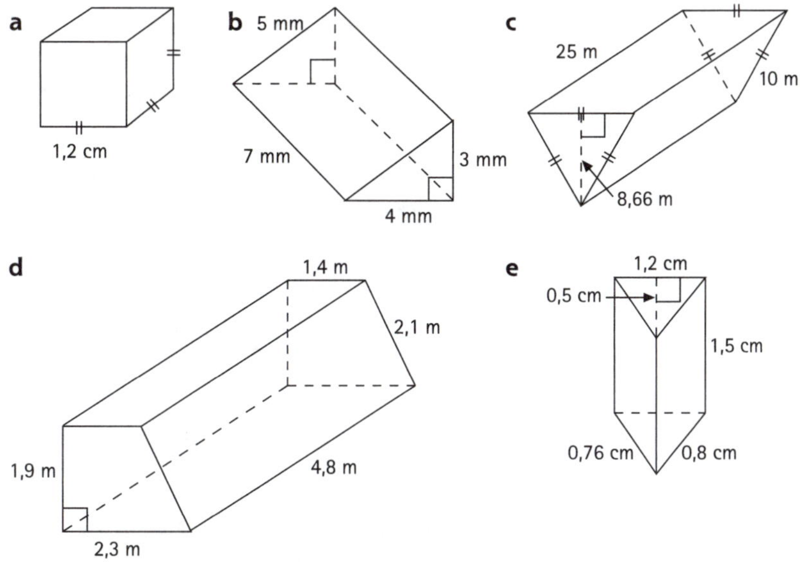

3 Find the total surface area of the outside of a pipe of radius 85 cm and length 4,5 m.

4 What is the minimum amount of paper required to wrap a box with dimensions 25 cm wide, 32 cm long and 20 cm high?

Volume of prisms and cylinders

The amount of space contained within the outside surfaces of a three-dimensional object is called the volume and is measured in cubic units.

Prisms

The number of cubic units at the end of a prism has the same numerical value as the area of the end of the prism, as shown here.

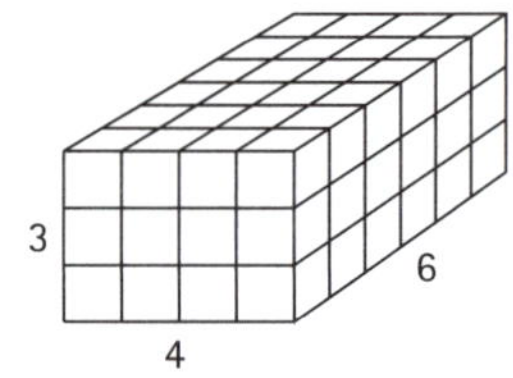

Number of cubic units (end) $= 4 \times 3$
$= 12$

Area (end) $= 4 \times 3$
$= 12 \text{ units}^2$

$\therefore$ Volume = area (end) $\times 6$
$= 12 \times 6$
$= 72 \text{ units}^3$

More generally, the volume of a prism is equal to the area of the constant cross-section (the area or A) multiplied by the height.

Consider this general case shown in the diagram on the right, if area $(A) = 5 \text{ cm}^2$:

Volume $= A \times h$
$= 5 \times 2$
$= 10 \text{ cm}^3$

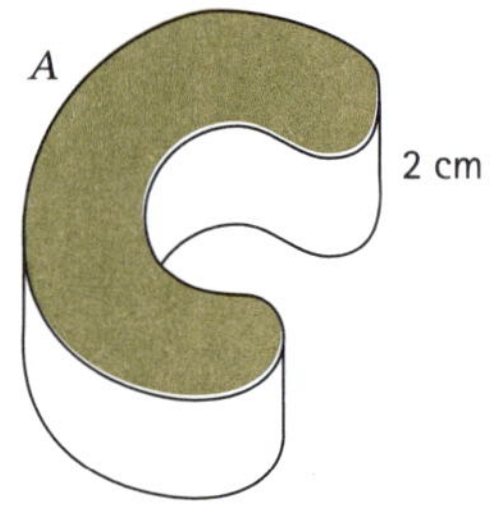

EXAMPLE

Find the volume of this cylindrical object.

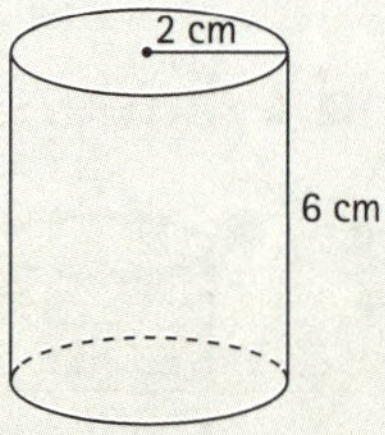

$$
\begin{aligned}
V &= Ah \\
&= \pi r^2 h \\
&= \pi (2)^2 \times 6 \\
&\approx 75{,}40 \text{ cm}^3
\end{aligned}
$$

PEER ASSESSMENT

See page 290 (rubric 5).

Investigation: Predicting volume

Work in groups.

1 Predict what will happen to the volume of a rectangular prism if one of its dimensions is multiplied by 5. Check your prediction by completing this table:

		Length	Breadth	Height	Volume
a	Original prism	4 cm	3 cm	5 cm	
b	Only length times 5	20 cm	3 cm	5 cm	
c	Only breadth times 5	4 cm	15 cm	5 cm	
d	Only height times 5	4 cm	3 cm	25 cm	

2 Generalise the results of question 1 by completing this table:

		Length	Breadth	Height	Volume
a	Original prism	l	b	h	
b	Only length times k	kl	b	h	
c	Only breadth times k	l	kb	h	
d	Only height times k	l	b	kh	

3 Predict what will happen to the volume of a closed cylinder if one of its dimensions is multiplied by 3. Check your prediction by completing this table (give your answers as multiples of π):

		Radius	Height	Volume
a	Original cylinder	2 cm	5 cm	
b	Only radius times 3	6 cm	5 cm	
c	Only height times 3	2 cm	15 cm	

4 Predict what will happen to the volume of a closed cylinder if one of its dimensions is multiplied by k. Check your prediction by completing this table (give your answers as multiples of π):

		Radius	Height	Volume
a	Original cylinder	r	h	
b	Only radius times k	kr	h	
c	Only height times k	r	kh	

5 Discuss your findings in your group. Write a short summary of your findings and present it to the class.

Exercise

Find the volume of these prisms. Round off your answers to 2 decimal places.

1

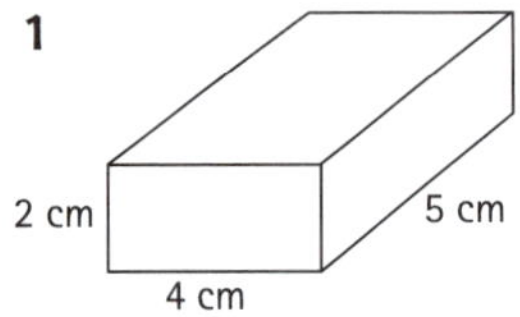

2

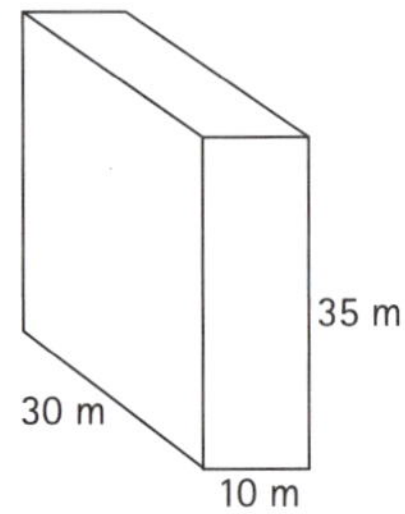

3

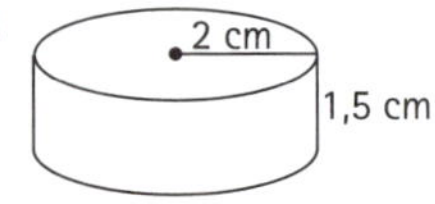

4

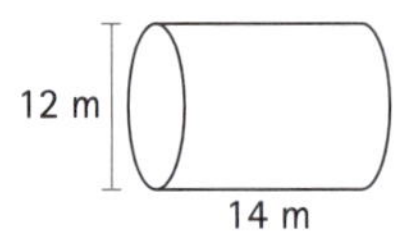

5

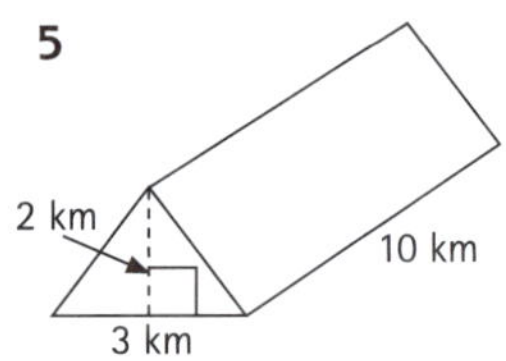

6

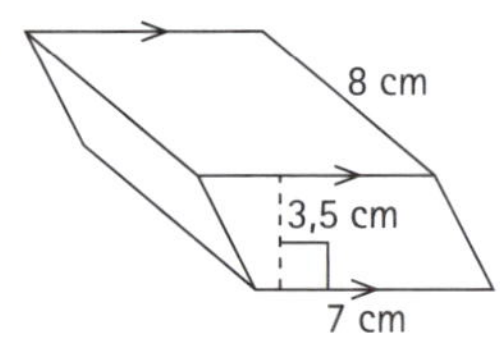

7

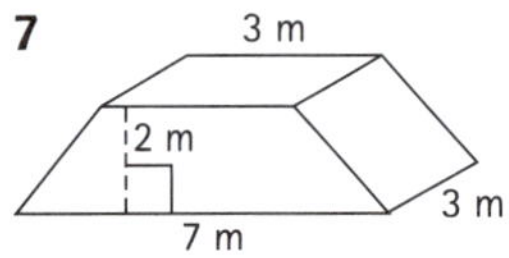

8

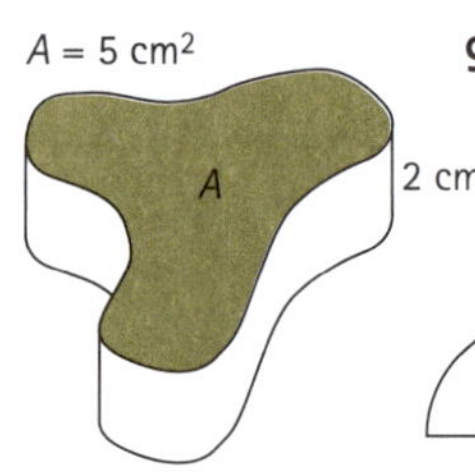

9

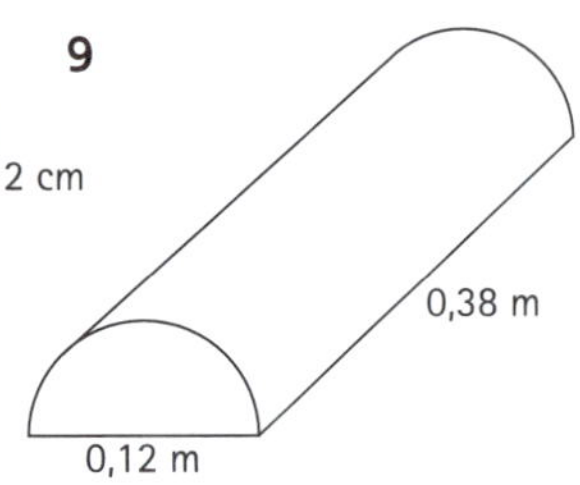

10

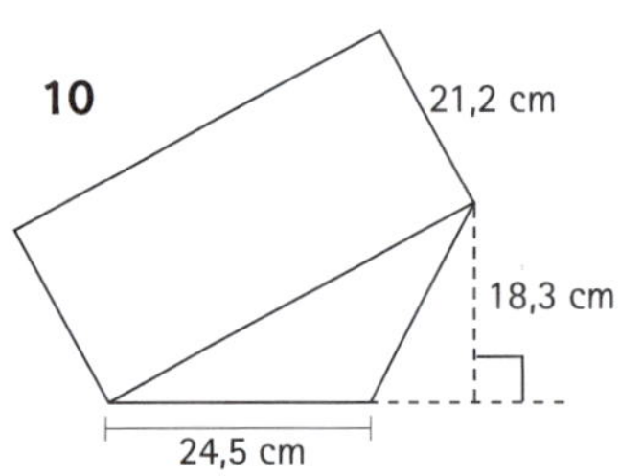

SELF-ASSESSMENT

See page 286 (rubric 1).

CHAPTER 2 COORDINATE GEOMETRY

KEY VOCABULARY

coordinate • Cartesian plane • y-axis • x-axis • Pythagoras' theorem • hypotenuse • line segment • mid-point • loci • gradient • rise • run • slope • horizontal line • vertical line

In this chapter, you will work with coordinates. A set of coordinates is a pair of numbers or letters, or a combination of numbers and letters, that tells you the position of a point on a grid. In mathematics, the first coordinate in a pair is always the x-coordinate (the position along the horizontal axis) and the second coordinate is the y-coordinate (the position along the vertical axis). You use coordinates in social sciences (geography) when you work with maps. When you work on a computer in a spreadsheet programme, the cells of the spreadsheet are named using a coordinate system consisting of letters for each column and numbers for each row (e.g. A1, B7, C9).

Exercise

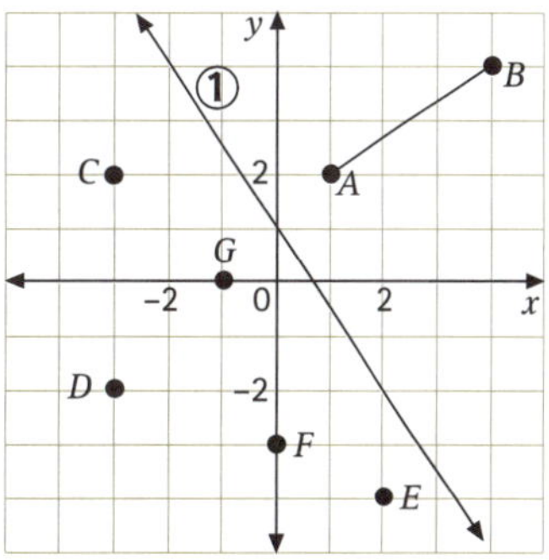

Give the coordinates (x; y) of the points:

1 A **2** E **3** C

4 D **5** F **6** G

What is the distance from:

7 C to A? **8** C to D?

9 F to the origin?

10 Find the gradient of AB.

11 For the line ①, what is the:

a gradient? **b** y-intercept? **c** equation?

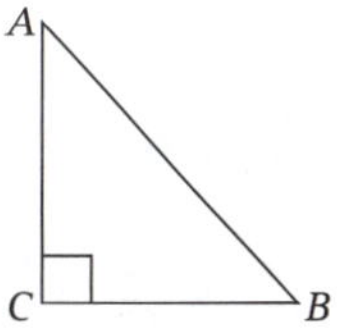

For ΔABC on the right:

12 Name the hypotenuse.

13 Complete the statement according to Pythagoras' theorem:

a $AB^2 = \ldots$ **b** $CB = \sqrt{\ldots}$

14 Find $AC^2 + BC^2$ if $AC = 16$ and $BC = 12$. Hence find AB.

Calculate:

15 $\frac{3 + 5}{2}$ **16** $\frac{-3 + 9}{2}$

17 the average of −6 and 2.

18 What number is halfway between −8 and 4?

19 Given that $x_M = \frac{x_1 + x_2}{2}$, evaluate x_M when $x_1 = 5$ and $x_2 = 9$.

20 Given that $y_M = \frac{y_1 + y_2}{2}$, evaluate y_M when $y_1 = -6$ and $y_2 = -3$.

The distance between two points

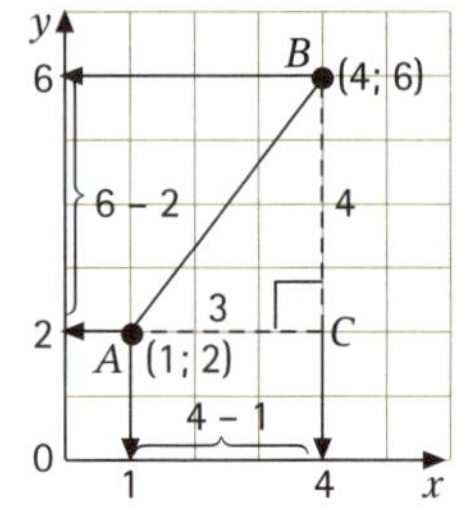

To find the distance from A to B, you will use Pythagoras' rule.

By drawing in the lines BC and AC, you create a right-angled triangle with **hypotenuse** AB, the required distance.

$BC = 6 - 2$ units $= 4$ units

That is, $BC = y_B - y_A$ (right-hand y-coordinate minus left-hand y-coordinate)

$AC = 4 - 1$ units $= 3$ units

That is, $AC = x_B - x_A$ (right-hand x-coordinate minus left-hand x-coordinate)

$$AB^2 = AC^2 + BC^2 \quad \text{(Pythagoras)}$$
$$AB = \sqrt{AC^2 + BC^2}$$
$$AB = \sqrt{(x_B - x_A)^2 + (y_B - y_A)^2}$$
$$= \sqrt{3^2 + 4^2}$$
$$= \sqrt{25}$$
$$= 5 \text{ units}$$

GROUP ASSESSMENT

See page 291 (rubric 7).

Investigation: The distance between two points

Work in groups.

Investigate why it is important for people in these jobs to measure the distance between two points:

- air traffic controllers
- anti-poaching officials
- architects
- tree surgeons.

If you are unsure what any of these jobs involve, do some research to find out. Present your findings to the class.

Using a formula

The distance d between two points $(x_1; y_1)$ and $(x_2; y_2)$ is found using the formula:

$$d = \sqrt{(x_2 - x_1)^2 + (y_2 - y_1)^2}$$

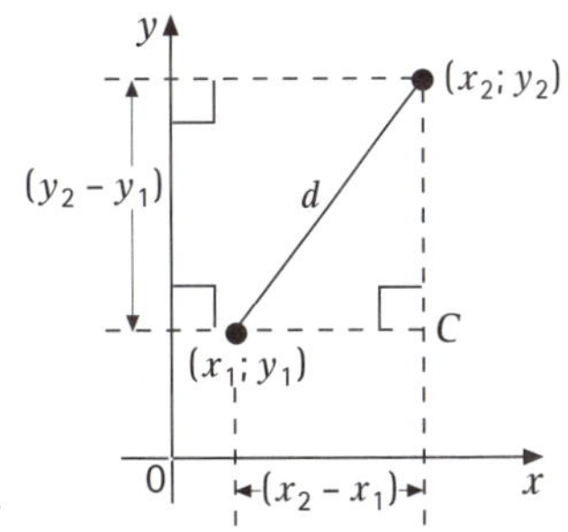

Find the difference between the x-coordinates (right to left), then square it. Similarly, find the difference between the y-coordinates and square it. Finally, calculate the square root of the sum.

EXAMPLES

1 Find the distance between the points (2; 3) and (7; 15).

Make a sketch.

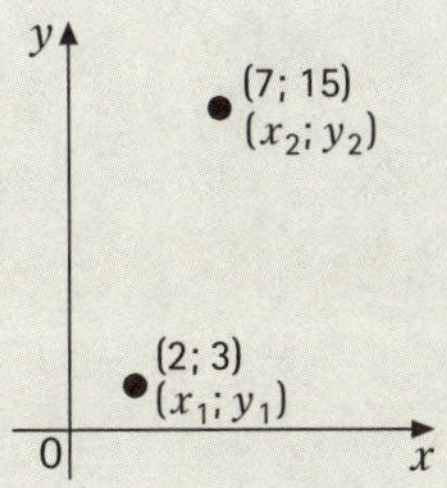

$d = \sqrt{(x_2 - x_1)^2 + (y_2 - y_1)^2}$	Write the formula.
$= \sqrt{(7 - 2)^2 + (15 - 3)^2}$	Substitute with $(x_2; y_2)$ as right-hand coordinates (7; 15).
$= \sqrt{5^2 + 12^2}$	
$= \sqrt{25 + 144}$	For $(x_1; y_1)$, use (2; 3).
$= \sqrt{169}$	5 x^2 + 1 2 x^2 = √ 1 3
$= 13$	25 144 169

2 How far is it from (–3; 0) to (6; –5), to 1 decimal place?

Sketch the points.

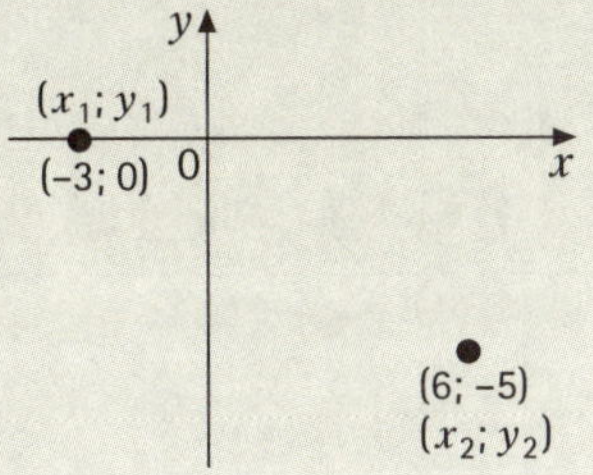

$d = \sqrt{(x_2 - x_1)^2 + (y_2 - y_1)^2}$ Write the formula.

$= \sqrt{(6 - [-3])^2 + (-5 - 0)^2}$ Substitute.

$= \sqrt{9^2 + (-5)^2}$ Calculate. (Remember to use the

$= \sqrt{81 + 25}$ ± key to generate –5.)

$= \sqrt{106}$

$= 10{,}29563$ Round off to 1 decimal place.

$= 10{,}3$ (to 1 decimal place) (If no guidance is given, leave answers in $\sqrt{\ }$ form.)

3 Show that the points (–2; 0), (1; –4) and (2; 3) form the vertices of an isosceles triangle.

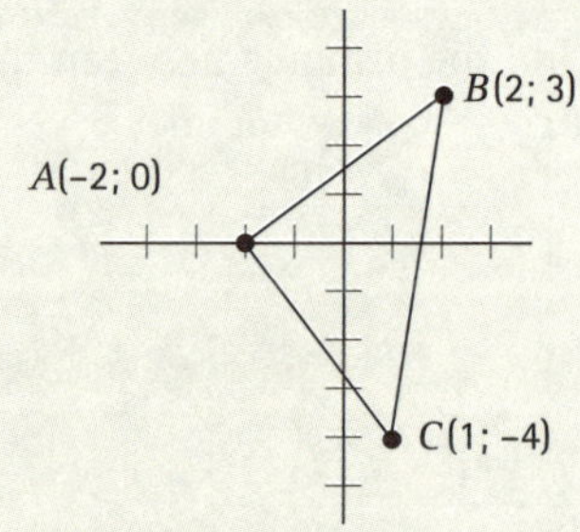

Make a sketch, using rough scaling to locate points.
Name each point for easy reference.
Isosceles triangles have **two equal sides.**
The sketch suggests that they are *AB* and *AC*.
Calculate distances *AB* and *AC*.
If they are equal, the triangle is isosceles.

$d_{AB} = \sqrt{(x_2 - x_1)^2 + (y_2 - y_1)^2}$

$= \sqrt{(2 - [-2])^2 + (3 - 0)^2}$

$= \sqrt{4^2 + 3^2}$

$= 5$

Similarly, $d_{AC} = 5$.

$\therefore \Delta ABC$ is isosceles.

Note: Is it necessary to find d_{BC}? Discuss this in your groups.

Exercise

Use Pythagoras' theorem to find the length of each line segment in this number plane. (Leave answers in square root form.)

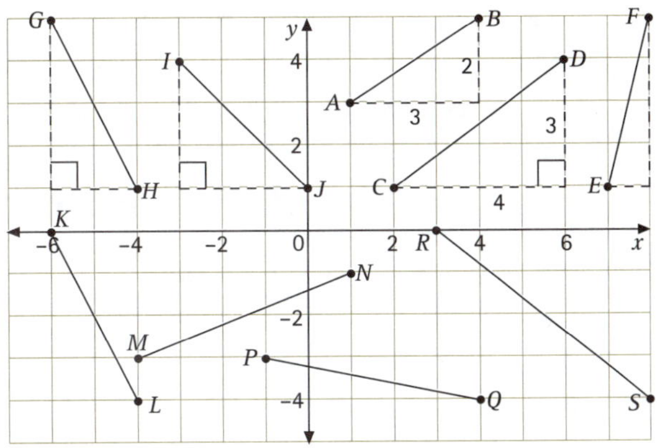

1 *AB*	**2** *CD*	**3** *EF*	**4** *GH*	**5** *IJ*
6 *KL*	**7** *MN*	**8** *PQ*	**9** *RS*	**10** *GC*
11 *HA*	**12** *LE*	**13** *QF*	**14** *LJ*	**15** *IQ*

Use the formula $d = \sqrt{(x_2 - x_1)^2 + (y_2 - y_1)^2}$ to find the distance between each pair of points.

16 (2; 3) and (5; 7) **17** (0; 1) and (6; 9) **18** (0; 0) and (5; 12)

19 (−4; 8) and (2; 0) **20** (−3; 0) and (0; 4) **21** (0; −1) and (2; −4)

22 (2; 5) and (1; −3) **23** (4; 9) and (−1; −3) **24** (5; −3) and (−2; −7)

25 (2; −3) and (−2; 3) **26** (0; 0) and $(-2\frac{1}{2}; 6)$

27 (−4,5; 4,5) and (−3,5; 3,5)

28 Calculate the distance from the origin to (−4; −7).

29 Which of the points (3; 5) and (−4; −4) is closer to the origin?

30 Show that the points (3; 6) and (−1; 4) are equidistant from (2; 3).

31 A triangle has vertices (−4; 0), (0; 3) and (0; −3). Show that it must be isosceles.

32 A quadrilateral has vertices (−1, −1), (3, 1), (1, −3) and (−3; −5).

- **a** Find the lengths of its sides. What sort of quadrilateral is it?
- **b** Find the lengths of its diagonals.

33 Given that *A* is (−2; 2), *B* is (3; 2), *C* is (1; −2) and *D* is (−4; −2), show that the opposite sides of parallelogram *ABCD* are equal.

34 Prove that a circle can be drawn with centre (–2; 1) and passing through the points (–5; 3), (–4; –2) and (1; 3).

35 **Challenge:** $(x; y)$ are the coordinates of **any** point on the number plane 4 units from the origin.

a Form an algebraic equation using the distance formula.

b Simplify the equation by eliminating square root signs.

c What shape is represented by the equation?

The mid-point of a line segment

PEER ASSESSMENT

See page 290 (rubric 6).

Investigation: Develop a formula

Work with a partner.

The mid-point of a line segment is its **halfway** point. Is there a pattern for calculating the coordinates of the mid-point of **any** line segment, so that you don't need to draw diagrams each time?

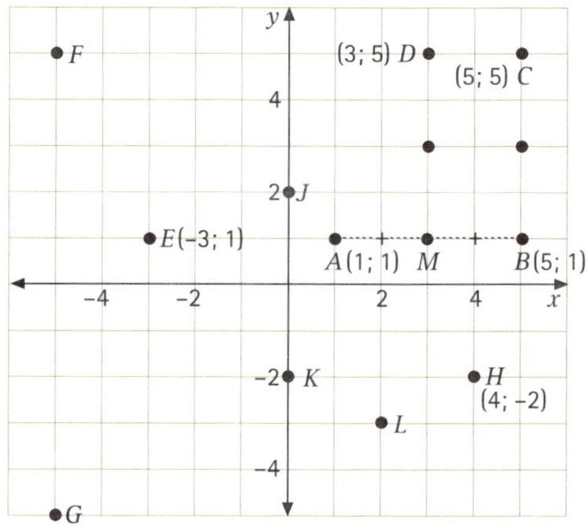

1 From the diagram, find the coordinates of the mid-point of these line segments:

a *AB* **b** *BC* **c** *AC* **d** *AD* **e** *AE*

f *FE* **g** *EG* **h** *FG* **i** *KJ* **j** *EK*.

2 For each line segment in question 1, list the coordinates of the end-points, then those of the mid-point. Look for a number pattern in the x-coordinates, then in the y-coordinates. Remember that you are investigating **halfway** points.

3 Use your discovery to **predict** the coordinates of the mid-point of:

a *DC* **b** *JK* **c** *JD*.

4 From the diagram, find the location of the other end-point of the line segment *H* ... for which:

a *L* is the mid-point **b** *M* is the mid-point.

5 Develop an algebraic formula for x_M, the x-coordinate of any mid-point, and another for y_M, the y-coordinate of any mid-point. Write instructions on how to use your formula.

The mid-point formula

If M is the **mid-point** of AB, it is **halfway** between A and B.

The x-coordinate of M (x_M) is halfway between 1 and 5.
That is, it is the **average** of 1 and 5:

$$\frac{1 + 5}{2} = 3$$

So: $x_M = \frac{x_A + x_B}{2}$

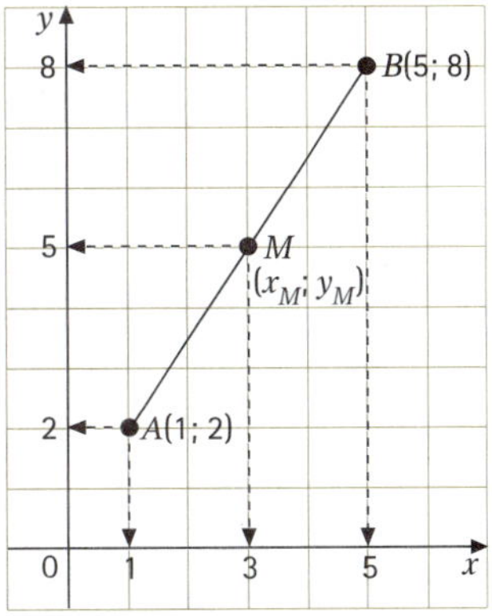

Similarly, the y-coordinate of M (y_M) is halfway between 2 and 8.
That is, it is the **average** of 2 and 8:

$$\frac{2 + 8}{2} = 5$$

So: $y_M = \frac{y_A + y_B}{2}$

Therefore the coordinates of the mid-point of a line segment are:

$$x_M = \frac{x_A + x_B}{2} \quad \text{and} \quad y_M = \frac{y_A + y_B}{2}$$

EXAMPLES

1 Find the mid-point of the line segment joining $A(2; 3)$ to $B(5; 7)$.

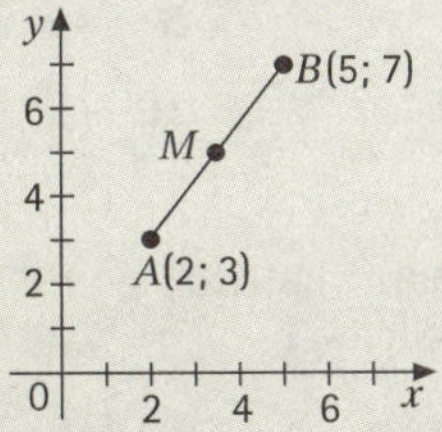

Sketch and label a diagram.

$x_M = \frac{x_1 + x_2}{2}$	$y_M = \frac{y_1 + y_2}{2}$	State the formula.
$= \frac{2 + 5}{2}$	$= \frac{3 + 7}{2}$	Substitute.
$= 3{,}5$	$= 5$	Simplify.
$M = (3{,}5; 5)$		Give the coordinates as an ordered pair. Check against the diagram for reasonableness of values.

2 (1; –2,5) is the mid-point of the line segment from (t; –4) to (4; –1). Find t.

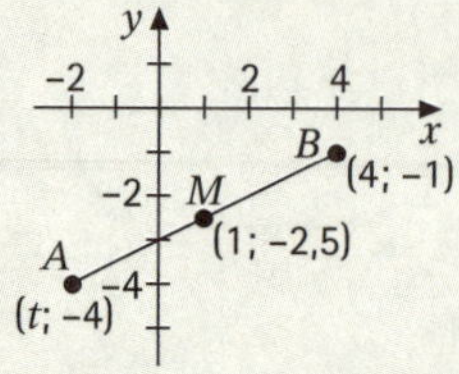

Sketch and label a diagram.

The x-coordinate of A is required.

$x_M = \frac{x_1 + x_2}{2}$ State the x-formula.

$1 = \frac{t + 4}{2}$ Substitute and solve.

$2 = t + 4$

$-2 = t$

Exercise

Find the mid-point of the line segment joining the points:

1 (1; 4) and (3; 6)

2 (3; 7) and (5; 9)

3 (0; 4) and (6; 6)

4 (2; 4) and (3; 5)

5 (7; 2) and (5; 3)

6 (1; 6) and (4; 2)

7 (0; 0) and (–2; –4)

8 (–2; –3) and (–4; –5)

9 (–3; –1) and (–5; –5)

10 (–3; –4) and (5; 6)

11 (0; –8) and (–6; 0)

12 (3; –4) and (–3; 4)

Supplies must be dropped at halfway points between the following grid references on a map. Locate each drop zone for the pilot:

13 (300; 52) and (100; 38)

14 (49; –20) and (33; –10)

15 (–71; 52) and (38; –19)

Find the point of bisection of AB:

16 $A(-3; -4)$, $B(-8; -3)$

17 $A(3a; 2t)$, $B(5a; 4t)$

18 $A(4p; -2n)$, $B(2p; 6n)$

19 $A(a; 3b)$, $B(-a; -3b)$

20 $A(\frac{1}{3}; \frac{1}{4})$, $B(\frac{1}{2}; \frac{2}{3})$

21 $A(2n^2; -3p)$, $B(n^2; -p)$

22 a Show that the mid-point of the line segment joining (–4; –4) and (4; 0) lies on the y-axis.

b How far from the origin is this mid-point?

23 Show that the x-axis bisects the line segment joining (–6; 3) to (–8; –3).

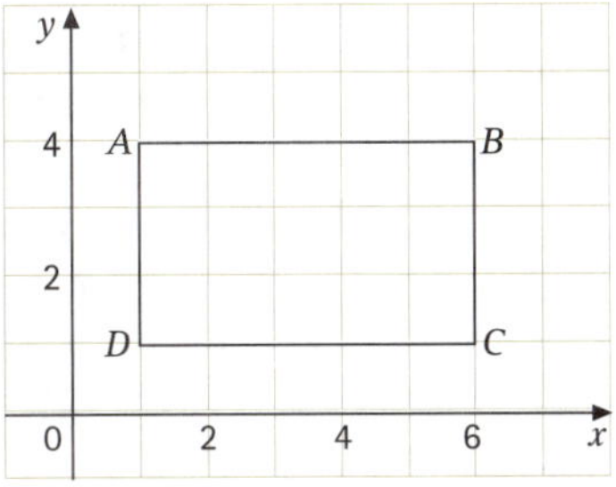

24 a Find the mid-point of AC.

b Find the mid-point of BD.

c Are **a** and **b** the same point?

d What property of the diagonals of a rectangle does this show?

25 a Plot the four corner points of any square in the first quadrant.

b Use the coordinates of the points to show that the diagonals of your square bisect each other, and that the diagonals are equal.

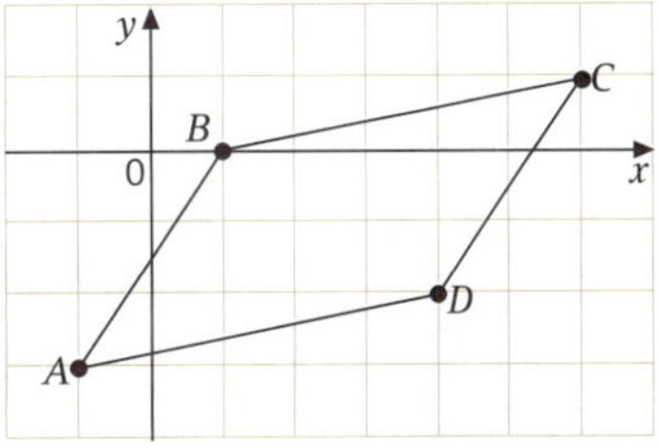

26 a Show that the diagonals of $ABCD$ bisect each other.

b What sort of quadrilateral could it therefore be?

c Show that the opposite sides of $ABCD$ are equal.

d Find the gradients of AB and CD and of BC and AD.

e What does the result of **d** show figure $ABCD$ to be?

27 The diameter of a circle, centre C, stretches from $A(1; 1)$ to $B(5; 7)$.

a Find the coordinates of C.

b Find the radius of the circle.

c Show that $D(5; 1)$ lies on the circle.

In each of the following, M is the mid-point of AB. Find the unknown coordinates:

28 $M(3; 5)$, $A(0; 4)$, $B(6; t)$

29 $M(1; 5\frac{1}{2})$, $A(-4; 6)$, $B(p; 5)$

30 $M(2; 3)$, $A(-3; 1)$, $B(p; t)$

31 $M(8; 5)$, $A(a; b)$, $B(10; 2)$

Challenge!

32 If the mid-point of a line segment is (1; 4), what could the end-points be?

33 Three vertices of a parallelogram are (10; 2), (2; 6) and the origin. Use the property of diagonals of a parallelogram bisecting each other to find the fourth vertex. Is there more than one possible answer?

The gradient of a line segment

This grid shows two sets of coordinates: point $A(1; 1)$ and point $B(3; 4)$. The line segment AB joins the two points. To get from point A to point B you have to move 2 places across (along the x-axis) and 3 places up (along the y-axis). You can express the gradient of the line segment as $\frac{\text{run}}{\text{rise}} = \frac{2}{3}$.

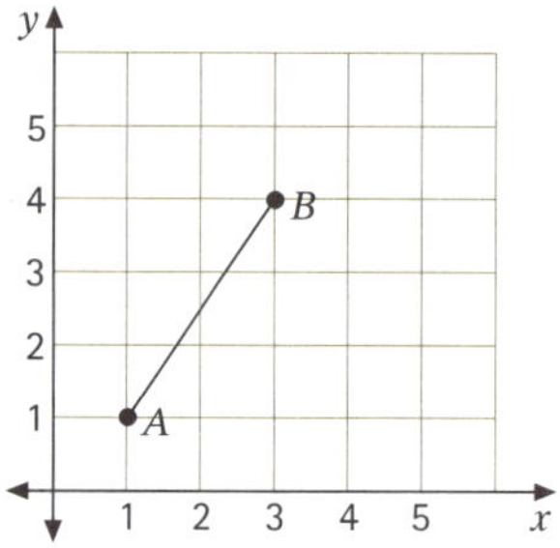

Investigation: A wheelchair ramp

Work in a group.

A wheelchair ramp is an example of a slope with a constant gradient. Each group should try to find a different wheelchair ramp (at the school, at a shopping centre or at other public buildings).

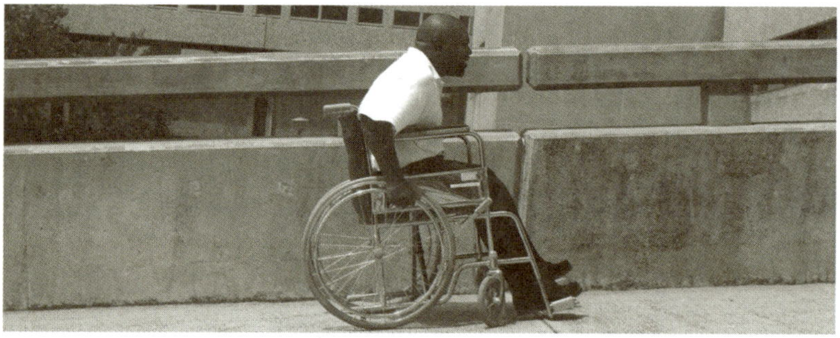

1 Measure the length and height of the ramp you have chosen.
2 Then draw a grid like the one shown above and work out the gradient of the ramp.
3 Compare your answer with other groups' answers. Do you think that all wheelchair ramps have the same gradient? Discuss.
4 Does your school provide access for disabled people, e.g. wheelchair ramps? If not, do you think the school should? Discuss this as a class.

Exercise

1 Complete the statement:

gradient = $\frac{\text{rise}}{\dots} = \frac{\dots\text{ change}}{\text{horizontal change}} = \frac{\text{change in } y}{\dots}$

Are the statements in questions 2 to 5 true or false?

2 The variable m is used for gradient.

3 Parallel lines have the same gradient.

4 The gradient between any two points on a line is the same.

5 A gradient of $-\frac{3}{5}$ means a line **falls** 3 units for every 5 units to the **right**.

Find the gradient of the line segment:

6 AB **7** CD **8** EF **9** BD

10 Is $AB \parallel EF$? Give a reason.

11 A line segment starting at E has gradient 0,4.

Give the coordinates of any point T lying on the line segment.

12 A line segment starting at the origin has a gradient of −0,25.

Find the coordinates of any point N on the line segment.

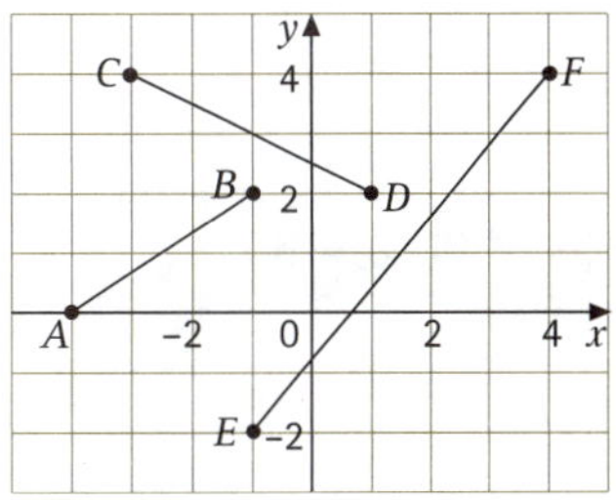

13 Which of AB and EF has the steeper slope? Give a reason.

14 Now look at the graph showing the speed of three long-distance swimmers, Lisa, Rodney and Bheki, who completed a 5-kilometre swim.

Work out the gradient of each swimmer's line in order to work out their race pace.

Speed of long-distance swimmers

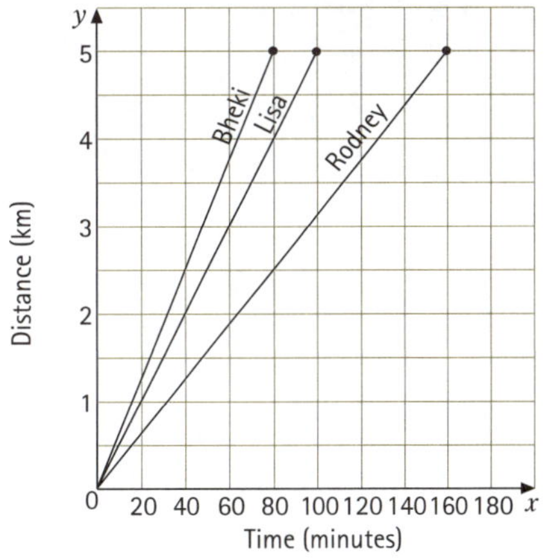

The gradient formula

The gradient of a line segment joining any two points A and B in the plane is represented by the symbol m_{AB}.

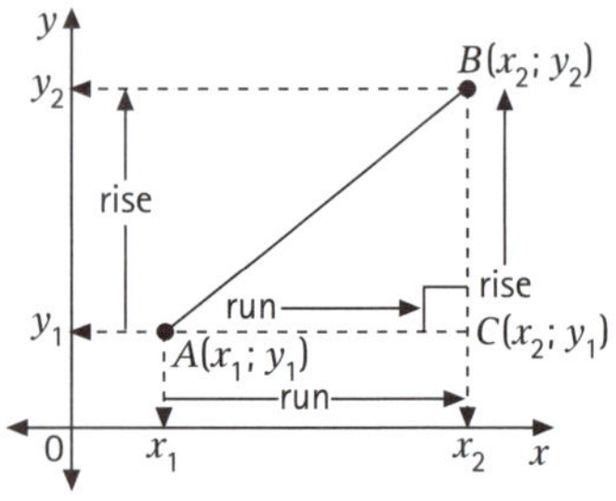

In this diagram:
rise is CB, which equals $y_2 - y_1$
run is AC, which equals $x_2 - x_1$.

So $m_{AB} = \frac{\text{rise}}{\text{run}} = \frac{y_2 - y_1}{x_2 - x_1}$

The gradient of any line segment

$$= \frac{\text{rise}}{\text{run}} = \frac{\text{change in } y}{\text{change in } x} = \frac{\text{difference of } y\text{-coordinates}}{\text{difference of } x\text{-coordinates}}$$

$$m = \frac{y_2 - y_1}{x_2 - x_1}$$

EXAMPLE

Find the gradient of the line joining $P(-2; 3)$ to $Q(3; -5)$.

Sketch and label a diagram.

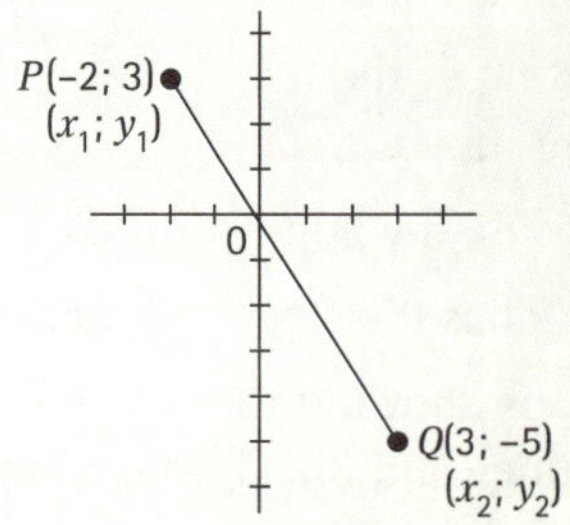

The right-hand point is $(x_2; y_2)$: Q.
The left-hand point is $(x_1; y_1)$: P.

$m_{PQ} = \frac{y_2 - y_1}{x_2 - x_1}$ — State the formula. Note that the order of the coordinates is from right to left.

$= \frac{-5 - 3}{3 - (-2)}$

$= \frac{-8}{5}$ — Substitute and simplify.

DISCUSS

Does it change the answer if strict right-to-left order is not followed? Check and discuss your findings with a partner.

Exercise

Use the formula to find the gradient of the line segment joining the points:

1 (2; 3) and (3; 5)
2 (2; 2) and (4; 3)
3 (3; 5) and (6; 7)
4 (0; 0) and (4; 6)
5 (1; 2) and (9; 8)
6 (1; 7) and (5; 2)

7 (6; 1) and (2; −1)

8 (0; 3) and (−3; 2)

9 (−1; −1) and (−4; −1)

10 (−5; −3) and (−2; −5)

11 (0; −2) and $(3; -\frac{1}{2})$

12 (7,5; 4) and (−4,5; 1,5)

13 $(t; t)$ and $(3t; 5t)$

14 $(n; -n)$ and $(4n; 0)$

15 $(-t; 2t)$ and $(5t; 0)$

16 What is the gradient of the line passing through the origin and (−3; 2)?

Refer to this diagram to find the gradient of:

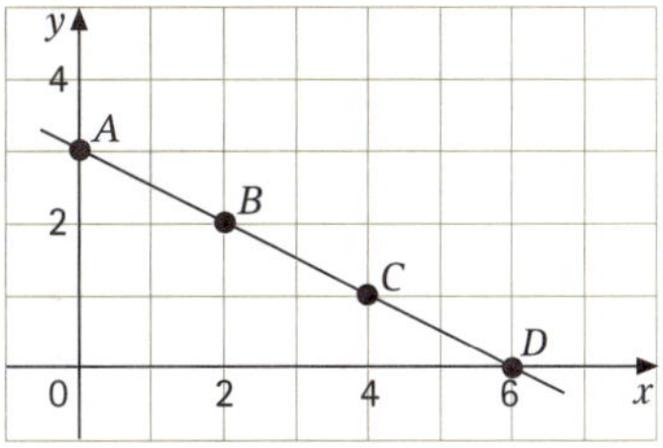

17 AB

18 BC

19 CD

20 AC

21 BD

22 AD

Consider the results of questions 17 to 22 to decide if the following statements are true or false:

23 Any two points on a straight line can be used to find its gradient.

24 The gradient of a line always remains the same.

25 If A, B and C are three points such that $m_{AB} = m_{BC}$, the points must be **collinear.** (Note that a common point B is used in both gradients.)

Gradients of horizontal and vertical lines

Horizontal lines

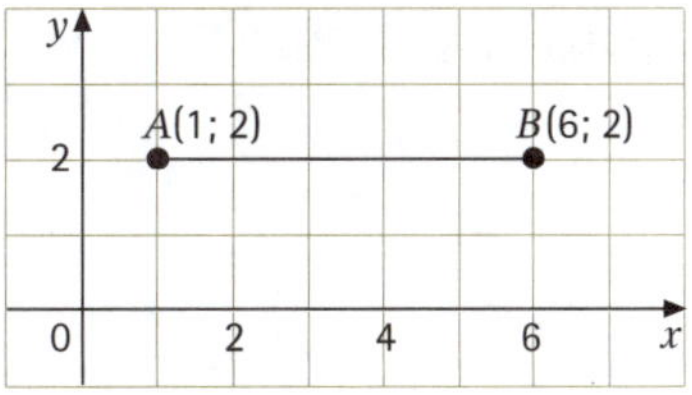

$$m_{AB} = \frac{y_B - y_A}{x_B - x_A} = \frac{2-2}{6-1} = \frac{0}{5} = 0$$

For all horizontal lines, the difference in y-coordinates must equal zero.

Any fraction with numerator zero equals zero.

All **horizontal** lines (lines parallel to the x-axis) have gradient **zero.**

If $AB \parallel x$-axis, $m_{AB} = 0$.

Vertical lines

Consider the steeply sloping line AB_1:

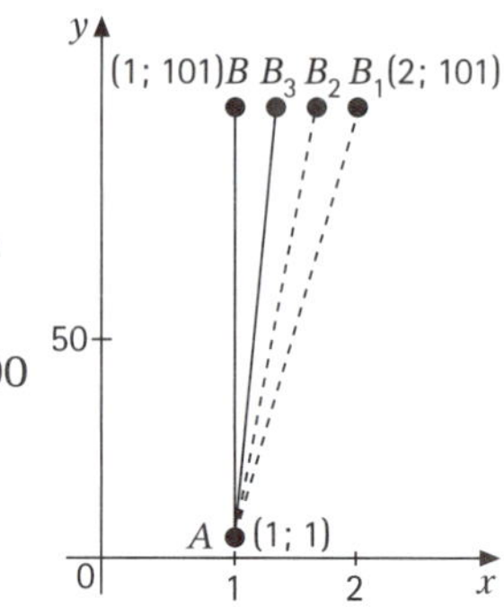

$$m_{AB_1} = \frac{101 - 1}{2 - 1} = \frac{100}{2} = 50$$

If B_2 is (1,5; 101): $m_{AB_2} = \frac{101 - 1}{1,5 - 1} = \frac{100}{0,5} = 200$

If B_3 is (1,1; 101): $m_{AB_3} = \frac{101 - 1}{1,1 - 1} = \frac{100}{0,1} = 1\,000$

As AB approaches the vertical, the gradient approaches **infinity** (symbol ∞).

Consider: $m_{AB} = \frac{101 - 1}{1 - 1} = \frac{100}{0} = ?$

All **vertical** lines (lines parallel to the y-axis) have gradient **infinity**.

If $AB \parallel y$-axis, $m_{AB} = \infty$.

Exercise

Write the gradient of each line (**1** to **6**) in the diagram.

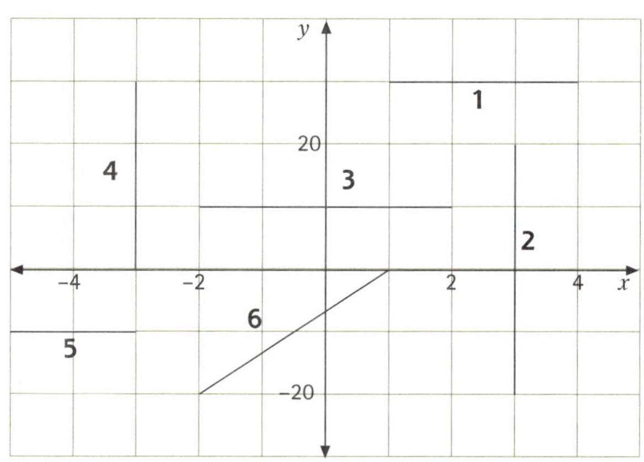

Find the gradient of:

7 the y-axis

8 the x-axis.

Draw a number plane with the same scale on both axes, with -4 to 5 on the x-axis and y-axis. You are given a point on the line and the gradient. Draw each line:

9 (1; 2), $m = 2$ **10** (3; 2), $m = \frac{1}{2}$ **11** (−2; 3), $m = -\frac{2}{3}$

12 (−4; 2), $m = 0$ **13** (−3; −4), $m = 1$ **14** (1; −4), $m = \infty$

15 Draw a line through the origin parallel to the line in question 11.

Calculate the slope of each line segment:

16 (–2; 1) to (1; 1) **17** (3; –4) to (3; 5) **18** (–5; –6) to (–5,1; 6)

19 Your local council decides that the municipal swimming pool needs a new coat of sealant. Look at the diagram.

a Work out the surface area of the sides and floor of the swimming pool.

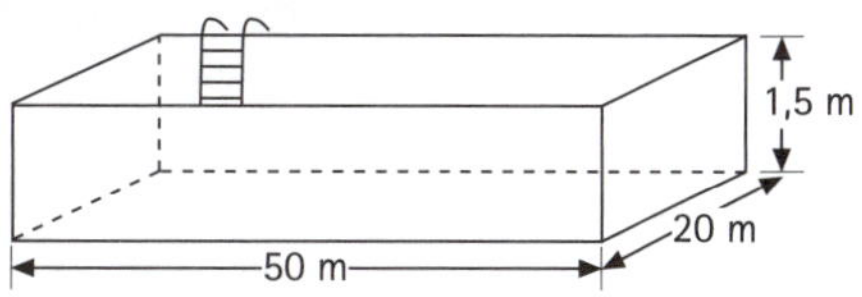

b A litre of sealant can cover five square metres. How many litres of sealant should the council buy?

c The sealant is available in 2-litre, 5-litre and 10-litre tins. The 2-litre tin costs R18, the 5-litre tin costs R40, and the 10-litre tin costs R70. How many tins of each should the council buy to keep costs down as much as possible?

d Based on your answer in **c**, work out how much the council has to pay for the sealant. Then work out how much the sealant costs the council, on average, per litre.

e The graph shows how long it takes for the water to drain out of the swimming pool. Calculate the gradient of the line, and use this to express the speed at which the water leaves the pool.

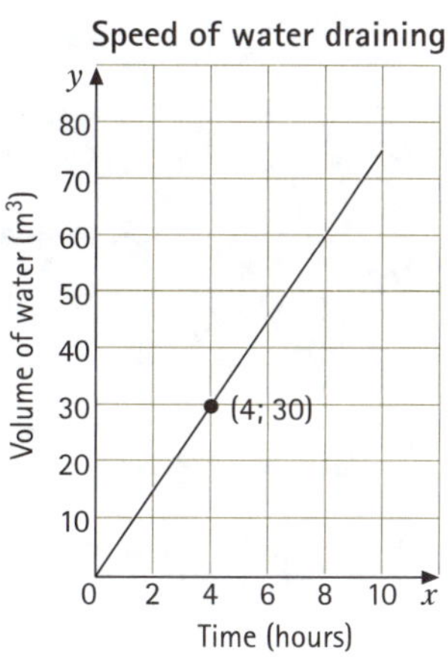

SELF-ASSESSMENT

See page 286 (rubric 1).

f Use the gradient you found to work out how long it will take to empty the whole swimming pool.

CHAPTER 3 TRANSFORMATION GEOMETRY

KEY VOCABULARY

transformation • translation • reflection • rotation • coordinate • image

Transformation geometry is the field of mathematics that deals with moving shapes into different positions. Look at the two pictures of the draught board below. Can you think of ways of using coordinates to describe the movement of the draughts from Figure 1 to Figure 2?

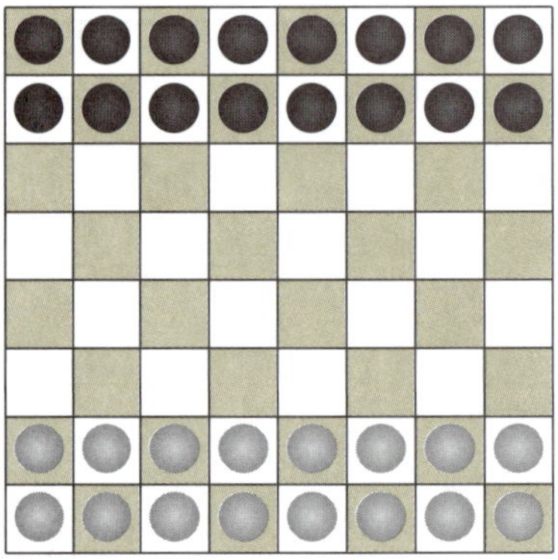

Figure 1

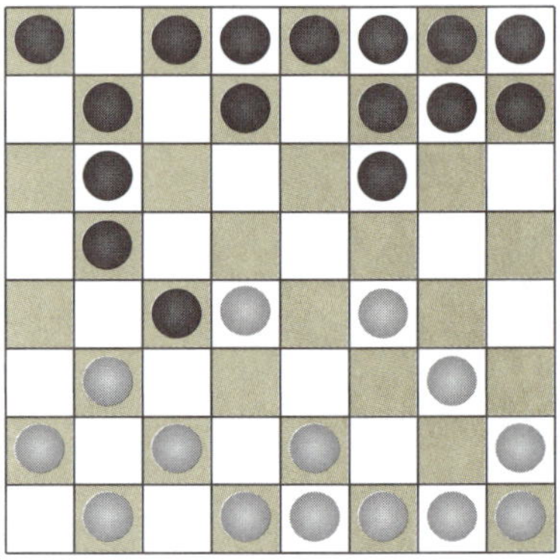

Figure 2

Transformations are an important aspect of design and technology. Look at these fabrics. The designers have repeated shapes by moving them at regular distances and angles.

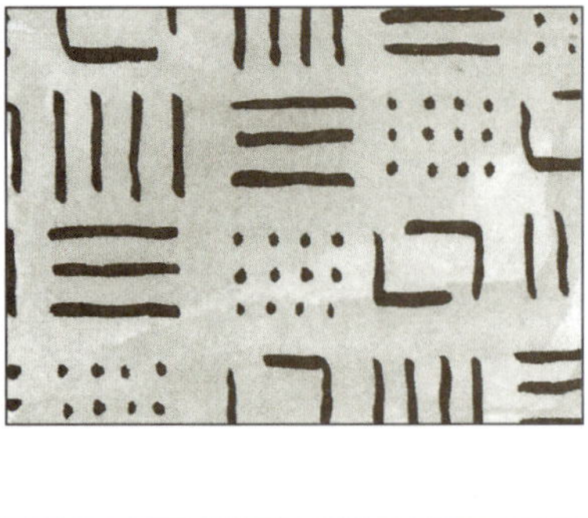

Transformations

If two figures are **congruent**, one figure can be moved to fit **exactly** onto the other. Moving a shape is called **transformation.**

The following transformations leave the size and shape of the figure unchanged, and so the two figures are congruent:

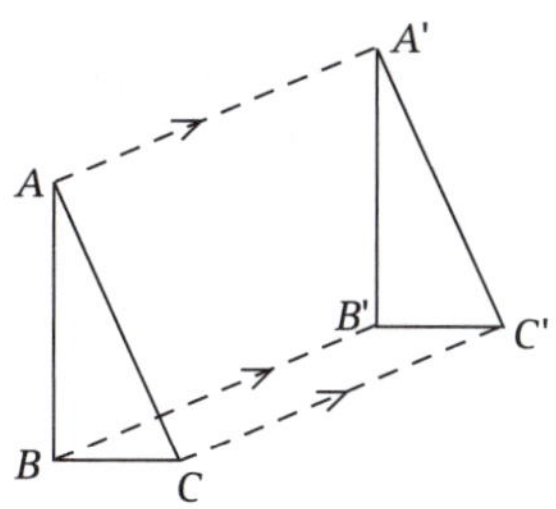

Translation

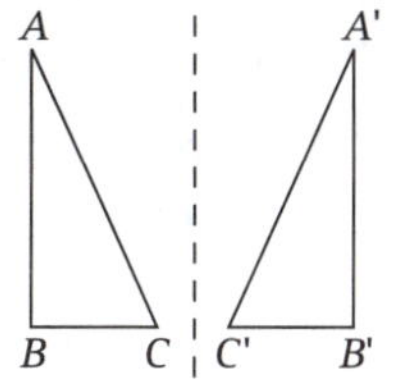

Reflection

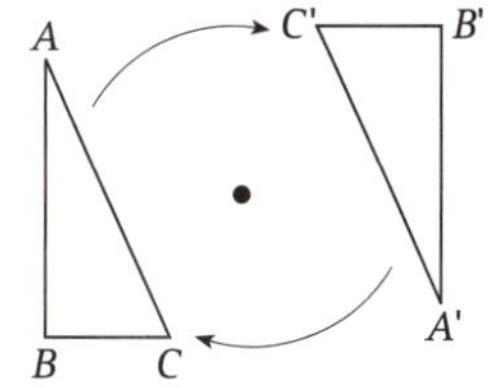

Rotation

The new figure that is produced is called the **image**, and the corners of the image figure are labelled $A'B'C'$ etc. A' is the new position of point A, B' is the new position of point B, and so on.

EXAMPLES

1 What transformations might have been performed to obtain the image figure from the original?

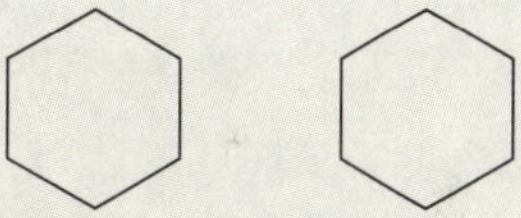

2 What additional information would you need to determine which of these transformations were performed?

1 The transformation could have been a translation:

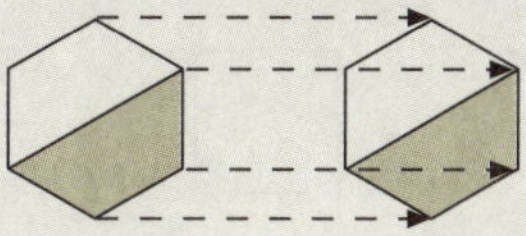

or it could have been a rotation:

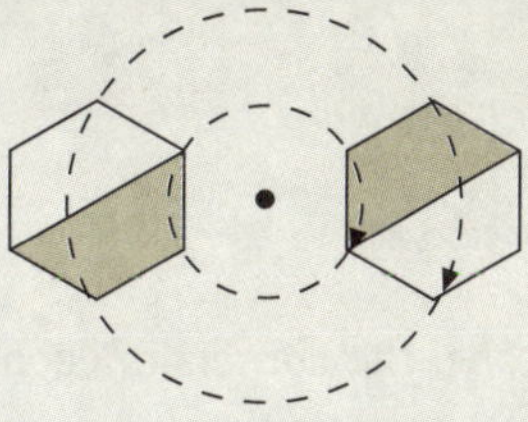

or it could have been a reflection:

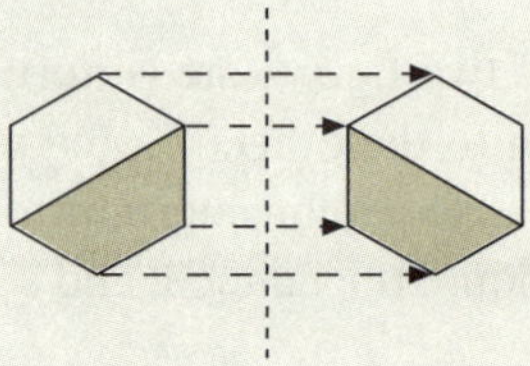

2 To determine which transformation was performed, the positions of the original corners need to be shown on the image figure, for example:

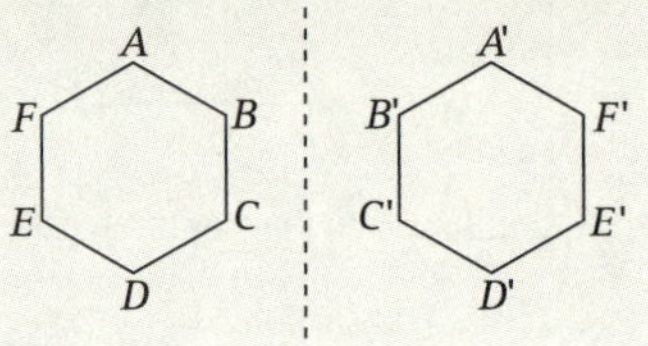

This would show that a reflection was performed.

Exercise

For each of these figures, determine which transformations could have been performed to obtain the image (in colour):

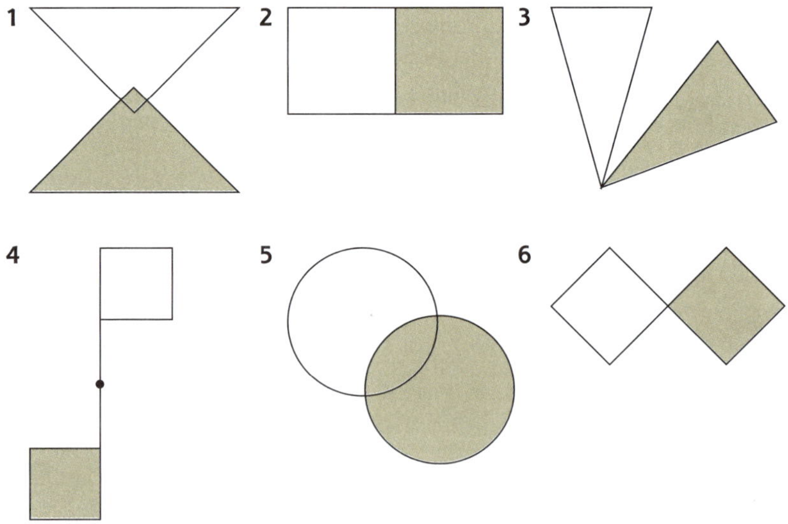

Label the corners of each image to show that the stated transformation has been performed in questions 7 to 12 on the next page:

7

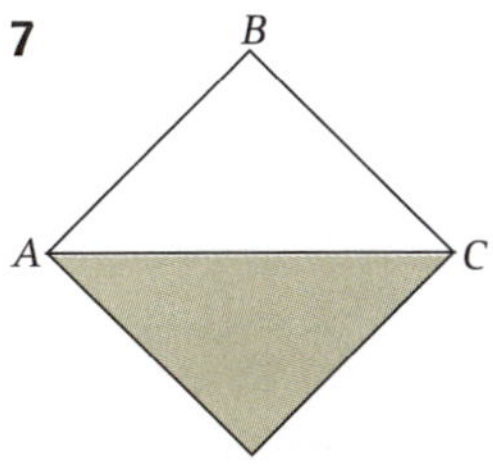

Rotation

8

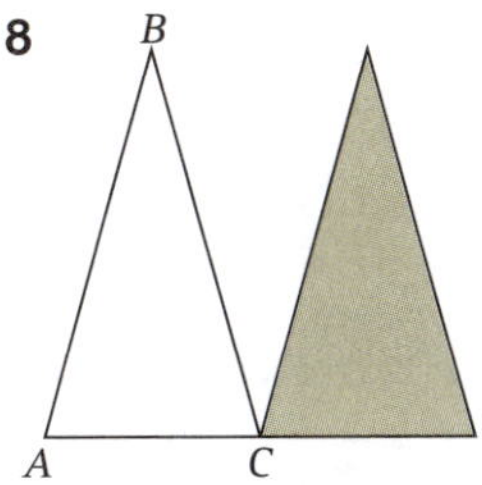

Translation

9

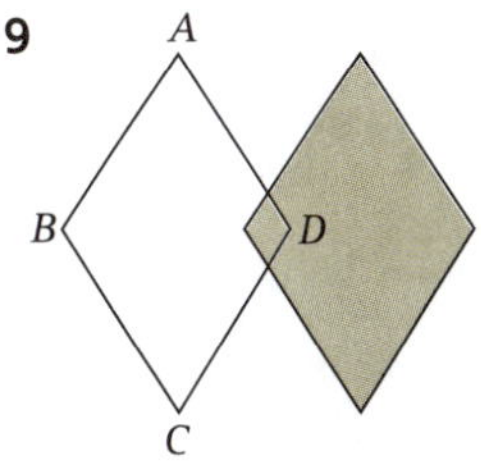

Reflection

10

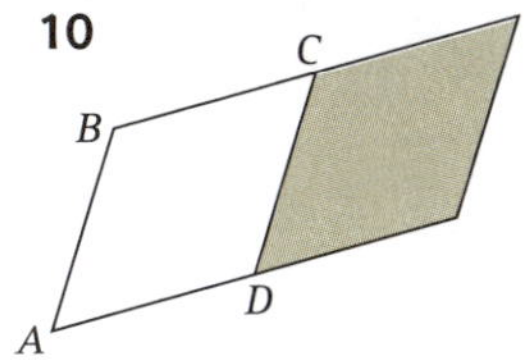

Translation

11

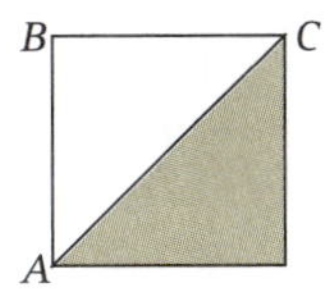

Reflection

12

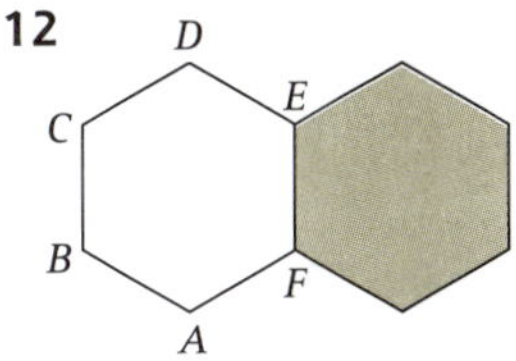

Rotation
(centre of rotation at F)

Copy Figures **13** to **16** into your book, and draw the image after performing the transformations stated below the grid. Give the coordinates of A' after each transformation.

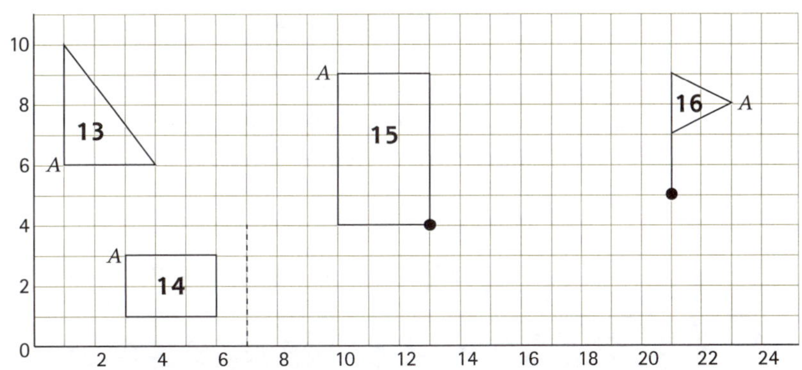

13 A translation 3 units to the right.

14 A reflection in the vertical dashed line indicated.

15 A 90° clockwise rotation about the point (13; 4).

16 A 180° anticlockwise rotation about the point (21; 5).

Challenge:

Each of the shapes numbered **17** to **19** on page 94 has undergone two transformations.

Describe the possible combinations of transformations used to get from the original shape to its image.

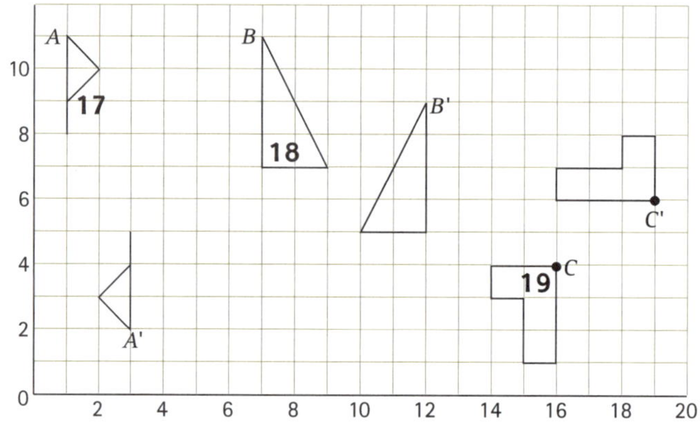

Investigation: Patterns

Fabric designs

Many African fabric patterns demonstrate excellent congruence transformations. Look at the examples on page 90 again.

Collect come examples of fabric patterns that show transformation. You may be able to obtain samples from a decorator's supplier, or you may find some leftover pieces at home.

Identify the part of the pattern that is repeated, and draw a line around it. Decide what types of transformations are shown. (There will probably be more than one.)

Patterns from other cultures

Identify the transformations that are present in these designs:

Batik work – Java

Dali tie-dye cloth – China

Islamic marble window grille (detail)

Find some examples of designs or patterns that originate from different African cultures. Look for the congruence transformations, and list those that are present. Copy or trace the designs, and colour them to look as authentic as possible.

Investigating transformations using coordinates

Investigation: Transformations

Translations

Work with a partner or in a group.

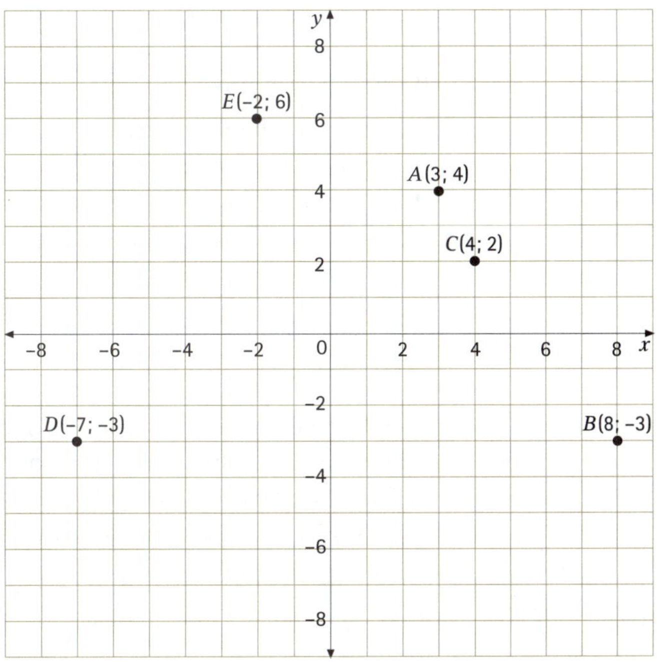

1 Copy the graph on squared paper or graph paper. Perform these translations and write down the coordinates of the images A', B', C', D' and E':

On point $A(3; 4)$: a translation of +4 units horizontally.

On point $B(8; -3)$: a translation of −3 units vertically.

On point $C(4; 2)$: a translation of +3 units horizontally and +6 units vertically.

On point $D(-7; -3)$: a translation of +9 units horizontally and -3 units vertically.

On point $E(-2; 6)$: a translation of -4 units horizontally and -10 units vertically.

2 Use your experience from doing question 1 to perform these **translations** and write down the coordinates of the images F', G' and H':

On point $F(a; b)$: a translation of c units horizontally.

On point $G(e; f)$: a translation of g units vertically.

On point $H(x; y)$: a translation of p units horizontally and q units vertically.

Discuss your findings with your partner or group.

Reflection in the x-axis

3 Copy the graph shown on page 95 on squared paper or graph paper again. Perform a **reflection in the x-axis** on each point and write down the coordinates of the images A', B', C', D' and E'.

4 Perform a **reflection in the x-axis** on the point $F(x; y)$ and write down the coordinates of the image F'.

Discuss your findings with your partner or group.

Reflection in the y-axis

5 Copy the graph shown on page 95 once more on squared paper or graph paper. Perform a **reflection in the y-axis** on each point and write down the coordinates of the images A', B', C', D' and E'.

6 Perform a **reflection in the y-axis** on point $F(x; y)$ and write down the coordinates of the image F'.

Discuss your findings with your partner or group.

Reflection in the line $y = x$

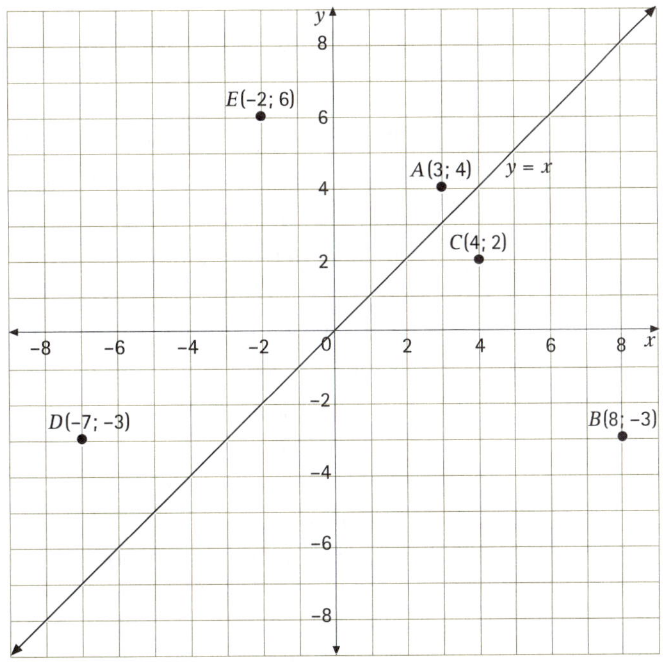

7 Copy the graph on squared paper or graph paper again. Add line $y = x$ as shown. Perform a **reflection in the straight line $y = x$** on each point and write down the coordinates of the images A', B', C', D' and E'.

8 Perform a **reflection in the straight line $y = x$** on the point $F(x; y)$ and write down the coordinates of the image F'.

Again, discuss your findings with your partner or group.

The investigation should have shown:

- Under **translation** of p units horizontally and q units vertically on point $H(x; y)$ the image is $H'(x + p; y + q)$.
- Under **reflection in the x-axis** the image of the point $F(x; y)$ is $F'(x; -y)$.
- Under **reflection in the y-axis** the image of the point $F(x; y)$ is $F'(-x; y)$.
- Under **reflection in the line $y = x$** the image of the point $F(x; y)$ is $F'(y; x)$.

Exercise

1 Perform these translations and write down the coordinates of the images A', B', C', D' and E':

On point $A(-5; 4)$: a translation of +3 units horizontally.

On point $B(-7; -12)$: a translation of −12 units vertically.

On point $C(-5; -12)$: a translation of +3 units horizontally and +6 units vertically.

On point $D(7; -3)$: a translation of +3 units horizontally and −3 units vertically.

On point $E(-12; 2)$: a translation of −4 units horizontally and −1 unit vertically.

Then plot these points on squared paper or graph paper to check your answers.

2 ΔABC has vertices $A(2; 3)$, $B(-4; -9)$ and $C(3; -7)$. A translation of −5 units horizontally and 4 units vertically is performed on the triangle. Write down the coordinates of the vertices of the image $\Delta A'B'C'$. Then plot these points on squared paper or graph paper to check your answers.

3 The points $A(2; 3)$, $B(-4; -9)$ and $C(3; -7)$ are given. Perform a reflection in the x-axis on each point and write down the coordinates of the images A', B' and C'.

4 Quadrilateral $ABCD$ has vertices $A(-2; 3)$, $B(-4; -9)$, $C(-3; -7)$ and $D(-2; -3)$. The quadrilateral is reflected in the y-axis. Write down the coordinates of the vertices of quadrilateral $A'B'C'D'$. Then plot these points on squared paper or graph paper to check your answers.

5 Quadrilateral $ABCD$ has vertices $A(-5; -6)$, $B(-4; -9)$, $C(-3; -7)$ and $D(-2; -3)$. The quadrilateral is reflected in the straight line $y = x$. Write down the coordinates of the vertices of quadrilateral $A'B'C'D'$. Then plot these points on squared paper or graph paper to check your answers.

6 Thabile decides to rearrange the furniture in her living room. She draws a floor plan of the room, showing the furniture in its existing position. Use graph paper to draw the new positions of the furniture if these are the transformations:

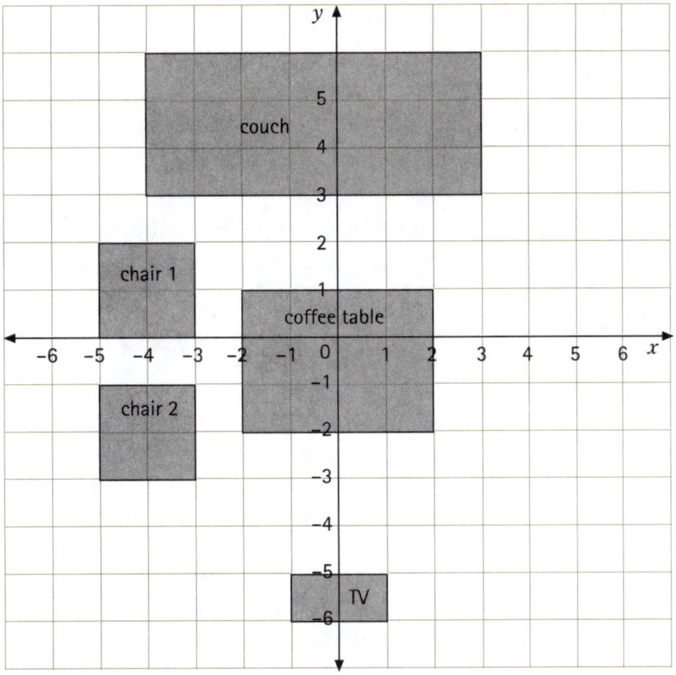

a Chair 1 is reflected in the y-axis.

b Chair 2 is translated 7 units horizontally and −2 units vertically.

c The coffee table is translated 1 unit vertically and −3 units horizontally, then rotated 90° anticlockwise about the point (−1; 2).

d The TV is translated −4 units horizontally and 4 units vertically, and rotated 90° anticlockwise about the point (−5; −1).

SELF-ASSESSMENT

See page 286 (rubric 1).

e Write your own transformation for the couch. Give the coordinates and directions for the transformation, and draw the couch in the correct place once you have done the transformation.

CHAPTER 4 TRIGONOMETRY

KEY VOCABULARY

trigonometry • opposite side • adjacent side • hypotenuse • similar • ratio • proportion • tangent ratio • theta • angle of elevation • angle of depression • distance • clinometer • sine ratio • cosine ratio • degree • minute • second • revolution • scale drawing • map • in proportion • building plan • reduction • enlargement

Note: The word "trigonometry" comes from the Greek words for "triangle measurement".

Trigonometry is the area of mathematics in which you study the relationships between lines and angles of triangles. Mathematicians developed this field of mathematics to enable them to measure things that are too big or too far away to measure physically. The Egyptians and Babylonians used trigonometry to study the stars. Today, trigonometry is used in civil engineering, architecture and navigation. You can use it to measure the heights of mountains, and to measure the distance between the earth and stars.

Naming the sides of a triangle

In a right-angled triangle, the longest side is opposite the right angle. This side is called the **hypotenuse**.

The other sides are named in relation to one of the other angles. Here the right angle is at B. The angle at C is 52°.

- Side AC is the **hypotenuse**.
- Side AB is **opposite** the 52° angle.
- Side BC is **adjacent** to the 52° angle.

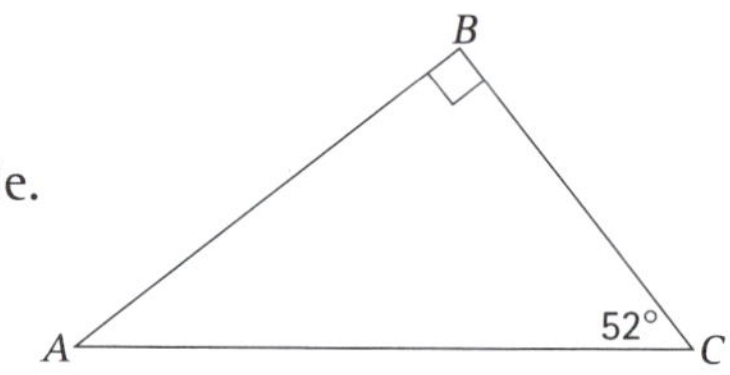

"Adjacent" means "next to". Sides AC and BC are both next to the 52° angle, but side AC has already been identified as the hypotenuse, and so BC is the only side to be called adjacent to angle 52°.

PEER ASSESSMENT

See page 288 (rubric 3).

Investigation: Ratio in similar triangles

Work with a partner.

The triangles on page 101 are all right-angled, and each has one angle of 53°.

1 What is the size of the third angle in each triangle?

The triangles are all **similar** – the **shapes** of the triangles are the same and all the triangles have the **same three angles**, but the **sizes** of the triangles are different.

2 For each triangle:

a measure the side opposite the 53° angle

b measure the side adjacent to the 53° angle

c calculate the ratio $\frac{\text{opposite}}{\text{adjacent}}$ to 1 decimal place.

What do you find? Discuss this with your partner.

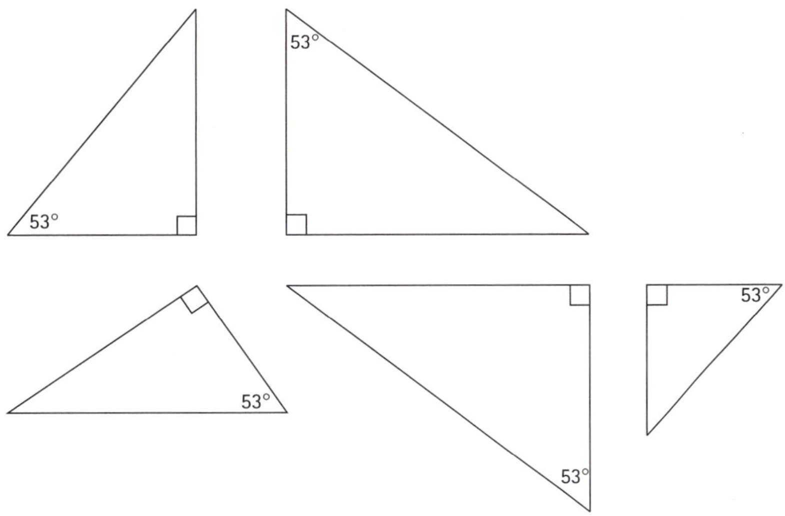

In any right-angled triangle with an angle of 53°, the ratio $\frac{\text{opposite}}{\text{adjacent}}$ is 1,3. This means that the side opposite the 53° angle is always 1,3 times the side adjacent to the 53° angle.

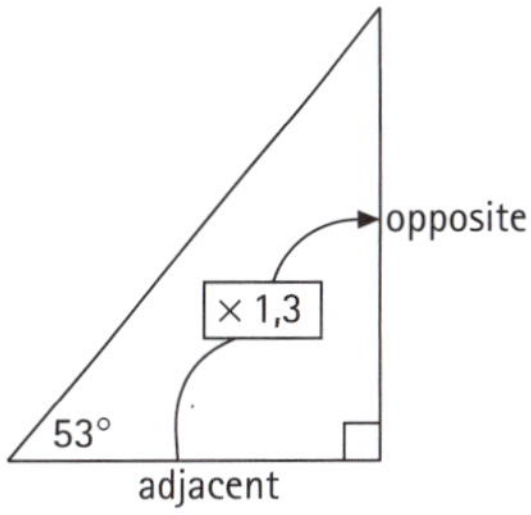

This fact can be used to calculate the length of the side opposite the 53° angle if you know the length of the adjacent side.

For example, if the adjacent side = 7 cm, the side opposite is $1{,}3 \times 7 = 9{,}1$ cm.

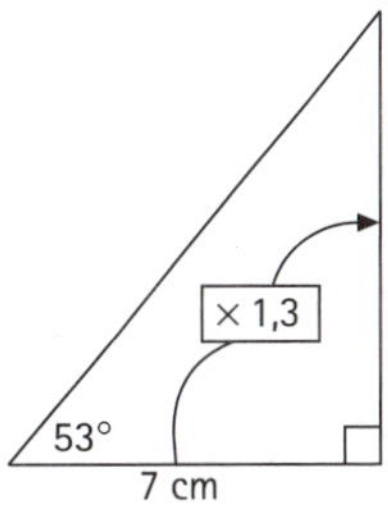

The tangent ratio

For a triangle with an angle of 53°, the ratio $\frac{\text{opposite}}{\text{adjacent}}$ is 1,3. This ratio is called the **tangent of 53°**. We say "the tangent of 53° is 1,3".

The tangent ratio is different for different angles. Every angle has its own tangent ratio.

Exercise

For each triangle:

a measure the side opposite the angle

b measure the side adjacent to the angle

c give the **tangent ratio** for the angle, correct to 2 decimal places.

1

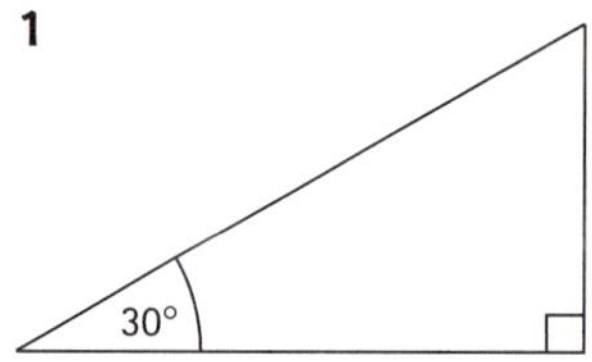

2

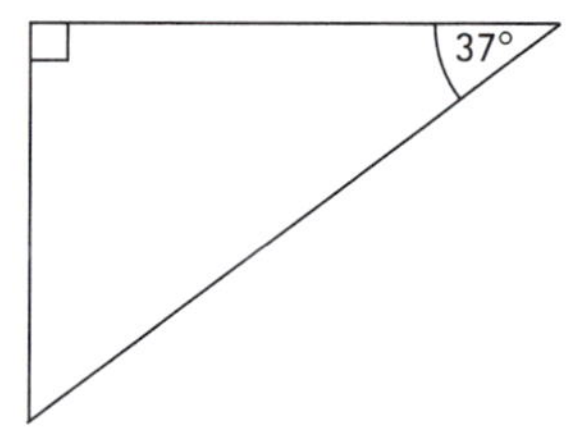

3

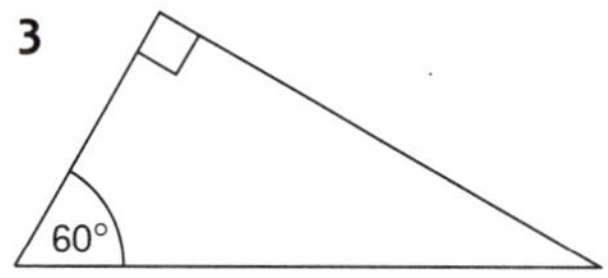

4

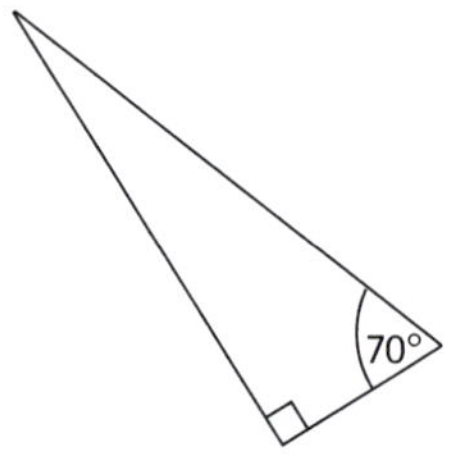

5

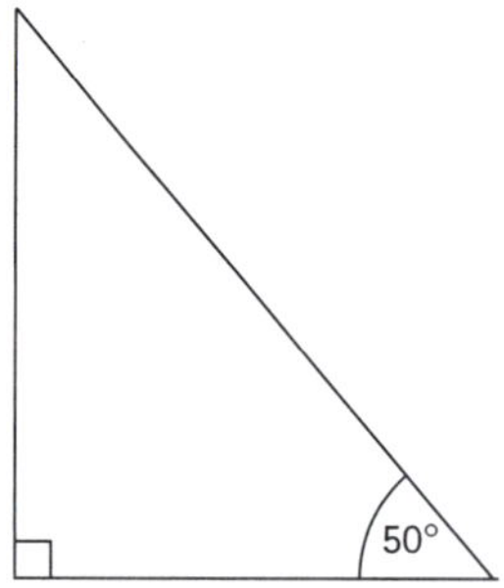

6

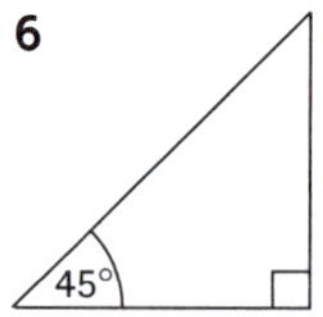

Using your calculator

Knowing the tangent ratio for an angle allows you to calculate side lengths in a triangle, but having to draw and measure a triangle every time is not very convenient.

Remember: Always check which method your calculator uses.

Fortunately, you can find the tangent ratio for any angle from your calculator. First, make sure your calculator is set to "degrees". It should show "D" or "deg" on the display. If your calculator is not set to degrees, press [MODE] [4]. Refer to the instruction manual if your calculator does not use this method, or ask your teacher for help.

To find the tangent of 53°, press [53] [tan]. (On some calculators, you will need to press [tan] [53].) (There is no need to press [=].) Your display should show 1.3270448 … . This confirms the measured ratio you found in the investigation for an angle of 53°, which was 1,3 to 1 decimal place.

Now check the tangent ratios for the angles in the triangles in the exercise on page 102. Do you get the same answers with your calculator?

Remember: Always write tan 53°, although some calculators swap the sequence.

You use the abbreviation **tan** for tangent. The tangent of 53° is written as tan 53°, and is entered into your calculator as [53] [tan] (or [tan] [53]).

Finding the side opposite the angle

You can use tangent ratios to calculate the lengths of sides of triangles.

EXAMPLE

Find the length of the side labelled x, correct to 1 decimal place.

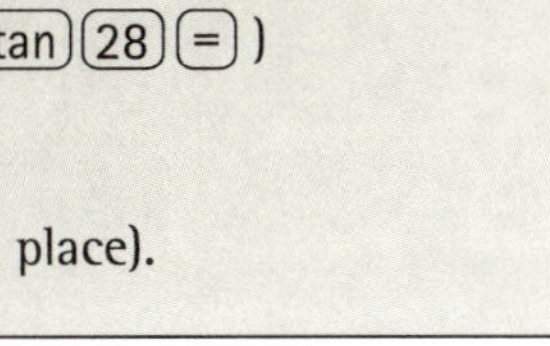

$$\tan 28° = \frac{\text{opposite}}{\text{adjacent}}$$

$$\tan 28° = \frac{x}{16}$$

$$x = 16 \times \tan 28°$$

$$= 8{,}5073509 \ldots$$

Press [16] [×] [28] [tan] [=]
(or [16] [tan] [28] [=])

The side labelled x is 8,5 cm (to 1 decimal place).

Exercise

Use your calculator to find the tangent ratio for each angle, correct to 2 decimal places:

1 25°	**2** 30°	**3** 45°	**4** 61°	**5** 38°
6 10°	**7** 85°	**8** 6°	**9** 52,5°	**10** $17\frac{1}{2}°$
11 $33\frac{1}{3}°$	**12** 1°	**13** 0°	**14** 89,5°	**15** 90°

Calculate the length of the side labelled d in each triangle, correct to 2 decimal places:

16

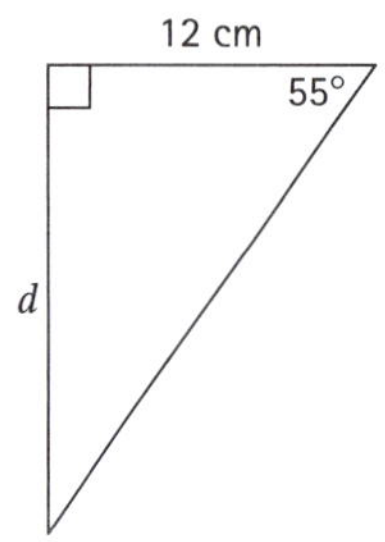

17

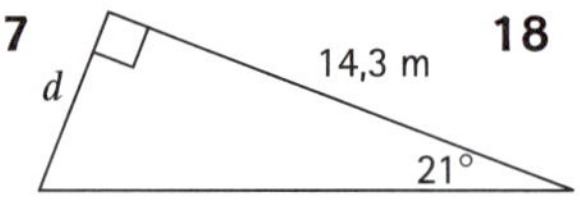

18

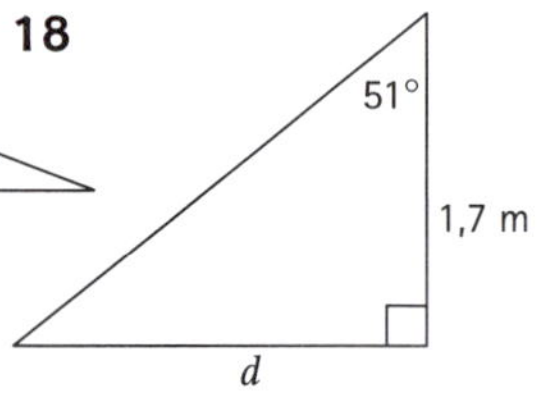

19

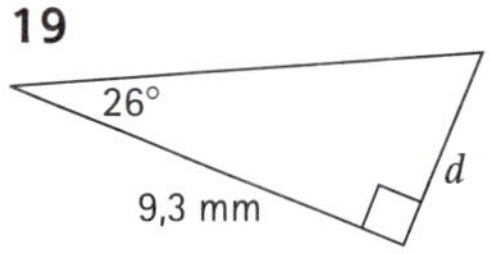

20

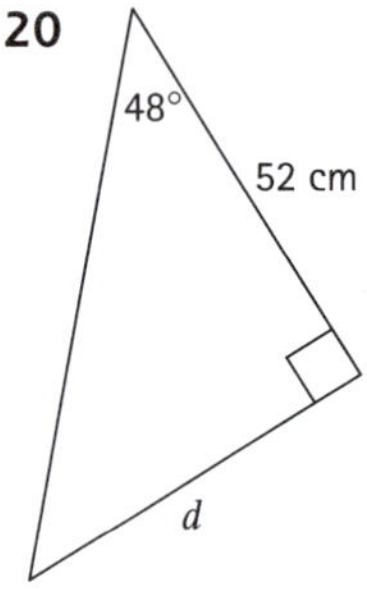

21

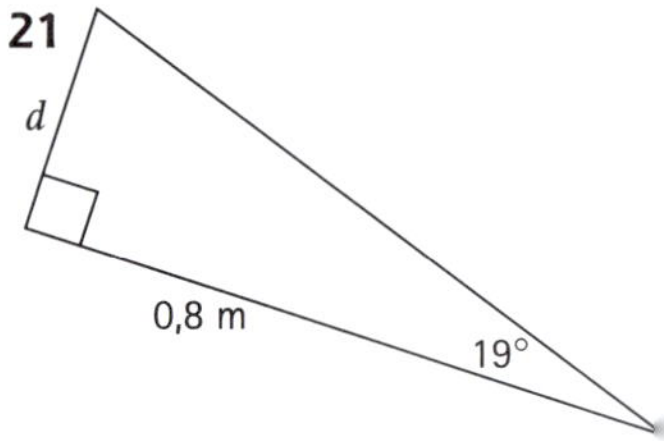

Finding the side adjacent to the angle

So far you calculated the **opposite** side when you knew the length of the **adjacent** side. But what if you know the length of the side opposite the angle, and want to find the length of the adjacent side?

EXAMPLE

Find the length of side PQ correct to 2 decimal places.

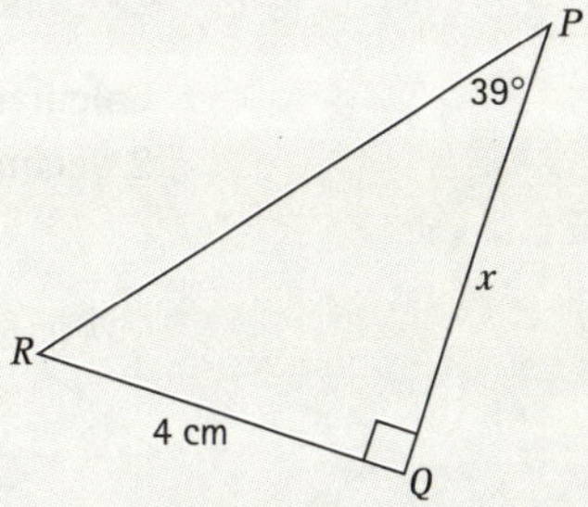

There are two possible ways of doing this. Have a look at both methods, then choose the one you prefer.

Method 1:

If you know that $\hat{P}$ is 39° and $\hat{Q}$ is 90°, you can find $\hat{R}$, as the three angles in a triangle must add up to 180°.

$$\hat{R} + 39° + 90° = 180° \qquad \text{(angle sum of a triangle)}$$
$$\hat{R} = 180° - 129°$$
$$= 51°$$

Now you have an angle for which you know the adjacent side and want to find the opposite side. This you can do in the same way as the questions in the last exercise, using the 51° angle.

$$\tan 51° = \frac{\text{opposite}}{\text{adjacent}}$$
$$\tan 51° = \frac{x}{4}$$
$$x = 4 \times \tan 51°$$
$$= 4{,}939588 \ldots$$
$$PQ = 4{,}94 \text{ cm (to 2 decimal places)}$$

Method 2:

If you know the adjacent side, you **multiply** by tan 39° to get the opposite side. So, if you know the opposite side, you must **divide** by tan 39° to get the adjacent side. That is:

$$\tan 39° = \frac{4}{x}$$
$$x \times \tan 39° = 4$$
$$x = \frac{4}{\tan 39°}$$
$$= 4{,}939588 \ldots$$
$$PQ = 4{,}94 \text{ cm (to 2 decimal places)}$$

Exercise

Calculate the length of the side labelled y in each diagram, correct to 2 decimal places:

1

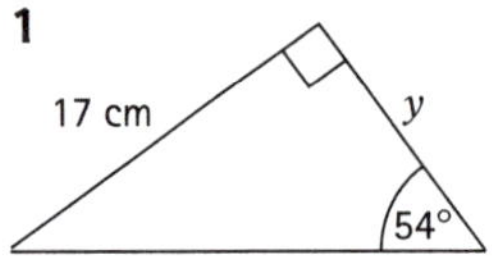

2

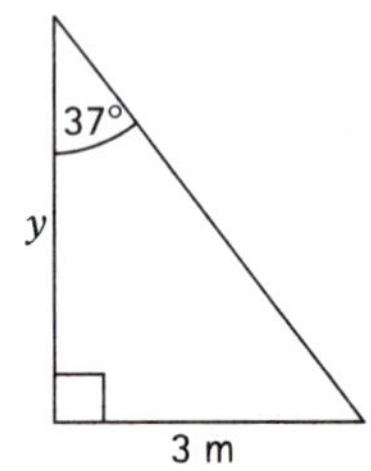

3

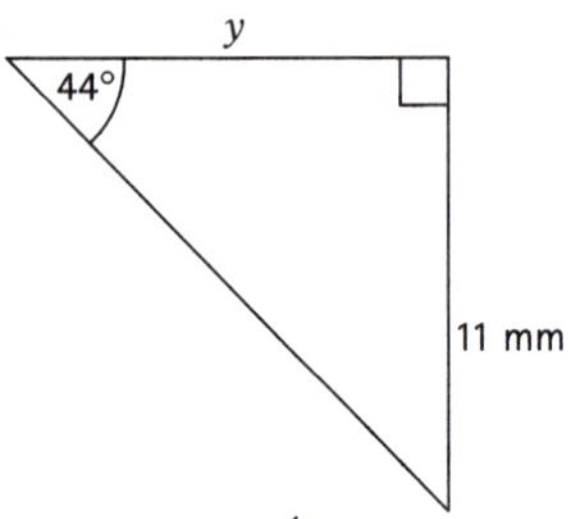

4

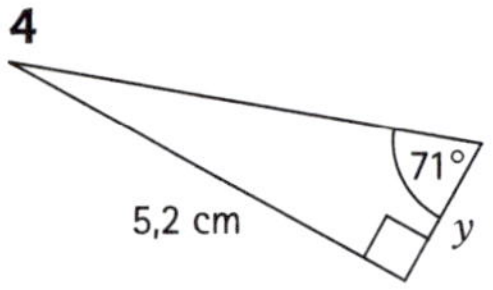

5

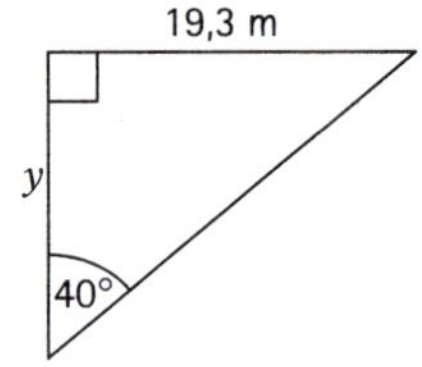

6

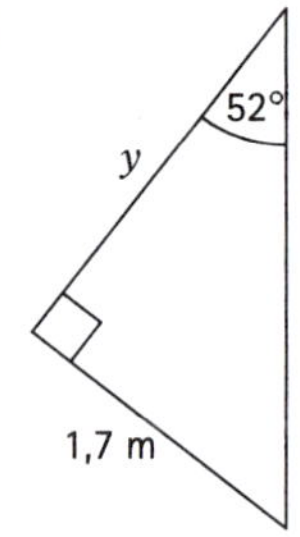

7

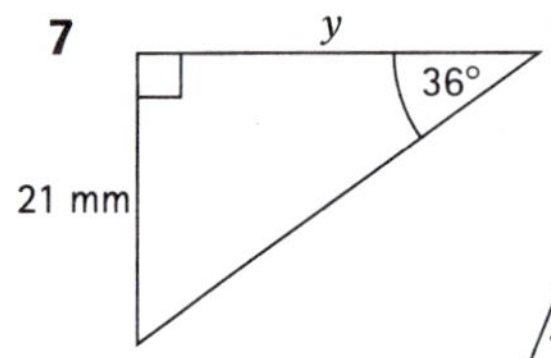

8

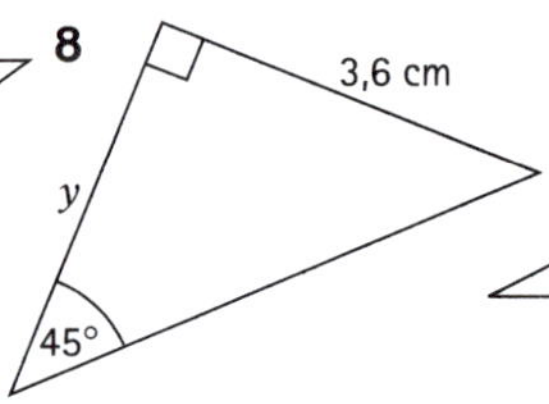

9

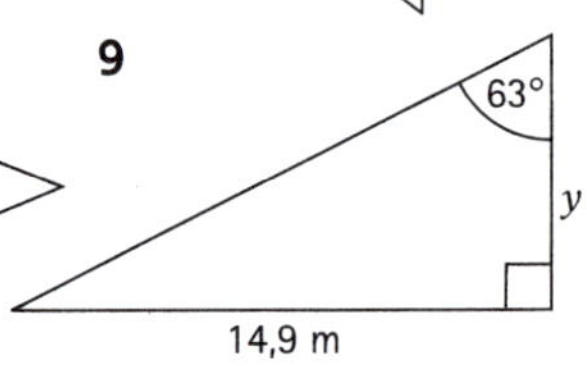

GROUP ASSESSMENT

See page 288 (rubric 3).

Remember:
$\tan\theta = \frac{\text{opposite}}{\text{adjacent}}$

Investigation: Estimating angles

Work in a group.

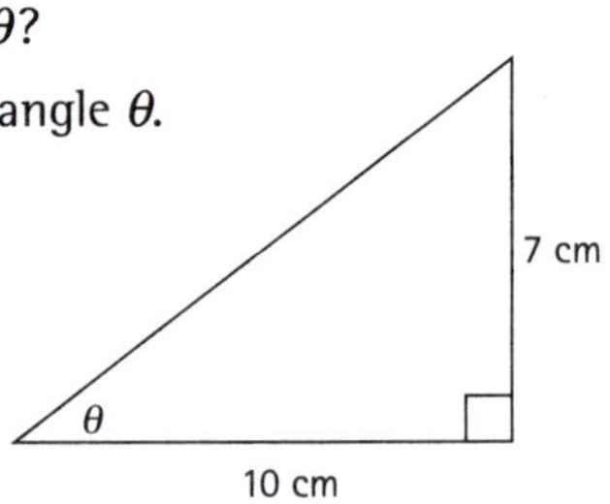

1. What is the tangent ratio for an angle θ?
2. Write down an estimate for the size of angle θ.
3. Use your calculator to find the tangent of your estimated angle.
4. Was the tangent of your angle close to the tangent for θ?
5. Try another estimate for θ and find the tangent. Were you closer this time?
6. Continue improving your estimate for θ until you get the same ratio as calculated in question 1 (accurate to 3 decimal places).

7 Compare your estimate for θ with those of other groups. Did you all find the same angle in the end?

Finding an angle

If you use the "trial-and-improvement" method, as in the investigation above, you will find the answer, but it is rather time-consuming. Fortunately, calculators can help.

If you start with an angle, you know how to use a calculator to find its tangent. Now you need to do the opposite: start with the **tangent** and find the **angle**.

Note: On some calculators you may have to press [INV] instead of [SHIFT].

For example, to find the angle whose tangent is 0,8, press [0.8] [SHIFT] [tan]. The label above the tan key is marked as "$\tan^{-1}$", which is the short way of writing "arc tan".

Your calculator display should show 38.659808 … . So the tangent of 38,7° is 0,8.

Check this by pressing [38.7] [tan]. You should get approximately 0,8. The answer won't be exact because you rounded off the angle to 1 decimal place.

EXAMPLES

1 What angle has a tangent of 2,1? Give your answer to the nearest degree.

Press [2.1] [SHIFT] [tan]. The display shows 64.536654 … . The angle whose tangent is 2,1 is 65°, rounded off to the nearest degree.

2 Given that $\tan \theta = 0{,}34$, find θ to 1 decimal place.

This means you need to find the angle θ whose tangent is 0,34. Press [0.34] [SHIFT] [tan]. The display shows 18.778033 … .

$\theta = 18{,}8°$ (to 1 decimal place)

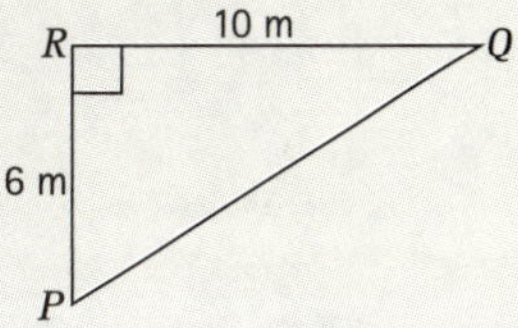

Remember: $\tan \hat{P} = \frac{\text{opposite side}}{\text{adjacent side}}$

3 Find the size of $\hat{P}$ to the nearest degree.

So $\tan \hat{P} = \frac{10}{6}$

$\tan \hat{P} = 1{,}66 \ldots$

The side opposite $\hat{P}$ is 10 m; the side adjacent to $\hat{P}$ is 6 m.

Now that you know the **tangent** of the angle, you can find the angle.
Leave the answer to 10 ÷ 6 on your calculator screen, then press (SHIFT)(tan). The calculator display shows 59.036243 ... .
$\hat{P} = 59°$ (to the nearest degree)

Exercise

Find the angle, to the nearest degree, whose tangent ratio is:

1 1,3	**2** 0,7	**3** 1	**4** 0,66
5 2,5	**6** 4	**7** 0,521	**8** 5,21
9 $\frac{1}{2}$	**10** $\frac{1}{4}$	**11** $\frac{3}{4}$	**12** $3\frac{2}{3}$

Find angle θ, correct to 1 decimal place:

13 $\tan \theta = 0,2$	**14** $\tan \theta = 6,1$	**15** $\tan \theta = 1,75$
16 $\tan \theta = 2,23$	**17** $\tan \theta = 14,5$	**18** $\tan \theta = 0,97$
19 $\tan \theta = \frac{2}{5}$	**20** $\tan \theta = 4\frac{3}{4}$	**21** $\tan \theta = 3\frac{1}{7}$

22 The foot of a ladder is 3 m from a wall, and the ladder reaches 6 m up the wall. What angle does the ladder make with the ground, to the nearest degree?

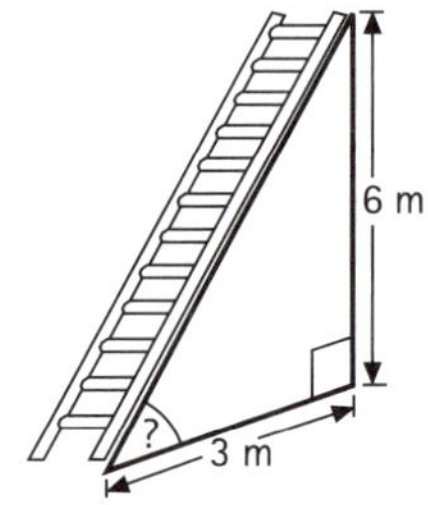

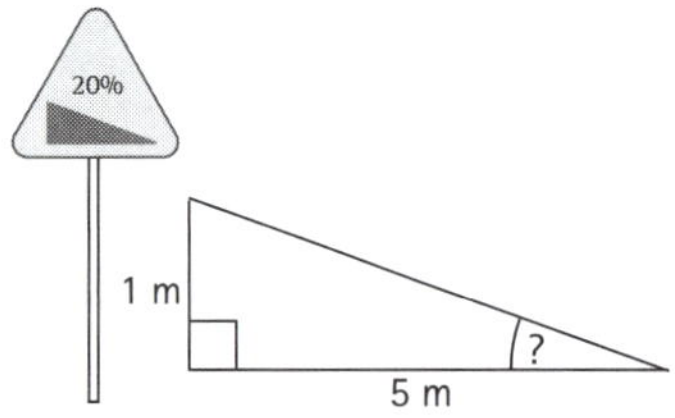

23 The road sign shows a slope of 20%. This means the road climbs 1 m for every 5 m travelled horizontally. What angle does the road make with the horizontal, to 1 decimal place?

24 The support wire for a 7 m flagpole is secured 2,5 m from the foot of the pole. What angle does the support wire make with the ground, to 1 decimal place?

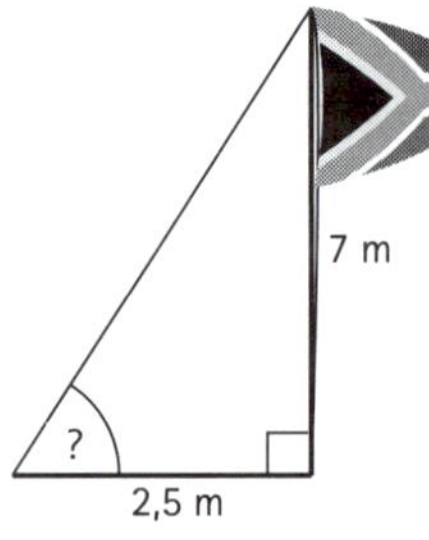

Angles of elevation and depression

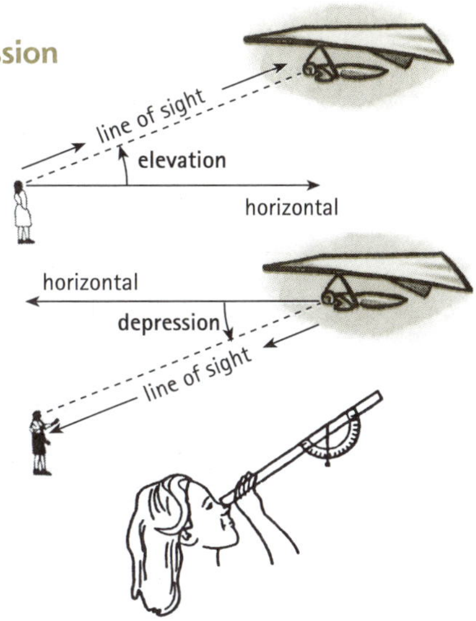

If you look **up** at an object, the angle between the horizontal and your line of sight is the **angle of elevation.**

If you look **down** at an object, the angle between the horizontal and your line of sight is the **angle of depression.**

Angles of elevation and depression can be measured using a **clinometer.**

EXAMPLES

1 A lighthouse is situated at the top of a vertical cliff, and the observation room is at a height of 120 m above sea-level. A ship is spotted at an angle of depression of 7°. How far out to sea is the ship, to the nearest metre?

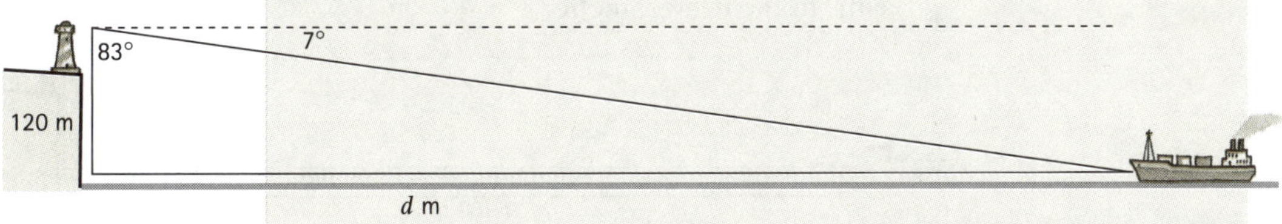

If the angle of depression is 7°, the angle **inside** the triangle is 90° – 7° = 83°.

Label the distance you are asked to find as d.

$$\text{Opposite} = \text{adjacent} \times \tan\theta$$
$$d = 120 \times \tan 83°$$
$$= 977{,}32157\ \dots$$

The ship is 977 m out to sea (to the nearest metre).

2 The angle of elevation of the top of a communications tower is 42° when measured at a point 76 m from the base of the tower. How high is the tower, correct to 1 decimal place?

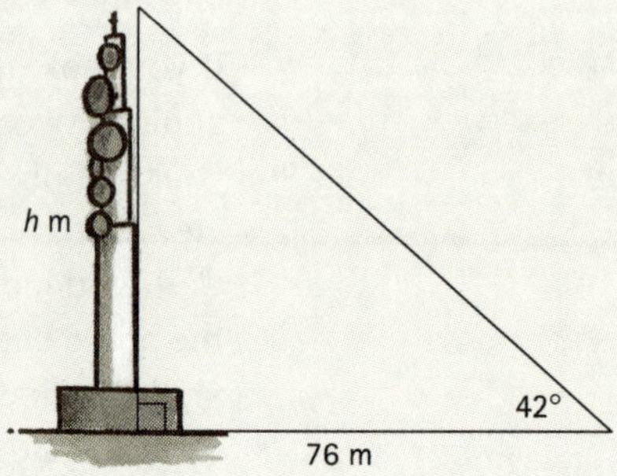

Label the height you are asked to find as h.

$$\text{Opposite} = \text{adjacent} \times \tan\theta$$
$$h = 76 \times \tan 42°$$
$$h = 68{,}43070 \ldots$$

The tower is 68,4 m high (to 1 decimal place).

Exercise

1 Sipho is at the top of a vertical cliff and measures the angle of depression of a boat 1 100 m out to sea to be 6°. How high is the cliff, to the nearest metre?

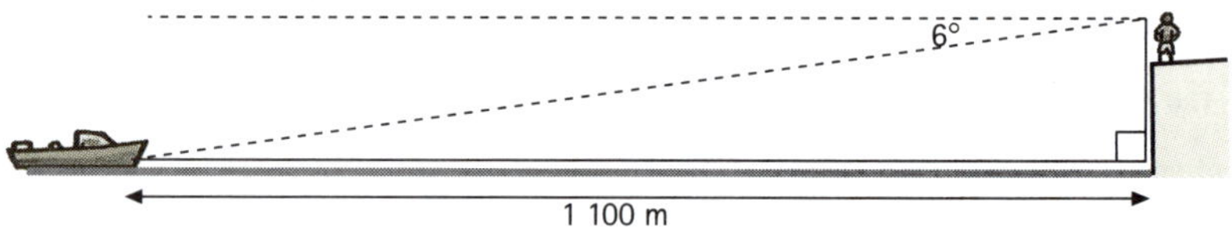

2 An aircraft flying at a height of 600 m has an angle of elevation of 4° measured from the runway. What horizontal distance must the aircraft cover before it reaches the runway, correct to 1 decimal place?

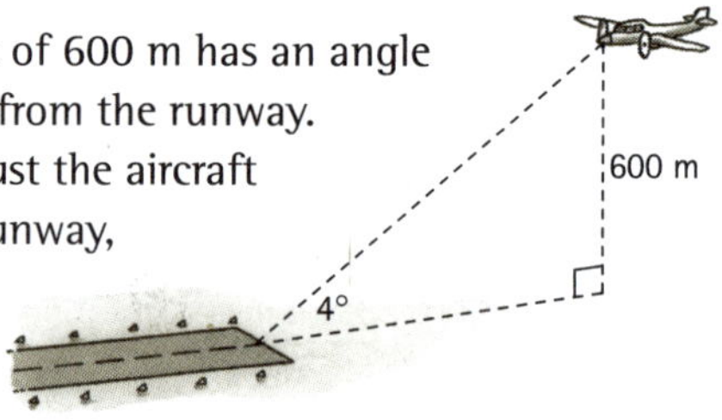

3 A boat detects a wreck at an angle of depression of 3°. If the wreck is at a depth of 76 m, how far must the boat sail to be directly above the wreck, to the nearest metre?

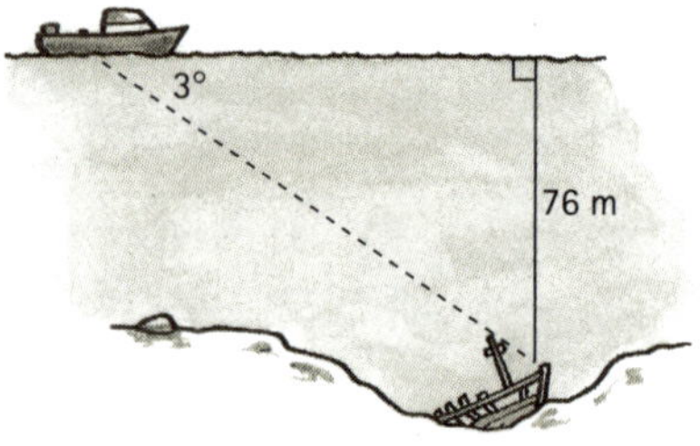

4 A ship's captain sees a lighthouse at an angle of elevation of 2,5°. The chart gives the height of the lighthouse as 136 m above sea-level. How far from the lighthouse is the ship, to the nearest metre?

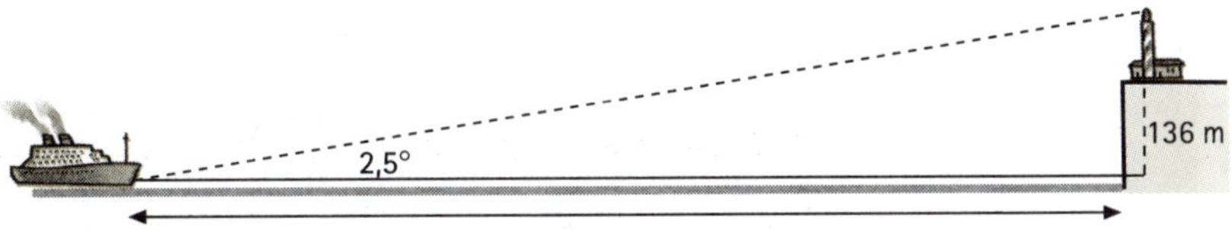

5 Vic is standing 53 m from the foot of a 14 m flagpole. What is the angle of elevation of the top of the pole from where Vic is standing, to 1 decimal place?

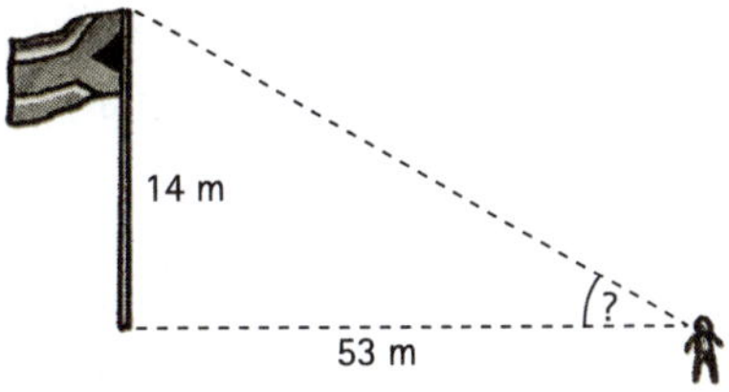

PEER ASSESSMENT

See page 288 (rubric 3).

Investigation: Calculating inaccessible heights

Work with a partner.

You will need: a clinometer, a tape measure.

You can calculate the height of an inaccessible object by using a clinometer and tape measure.

1 Find a tree, flagpole, building or other tall object in the school grounds.

2 Find a position that is on the same horizontal level as the base of the object, and from which you can clearly see the top of the object.

3 Measure the angle of elevation of the top of the object.

4 Measure the horizontal distance from your position to the base of the object.

5 Calculate the height of the object using the tangent ratio.

6 You have calculated the height from the position where you took the clinometer reading, which is your eye-level. Add on the distance from the ground up to your eye-level.

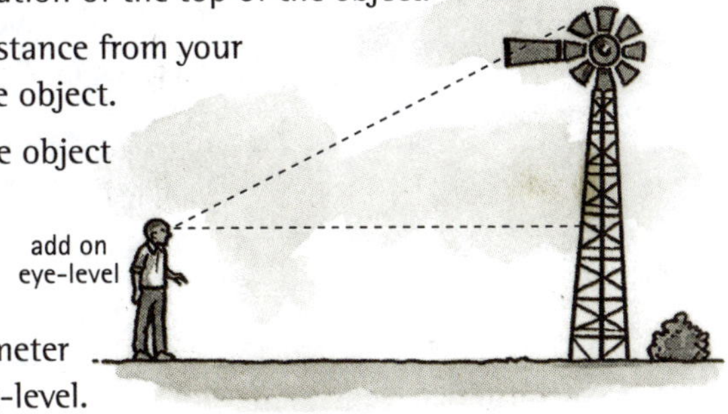

Swap your answers with another pair of learners and check each other's answers.

The sine ratio

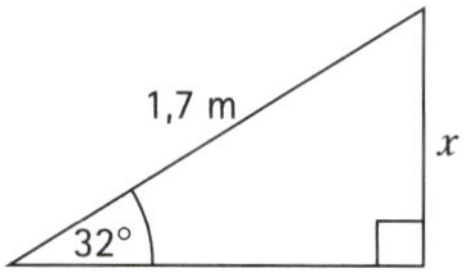

You will use the tangent ratio whenever an angle, its opposite side and its adjacent side are involved. If you know two of these values, you can calculate the third.

But what if you know an angle and the hypotenuse, and want to find another side?

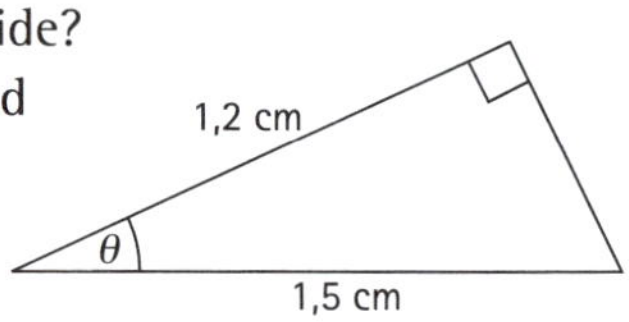

Or what if you need to find an angle, and only know the lengths of the side adjacent to it and the hypotenuse?

Two more ratios, like the tangent ratio, can help you with these problems.

Note: "sin θ" is pronounced "sine theta".

In a right-angled triangle with an angle θ, the ratio $\frac{\text{opposite side}}{\text{hypotenuse}}$ is called the **sine of θ**.

$$\sin\theta = \frac{\text{opposite}}{\text{hypotenuse}}$$

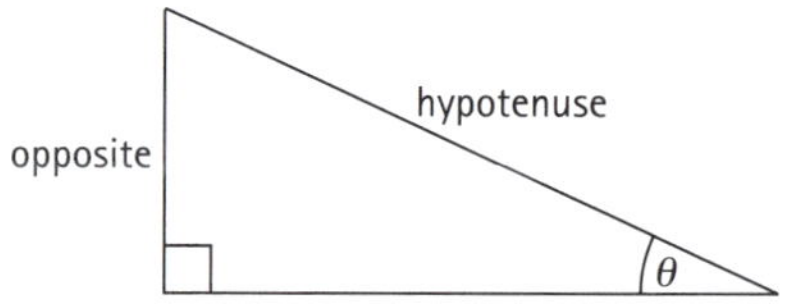

To find the sine of an angle using your calculator, enter the angle, then press the [sin] key (or press [sin] followed by the angle on direct logic calculators). Always make sure you use the correct order for **your** calculator.

EXAMPLES

1 What is the sine of 37° (to 2 decimal places)?

Press [37] [sin]. (Or press [sin] [37].)
The display shows 0.6018150 … .
The sine of 37° is 0,60 (to 2 decimal places).

2 Calculate the length of the side labelled x in this triangle (to 1 decimal place).

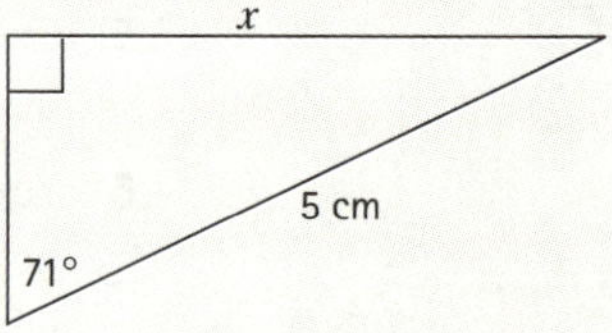

$$\sin 71° = \frac{\text{opposite}}{\text{hypotenuse}}$$
$$= \frac{x}{5}$$
$$x = 5 \times \sin 71°$$

Press [5] [×] [71] [sin] [=].
The display shows 4.7275928 … .
Side x is 4,7 cm long (to 1 decimal place).

3 Calculate the length of the hypotenuse in this triangle (to 1 decimal place).

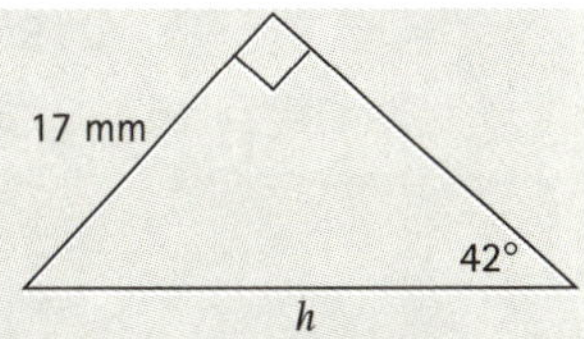

Call the hypotenuse h.

$$\sin 42° = \frac{17}{h}$$

$$h \times \sin 42° = 17$$

$$h = \frac{17}{\sin 42°}$$

Press [17] [÷] [42] [sin] [=].

The display shows 25.40610135 … .

The hypotenuse is 25,4 mm (to 1 decimal place).

Exercise

Find the sine of each angle correct to 2 decimal places:

1 30° **2** 10° **3** 78° **4** 45° **5** 23°

6 $25\frac{1}{2}°$ **7** 88,1° **8** $2\frac{1}{4}°$ **9** 0° **10** 90°

Calculate the length of the side labelled d in each triangle (correct to 1 decimal place):

11

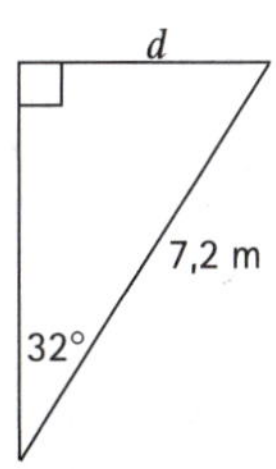

12

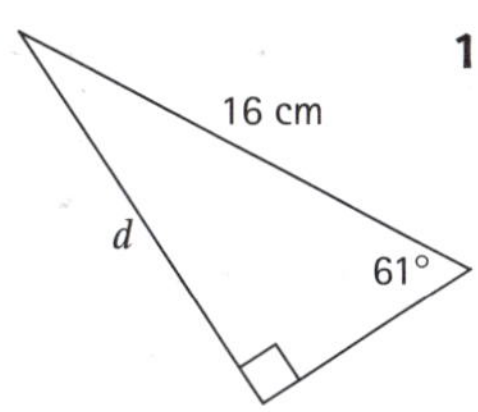

13

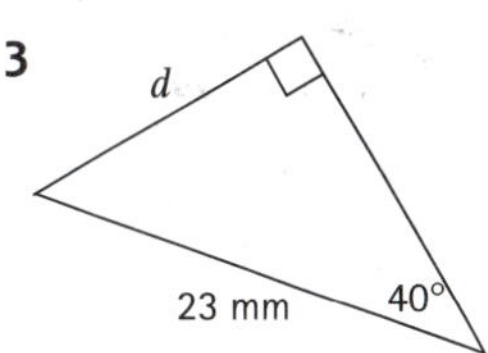

14

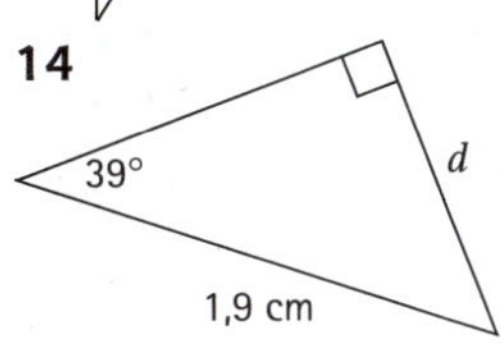

15

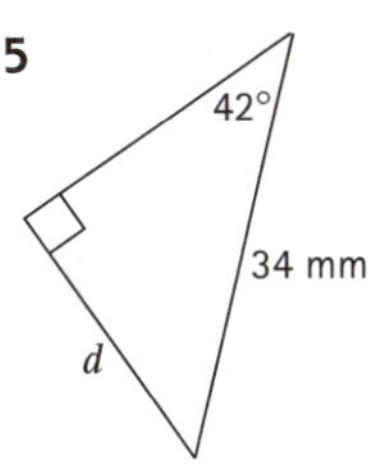

16

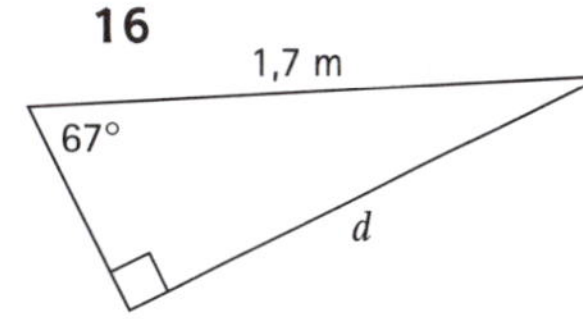

Find the length of the hypotenuse in each triangle, correct to 1 decimal place:

17

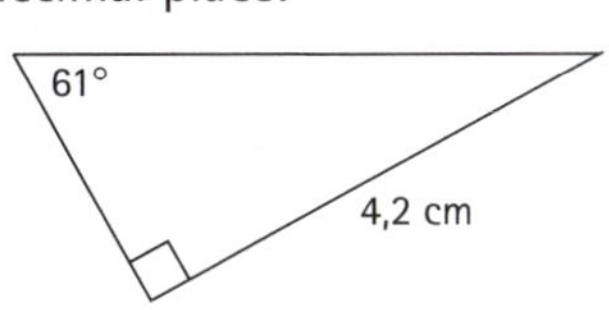

18

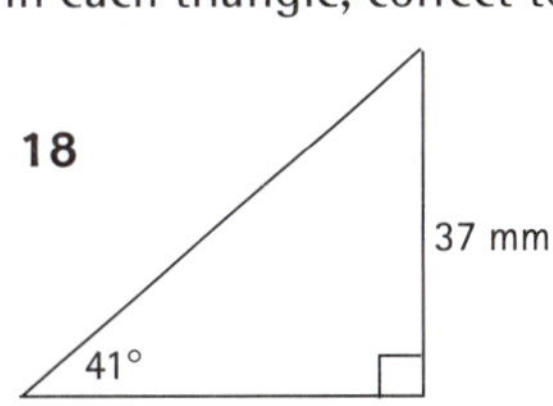

19
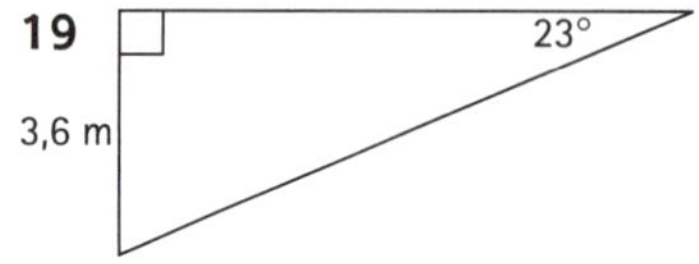

20
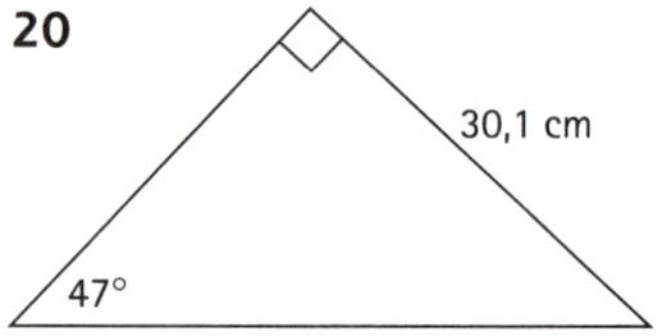

21
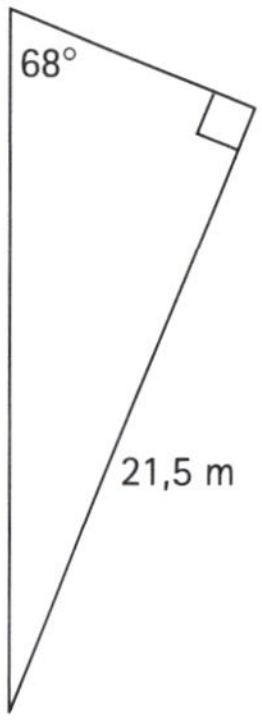

22
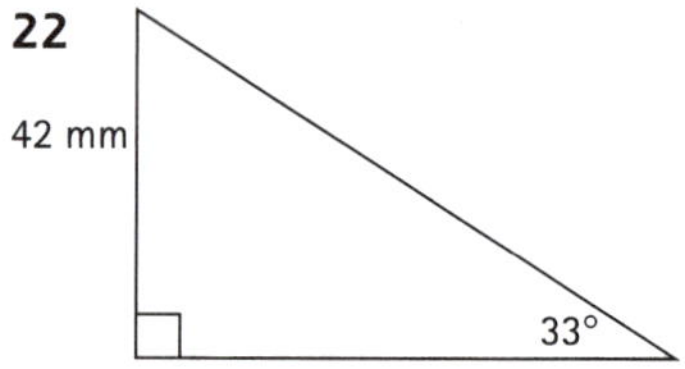

Finding an angle

To find an angle when you know the sine ratio, enter the ratio, then press [SHIFT] [sin].

EXAMPLES

1 What angle has a sine of 0,41? (Give your answer correct to the nearest degree.)

Press [0.41] [SHIFT] [sin]. The display shows 24.204834 ... .
The angle whose sin is 0,41 is 24° (to the nearest degree).

2 Find angle θ in this triangle, to the nearest degree.

$$\sin\theta = \frac{\text{opposite}}{\text{hypotenuse}}$$

$$\sin\theta = \frac{16}{23}$$

$$= 0{,}695652\ ...$$

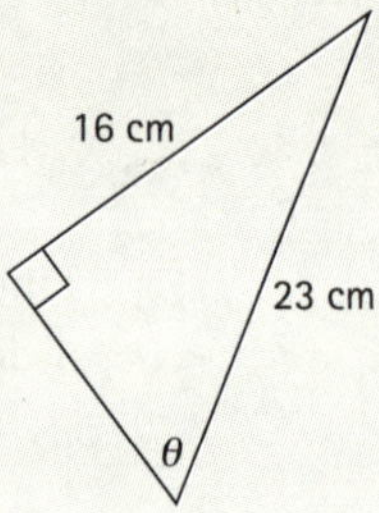

Leave this on the calculator screen, then press [SHIFT] [sin].
The display shows 44.079209 ... .
$\theta = 44°$ (to the nearest degree).

Exercise

Write the angle (to the nearest degree) that has a sine of:

1 0,1 **2** 0,3 **3** 0,95 **4** 0,355 **5** 0,719

6 $\frac{1}{4}$ **7** $\frac{3}{4}$ **8** $\frac{3}{7}$ **9** $\frac{12}{25}$ **10** $\frac{3}{16}$

Find the size of angle θ in each triangle (to the nearest tenth of a degree):

11

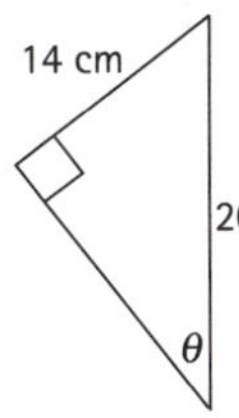

12

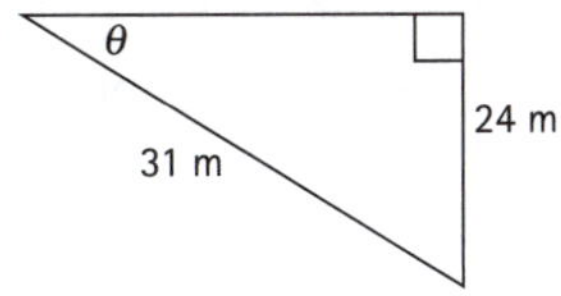

13

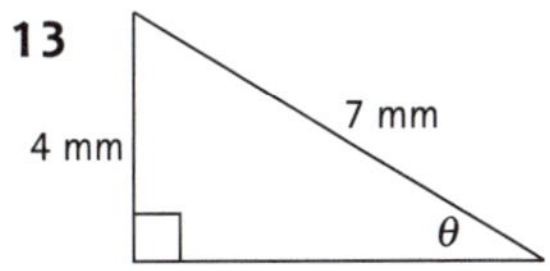

The cosine ratio

In a right-angled triangle with an angle θ, the ratio $\frac{\text{adjacent side}}{\text{hypotenuse}}$ is called the cosine of θ. The abbreviation for cosine is "cos".

$$\cos\theta = \frac{\text{adjacent}}{\text{hypotenuse}}$$

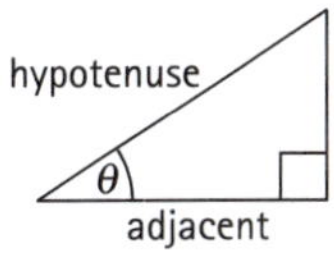

EXAMPLES

1 Find the cosine of 52° correct to 3 decimal places.

Press [52] [cos]. the display shows 0.615661475 .... .
The cosine of 52° is 0,616 (to 3 decimal places).

2 Find the length of side RS in this triangle, to the nearest centimetre.

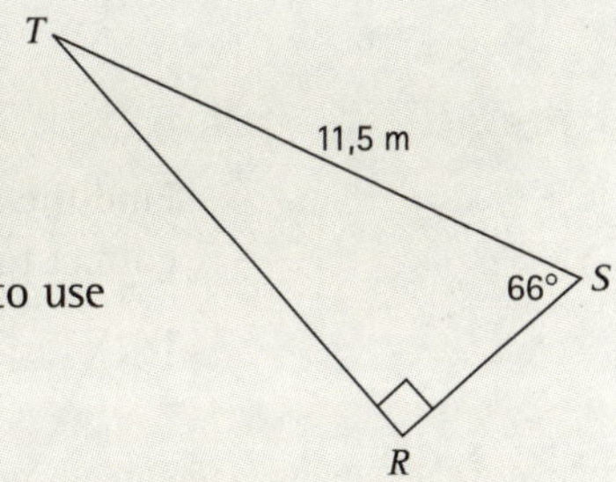

RS is the side adjacent to the 66° angle, and the 11,5 m side is the hypotenuse. This means you need to use the cosine ratio.

$$\cos\theta = \frac{\text{adjacent}}{\text{hypotenuse}}$$

$$\text{adjacent} = \text{hypotenuse} \times \cos\theta$$

$$RS = 11{,}5 \times \cos 66°$$

Press [11.5] [×] [66] [cos] [=].
The display shows 4.677471… .
Side RS is 4,68 m (to the nearest centimetre).

3 A ladder is placed 3 m from the foot of a wall. It makes an angle of 70° with the ground. How long is the ladder (to 2 decimal places)?

$$\cos 70° = \frac{3}{x}$$

$$x \times \cos 70° = 3$$

$$x = \frac{3}{\cos 70°}$$

Press [3] [÷] [70] [cos] [=].
The display shows 8.7714132… .
The ladder is 8,77 m long (to 2 decimal places).

Exercise

Find the cosine of each angle, to 3 decimal places:

1 20°	**2** 72°	**3** 45°	**4** 30°	**5** 60°
6 12°	**7** 5°	**8** 0°	**9** 85°	**10** 90°

Find the length of the side marked d in each triangle, correct to 2 decimal places:

11

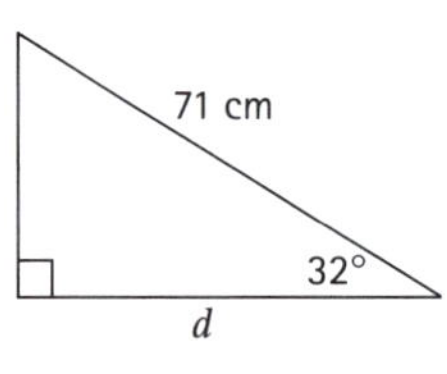

12

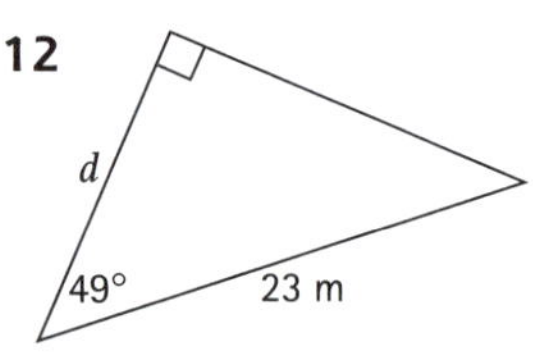

13

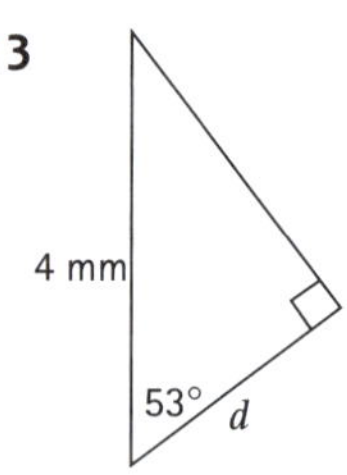

Find the length of the hypotenuse in each triangle, correct to 1 decimal place:

14

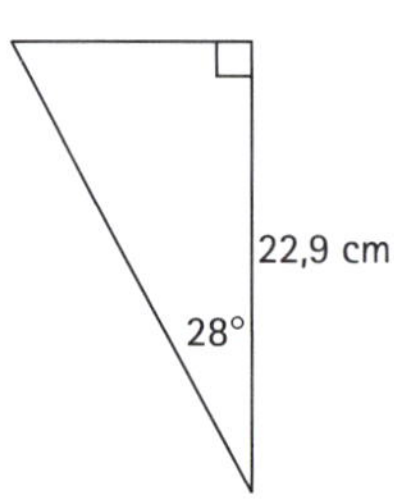

15

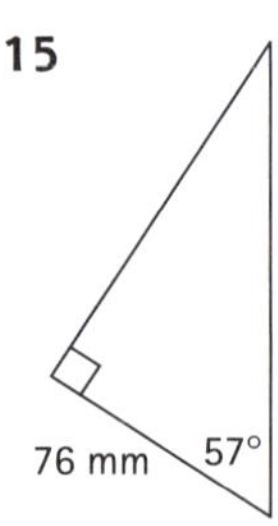

16

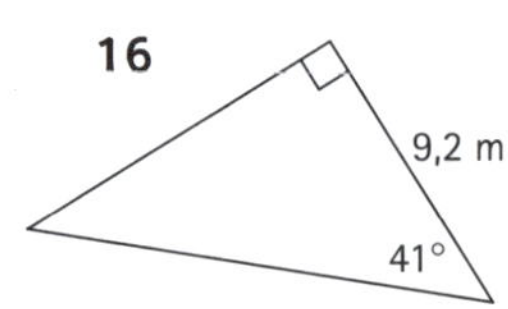

Finding an angle

To find an angle when you know the cosine ratio, you need to press [SHIFT] [cos].

EXAMPLES

1 Find the angle, to the nearest degree, whose cosine ratio is 0,7.

Press [0.7] [SHIFT] [cos]. The display shows 45.572996… .
The angle whose cosine ratio is 0,7 is 46° (to the nearest degree).

2 Find the size of angle θ, to 1 decimal place.

Adjacent side = 17 mm; hypotenuse = 21 mm.

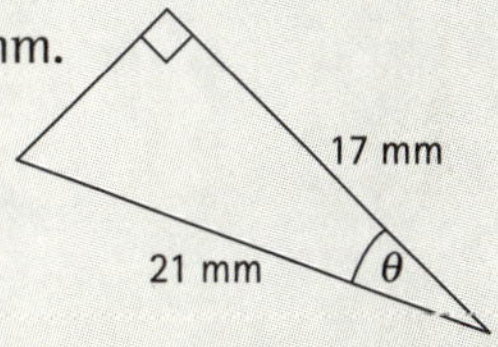

$$\cos\theta = \frac{\text{adjacent}}{\text{hypotenuse}}$$
$$= \frac{17}{21}$$

Press [17] [÷] [21] [=] [SHIFT] [cos].
The display shows 35.9505676… .
θ = 36,0° (to 1 decimal place).

Exercise

Find the angle, to the nearest degree, whose cosine ratio is:

1 0,3 **2** 0,16 **3** 0,85 **4** 0,5 **5** 0,05

6 0,418 **7** 0,297 **8** $\frac{1}{4}$ **9** $\frac{1}{3}$ **10** $\frac{7}{19}$

Find the size of angle θ in each triangle, to the nearest degree:

11

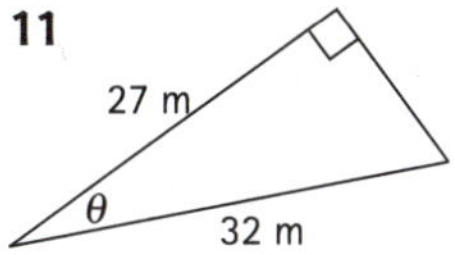

12

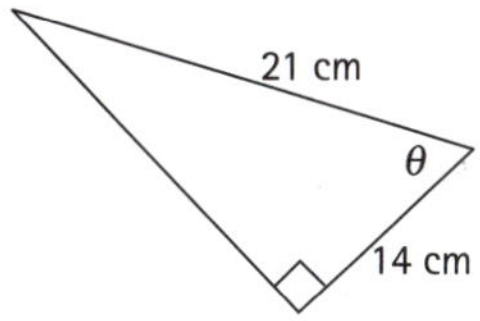

13

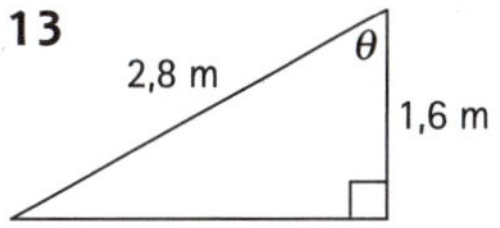

Sine, cosine or tangent?

When you solve problems with right-angled triangles, it is essential to choose the correct ratio. Each problem involves **two sides** and **one angle**. You will be asked to find either a side or the angle.

- If the opposite and hypotenuse are involved, use **sine**.
- If the adjacent and hypotenuse are involved, use **cosine**.
- If the opposite and adjacent are involved, use **tangent**.

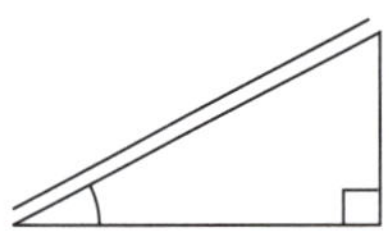
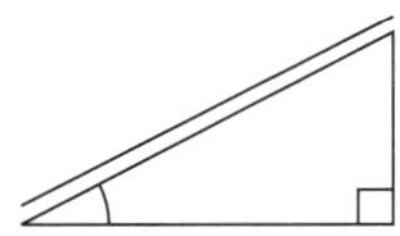
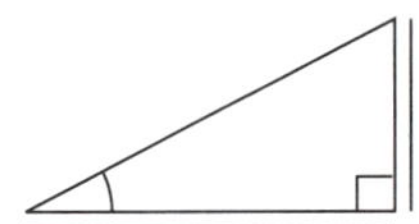

Remember these:

Sine $\theta = \frac{\text{opposite}}{\text{hypotenuse}}$ Cosine $\theta = \frac{\text{adjacent}}{\text{hypotenuse}}$ Tangent $\theta = \frac{\text{opposite}}{\text{adjacent}}$

In short:

SOH CAH TOA

EXAMPLES

a Which ratio should you use to find x?

b Find x, correct to 1 decimal place.

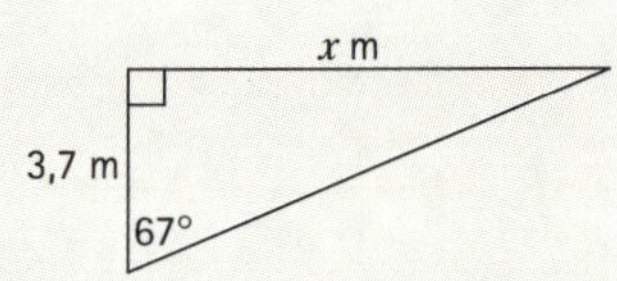

a Side x m is opposite the 67° angle; side 3,7 m is adjacent to the 67° angle;

so you use $\tan\theta = \frac{\text{opposite}}{\text{adjacent}}$

b $\tan 67° = \frac{x}{3{,}7}$

$x = 3{,}7 \tan 67°$ The display shows 8.7166537… .

$x = 8{,}7$ (to 1 decimal place)

Exercise

a Which ratio should be used to find x?

b Find x correct to 1 decimal place.

1

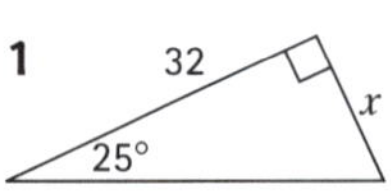

2

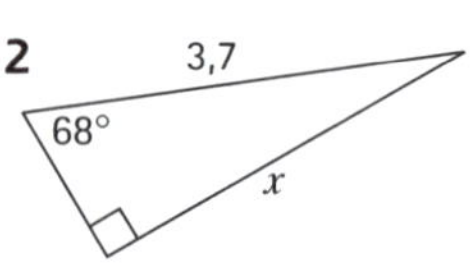

3

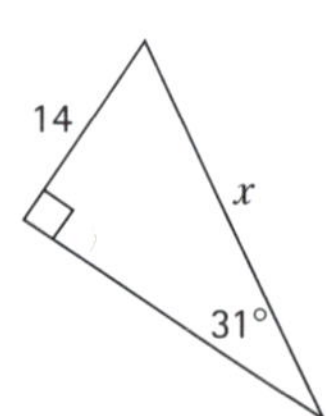

4

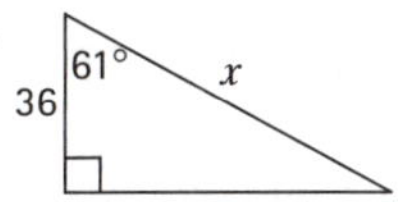

5

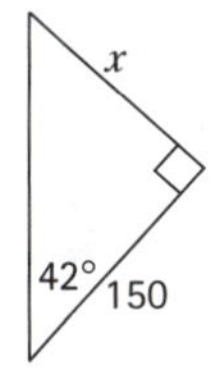

6

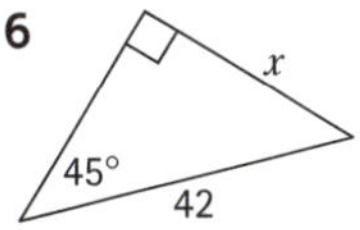

7

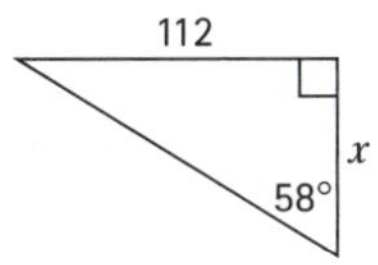

8

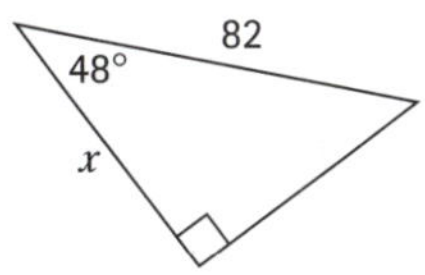

9

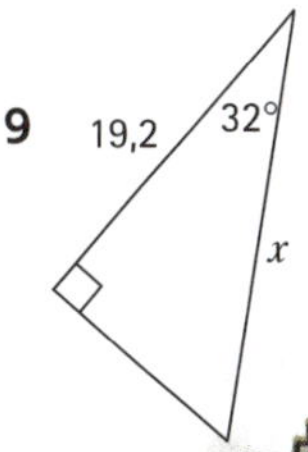

10 A cablecar takes tourists to the top of a mountain. The cable is 2,5 km long and makes an angle of 41° with the ground. What is the height of the mountain, to the nearest metre?

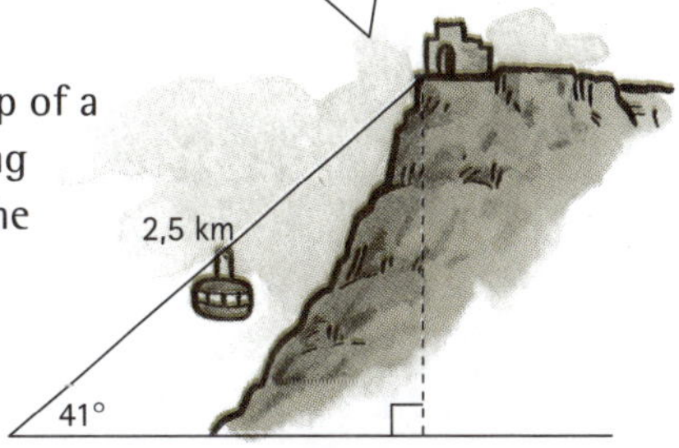

11 An escalator in a shopping centre rises 8 m from one floor to the next. If the escalator slopes at an angle of 30°, how long is the escalator?

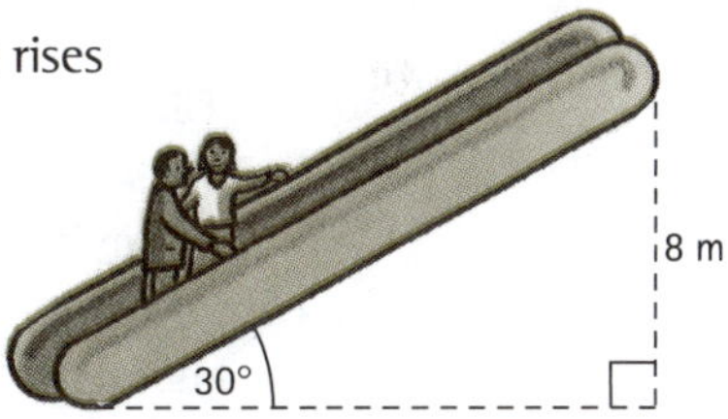

12 Zak is flying a kite on 15 m of string, and measures the angle of elevation of the kite to be 52°.

- **a** How high is the kite flying, to 1 decimal place?
- **b** Busisiwe is standing directly underneath the kite. How far apart are Busisiwe and Zak, to 1 decimal place?

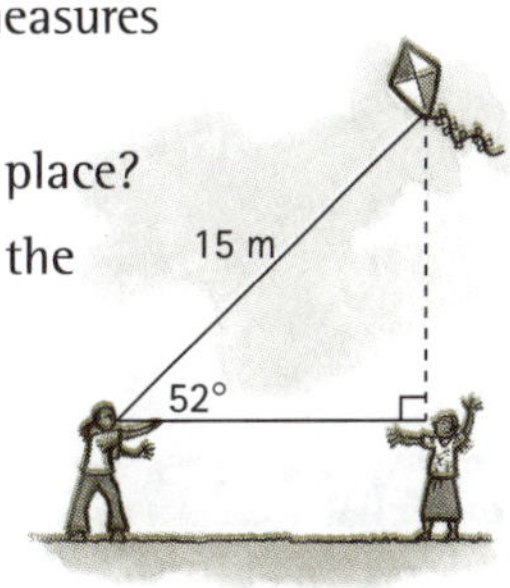

13 A boat is lying at anchor, and the length of the anchor cable is 13,5 m. If the angle between the anchor cable and the horizontal sea-bed is 37°, what is the depth of the water, to the nearest metre?

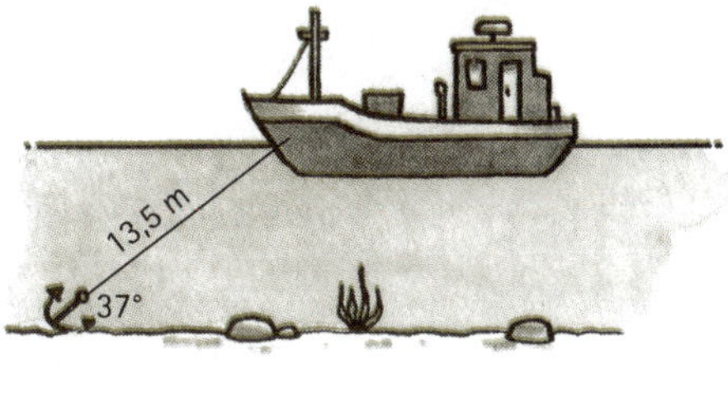

Degrees, minutes and seconds

If you need to measure land or buildings, you will need to use angles. Angles are measured in degrees, minutes and seconds. A complete revolution (turn) of a circle is 360 degrees (360°). A quarter turn of a circle is 90° or a right angle. This is the same angle as the corner of a square or a rectangle. By using right-angled triangles you will be able to measure angles and distances of many things, no matter how big they are.

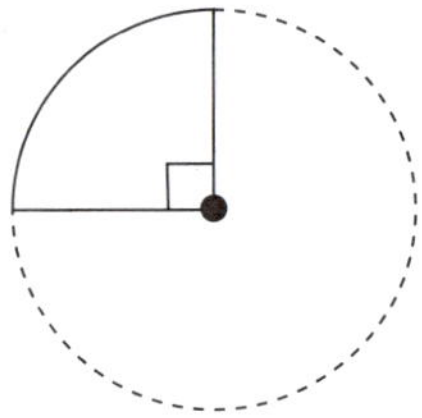

Remember:
One degree = 60 minutes
One minute = 60 seconds

One degree (1°) is the angle at the centre of the circle that describes an arc that takes up $\frac{1}{360\text{th}}$ of the circumference of the circle.

One minute (1') = $\frac{1}{60\text{th}}$ of a degree

One second (1") = $\frac{1}{360\text{th}}$ of a degree

For example, 26°25'16" means 26 degrees, 25 minutes and 16 seconds.

Trigonometry isn't the only field in which you will use degrees, minutes and seconds. When you locate places on maps, you also describe a position in degrees, minutes and seconds.

Angles can be entered into a calculator so that sin θ, cos θ and tan θ can be evaluated accurately. On some scientific calculators, you will need to convert angles to degrees only and enter them as a decimal. On most graphics calculators, angles can be entered directly, with degrees, minutes and seconds.

EXAMPLES

1 **a** Convert 15°35'24" to degrees only.
b Use a calculator to evaluate cos 15°35'24".

Hint: For a scientific calculator press [15.59], then [cos].

a $24'' = \frac{24}{60}$ and $35{,}4' = \frac{35{,}4}{60}$

$= 0{,}4'$ $\quad$ $= 0{,}59°$

$\therefore$ 15°35'24" = 15,59°

b cos 15°35'24" ≈ 0,9632

2 a Convert 36,18° to degrees and minutes, rounding to the nearest minute, then check your answer on a scientific or graphics calculator.

b Evaluate sin 36,18°.

Hint: Check your answer on a scientific or graphics calculator by pressing the key that directly converts to degrees, minutes and seconds.

a $0,18° = 0,18 \times 60$

$= 10,8'$

$\approx 11'$

$\therefore 36,18° \approx 36°\ 11'$

b $\sin 36,18° = \sin 36°10'48'' = 0,5903$

Exercise

1 Convert these angles in degrees and minutes to degrees only. Round off to 2 decimal places where necessary.

a 25°30' **b** 41°15' **c** 83°45' **d** 42°18'

e 59°13' **f** 1°14' **g** 23°52' **h** 58°22'

2 Use a calculator to evaluate these ratios by first converting to degrees only. Round off to 4 decimal places.

a sin 16°30' **b** cos 38°15' **c** tan 3°12'

d sin 83°48' **e** cos 19°13' **f** tan 43°37'

g sin 37°16' **h** cos 59°54'

3 Convert these angles to degrees and minutes. Round off to the nearest minute in each case.

a 18,5° **b** 24,25° **c** 5,037° **d** 83,2°

e 54,7° **f** 36,14° **g** 69,41° **h** 38,39°

4 Use a calculator to evaluate. Round off your answers to 4 decimal places.

a sin 37,43° **b** cos 27,94° **c** tan 36,73°

d sin 24,06° **e** cos 0,59° **f** tan 18,47°

g sin 23°24' **h** cos 20°21' **i** tan 49°10'

j sin 43°18' **k** cos 2°43' **l** tan 0,73°

Remember: Angle θ is called the **angle of elevation**, and it can be measured using a **clinometer**.

Other applications of ratio and proportion

Scale drawings

The height of a flagpole, tower or tall building can be found from a scale drawing if the following measurements are known:

- the length of the shadow
- the angle between the horizontal and the line of sight of the top of the building
- the eye-level of the person making the measurements.

EXAMPLE

In the figure above:

- θ is the angle of elevation. It is measured as 31°.
- The length of the shadow of the mosque is 18,4 m.
- The eye-level of the observer is 1,5 m.

From this information, you can construct a scale drawing.

A suitable scale here would be 1 cm to 2 m.

Draw in the 18,4 m shadow length.
(On a scale of 1 cm to 2 m, the length on the drawing is 9,2 cm.)

Measure the 31° angle, and draw in the line of sight.

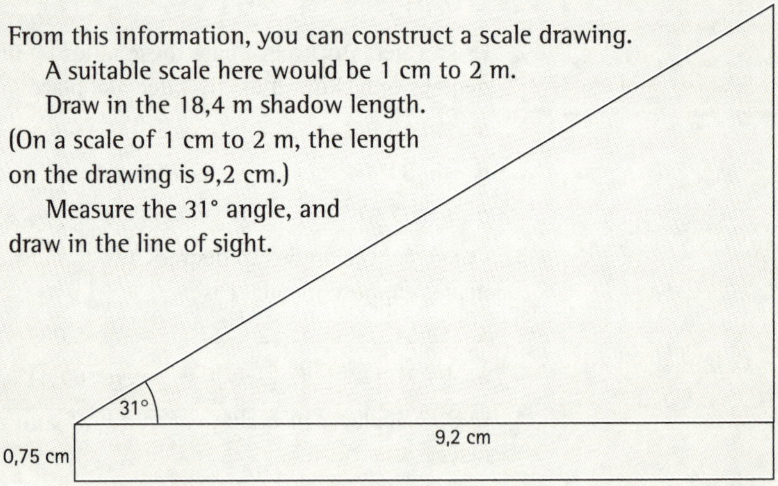

Add on the eye-level of the observer.

Draw in the height of the mosque, measure its distance on the drawing, and then calculate its actual height.

The height on the drawing is 6,2 cm. To calculate the actual height, use the scale of 1 cm to 2 m: 6,2 cm represents 12,4 m. So the height of the mosque is 12,4 m.

Exercise

Draw a scale drawing to represent each situation (use the suggested scale), and then calculate the height of the object:

1

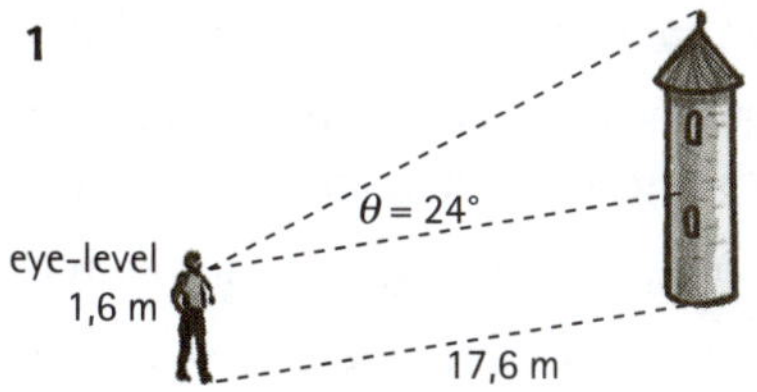

Use a scale of 1 cm to 2 m.

2

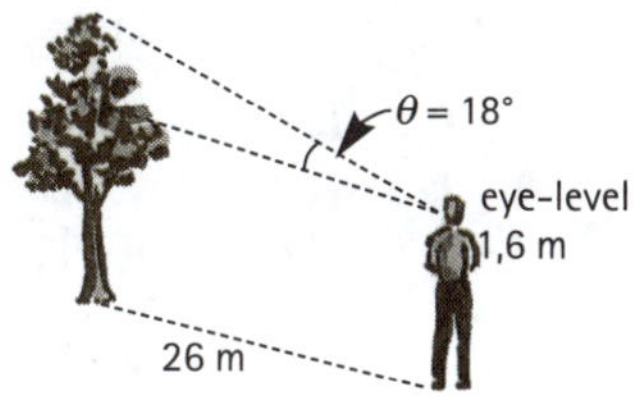

Use a scale of 1 cm to 4 m.

3

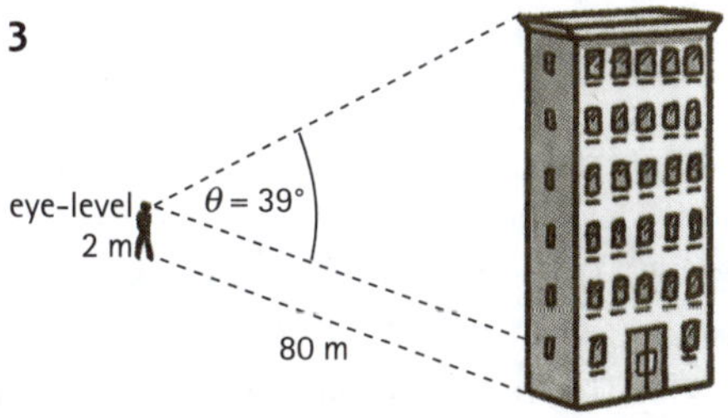

Use a scale of 1 cm to 10 m.

Investigation: A scale drawing

Work in groups of five.

You will need: a tape measure, a clinometer.

Find a tall building in the school grounds. Use the method of scale drawing to calculate the height of the building.

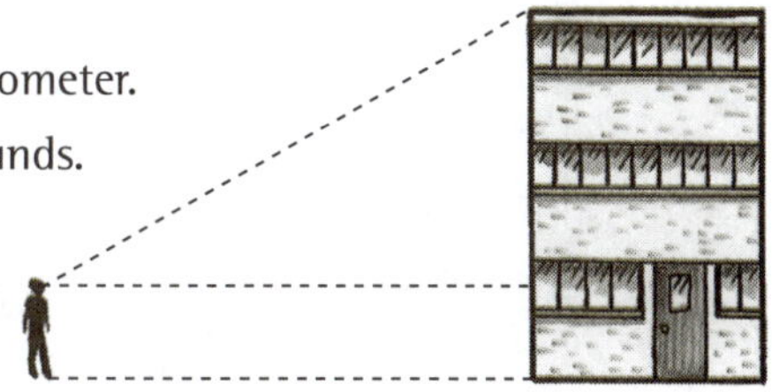

1 Draw a sketch like the one shown, and then mark on the measurements as you make them.

2 Measure the length of the building's shadow.

3 Stand at the end of the shadow and measure the angle of elevation to the top of the building.

4 Measure the eye-level of the person measuring the angle of elevation.

5 Select a suitable scale and construct a scale drawing.

6 Measure the height of the building on the drawing, and then use the scale to calculate its actual height.

Swap your scale drawing with someone else and check each other's work.

Investigation: Building a dog kennel

You have a piece of plywood that measures 280 cm by 150 cm, and your task is to design and build a dog kennel. The kennel should be as large as possible, does not require a floor and must not have a flat roof.

1 Use graph paper to draw a plan showing how the pieces would be cut from the plywood. (Remember to state the scale.)
2 Label each piece, describing its position in the completed kennel.
3 Make a three-dimensional sketch to show what the finished kennel will look like.

Maps

When you interpret maps, you also apply your knowledge of ratio and proportion. The scale given on a map is often expressed as a ratio, for example, 1:50 000, or 1:1 000 000. Look at the map on this page, which shows part of the Western Cape province. It has a scale of 1:1 000 000. This tells you that 1 cm on the map represents an actual distance of 10 km.

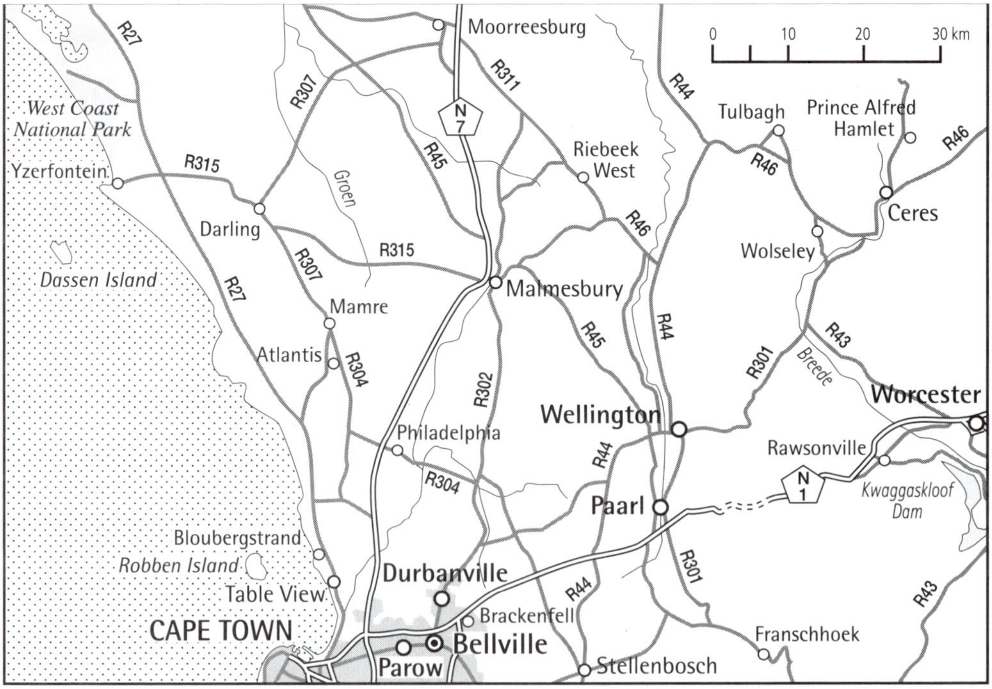

Scale 1 : 1 000 000

EXAMPLES

1 Use the map to find the distance in kilometres (as the crow flies) between Malmesbury and Ceres.

The map distance between Malmesbury and Ceres is 5,3 cm (found with a ruler).

actual distance = map distance in cm × 10 km
= 5,3 × 10 km
= 53 km

2 How should you interpret the scale in the top right-hand corner of the map?

The scale is 3 cm long, but it represents a distance of 30 km.

So 1 cm represents 10 km ($\frac{30}{3} = 10$), which is the same as 1 cm to 1 000 000 cm or 1:1 000 000.

3 The scale on a map is 1:42 000 000.

a What distance does 1 cm on the map represent in kilometres?

b Two capital cities on this map are 6,2 cm apart. What is the distance between them in kilometres?

c A major provincial centre is 672 km from its capital city. How far apart will the centre and the capital city be on the map?

a A scale of 1:42 000 000 means that
1 cm on the map represents an actual distance of 42 000 000 cm
∴ 1 km = 100 000 cm (1 km = 1 000 m = 1 000 × 100 cm)
∴ 1 cm on the map represents an actual distance of $\frac{42\ 000\ 000}{100\ 000} = 420$ km

b map distance = 6,2 cm
actual distance = 6,2 × 420 = 2 604 km

c actual distance = 672 km
map distance = $\frac{672}{420}$ = 1,6 cm

Photography

Cameras also work on the principle of ratios. Here is a simplified diagram that shows the path along which light travels from an object you want to photograph to its image on the film in your camera. In the diagram:

- h_0 represents the height of the object you are photographing
- d_0 represents the distance of the object from the lens of the camera
- h_i represents the height of the image on the film of the camera
- d_i represents the distance of the image from the lens of the camera.

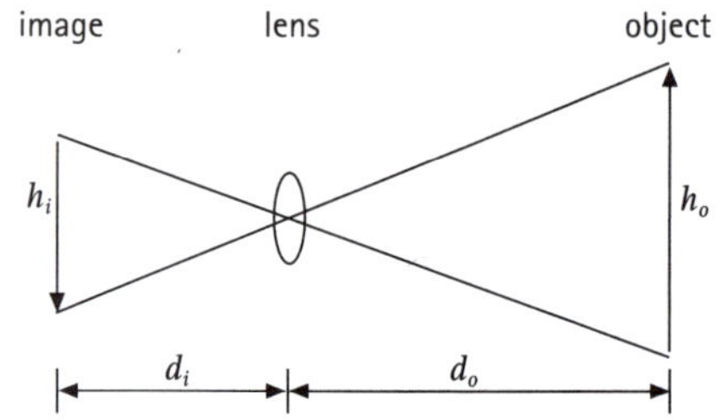

It can be shown that the ratios:

$$d_0 : d_i = h_0 : h_i$$

This means that the distances and heights are in proportion. Thus, if you know three of the four values, you can calculate the fourth, using the formula:

$$\frac{d_0}{h_0} = \frac{d_i}{h_i}$$

EXAMPLE

You are photographing a tree 150 m away. The height of the image of the tree on the film of your camera is 9 mm and the film is 45 mm from the lens. How tall is the tree?

From the information given:

$d_0 = 150$ m $\quad d_i = 45$ mm $\quad h_i = 9$ mm $\quad h_0 = ?$

If you substitute the information in the formula, you get:

$$\frac{150}{h_0} = \frac{45}{9} = 5$$

$$\therefore \quad h_0 = \frac{150}{5}$$

or $h_0 = 30$ m

Exercise

1 Use the map on page 124 to find, to the nearest kilometre, the shortest distance between:

a Wellington and Malmesbury

b Tulbagh and Wellington

c Wellington and Paarl

d Ceres and Paarl

e Brackenfell and Ceres.

2 These scales have all been taken from different maps. Use them to determine the scale of the map in terms of centimetres per kilometre.

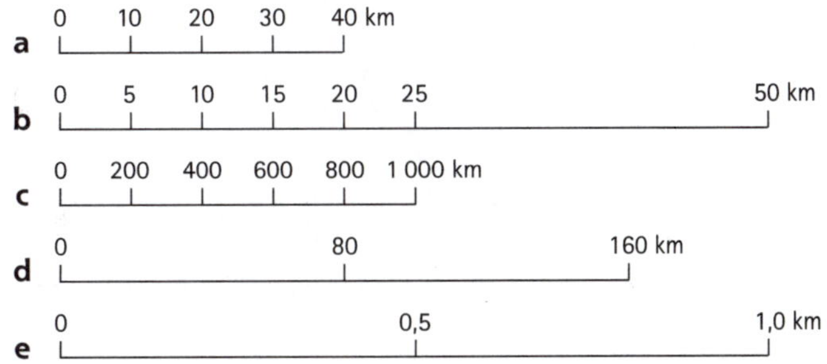

3 On a street map the distance between two points is 8 cm. If the actual distance between the points is 2 km, to what scale is the map drawn?

4 Neo is drawing a map of the school grounds to a scale of 1 cm to 10 m. What size should each of the following distances be on his drawing:

a the driveway, which is 30 m long?

b the front fence, which is 55 m long?

c the science block, which is 27 m long?

5 The scale on a map is 1:380 000.

a How far does 1 cm on the map represent in kilometres?

b Two towns on this map are 3,7 cm apart. What is the distance between them in kilometres?

c Two towns are 276 km apart. How far apart will they be on the map?

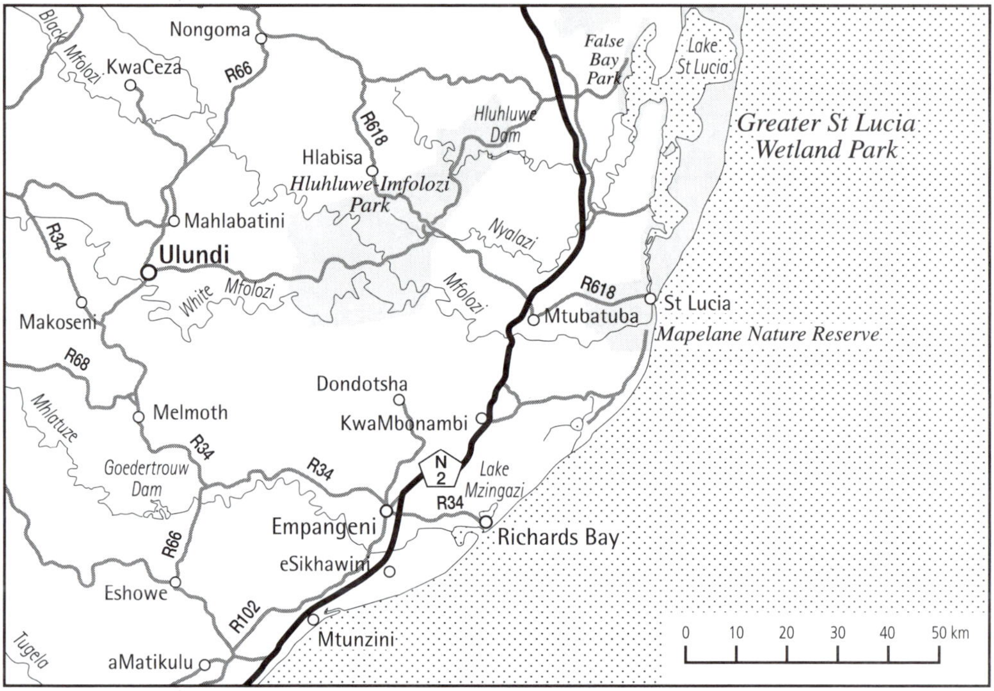

Scale 1 : 1 500 000

6 Use the scale information on this map of the eastern part of KwaZulu-Natal to determine to the nearest kilometre the shortest distance between:

- **a** Ulundi and Empangeni
- **b** Ulundi and Richards Bay
- **c** Mtubatuba and Hlabisa
- **d** Nongoma and St Lucia
- **e** Dondotsha and KwaCez a.

7 You are photographing a building 200 m away. The height of the image of the building on the film of your camera is 5 mm and the film is 50 mm from the lens. How tall is the building.

8 You are photographing a person who is 180 cm tall. The height of the image of the person on the film of your camera is 30 mm and the film is 10 mm from the lens. How far is the person from the camera? Give your answer to the nearest metre.

Models and building plans

Ratio and proportion are also applied when people make a scale model of an object. Architects and draughtspeople use a particular scale when drawing building plans.

Exercise

1 This model of an SAA 747 aeroplane was built to a scale of 1 cm to 20 m. How long is the real aeroplane?

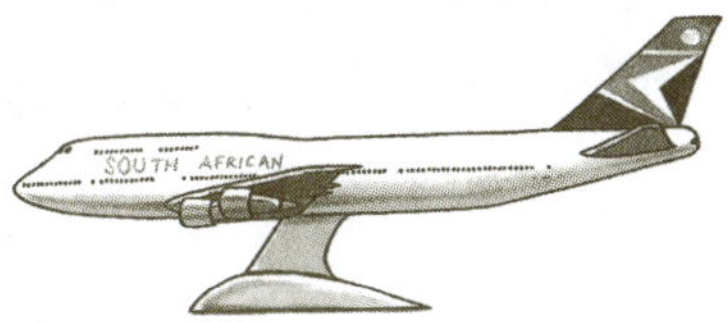

2 In 1999, Gladys received a house as part of a housing programme. This year, she built a new room onto her house. These two drawings show Gladys's house from two different positions.

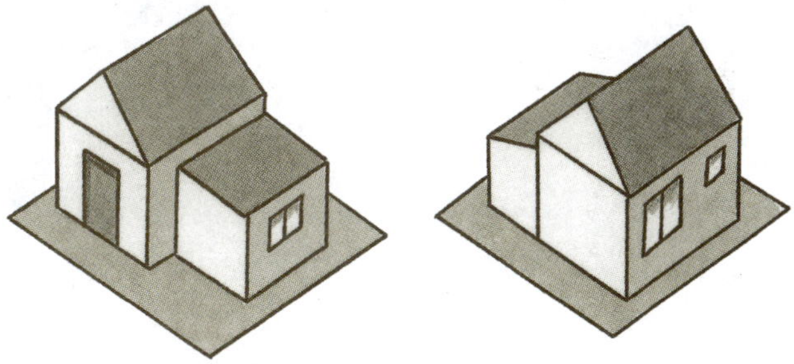

For each of the views shown in Figures 1, 2 and 3 on page 130, decide from which of the arrows A to H the view would be seen.

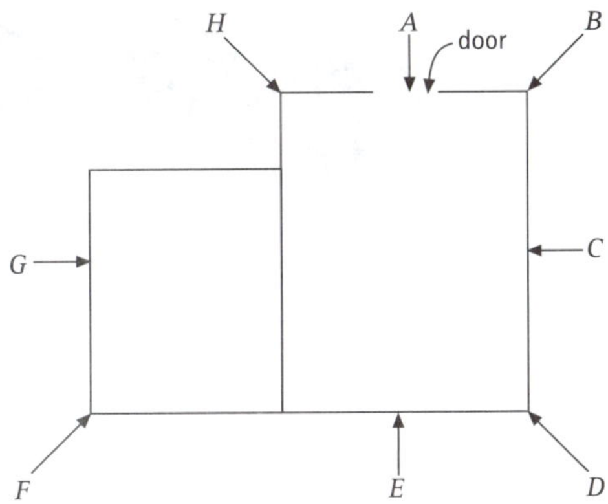

Figure 1

Figure 2

Figure 3

3 Draw the views from the other five arrows.

4 Draw the top view (or plan) of Gladys's house.

This plan shows part of a playground.

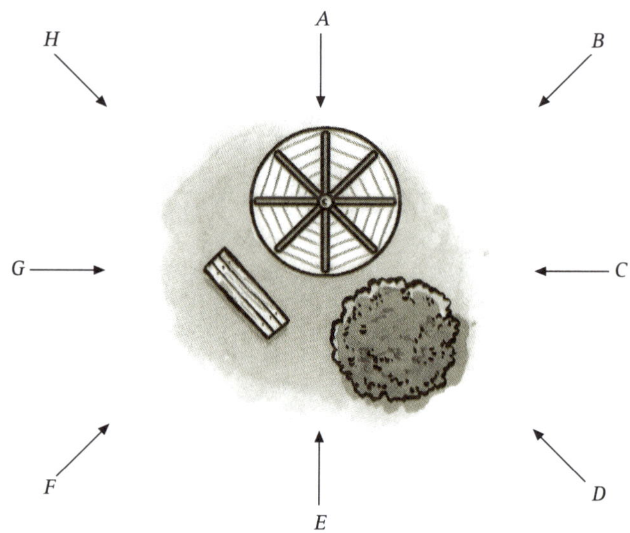

From which direction (A to H) would each of these views be seen?

5

6

7

GROUP ASSESSMENT

See page 288 (rubric 3).

Investigation: A test model

Work in groups.

You will need: a tape measure or metre wheel, square grid paper, stiff card, scissors, sticky tape.

Select a building or structure in the school grounds for constructing a scale model.

1 Make sketches of the structure from all sides, and draw a plan.

2 Make the necessary measurements, and mark them on your sketches.

3 Choose an appropriate scale, and construct scale drawings of all sides of the structure (front, back, left side, right side and top).

4 Cut out the pieces and make a test model. If everything looks correct, transfer the shapes onto the card and make the model. You might be able to put some of the pieces side by side on the card to reduce the number of joins. (This is like making a net of a solid. You may remember this from previous years.)

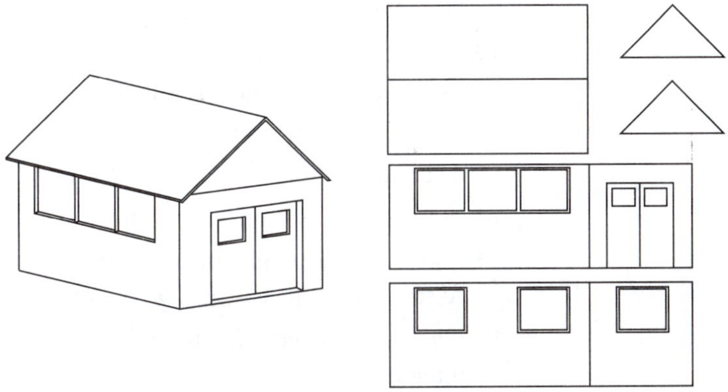

If your test model doesn't look right:

a Check that you've calculated the lengths correctly according to the scale you've chosen.

b Check that the original measurements you made were accurate.

c Ask another group or your teacher for help.

Language skills

clinometer	*parallelogram*	*collinear*	*quadrilateral*
distance	*rectangle*	*equilateral*	*reflection*
gradient	*regular*	*hypotenuse*	*square*
isosceles	*tangent*	*mid-point*	*theta*
parallel	*translation*		

Complete the following sentences using words from the list:

1 A triangle with exactly two sides equal is called an ... triangle.

2 An ... triangle is a triangle with three equal sides.

3 A ... is a plane figure with four sides.

4 A ... is a quadrilateral with both pairs of opposite sides parallel.

5 A parallelogram with one interior angle equal to 90° is called a ... .

6 A ... is a rhombus with one angle equal to 90°.

7 A ... polygon has all sides and all interior angles equal.

8 When two or more points are on the same straight line, they are ... .

9 State the gradient formula and explain how to apply it.

10 State the distance formula and explain how to apply it.

11 The point halfway between two points of a line segment is called the ... of the line segment.

12 A straight line with a gradient of zero is ... to the x-axis.

13 Under ... in the line $y = x$ the image of $A(x; y)$ is $A'(y; x)$.

14 Under ... of p units horizontally and q units vertically on point $A(x; y)$ the image is $A'(x + p; y + q)$.

15 What is the name given to the longest side of a right-angled triangle?

16 Which trigonometric ratio is found by dividing the opposite side by the adjacent side?

17 What instrument is used to measure angles of elevation?

18 What is the name of the Greek letter whose symbol is θ?

Answer these questions after discussing them with a partner.

19 Write a sentence explaining the difference between the **volume** and the **surface area** of a solid.

20 What plane shapes make up the net of a cylinder?

21 Give three examples of jobs that may require you to calculate the surface area of something.

22 Explain how to test whether two lines are parallel or perpendicular.

23 Explain how to find the y-intercept of a curve.

24 Write a sentence explaining to a Grade 10 learner what trigonometry is.

25 Draw a labelled diagram to explain angles of elevation and depression. List three jobs in which such information would be used.

SELF-ASSESSMENT

Also see page 293 (rubric 9).

Check yourself

Make a mind map of the key facts and formulas in this module under the four headings listed here. It will help if you look at the key skills given at the beginning of this module (see page 53).

Polygons, prisms and cylinders
Coordinate geometry
Transformation geometry
Trigonometry

Then complete the Review. The questions in the Review consolidate your knowledge and understanding of the topics in this module.

MODULE 2

REVIEW

1 Find the value of the variable, giving reasons for your answers:

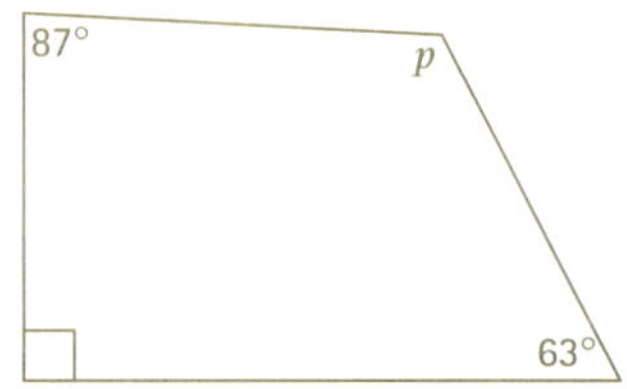

2 What is the interior angle sum of this polygon?

3 What is the size of each exterior angle of a regular hexagon?

4 Find the area of these shapes to 2 decimal places:

a

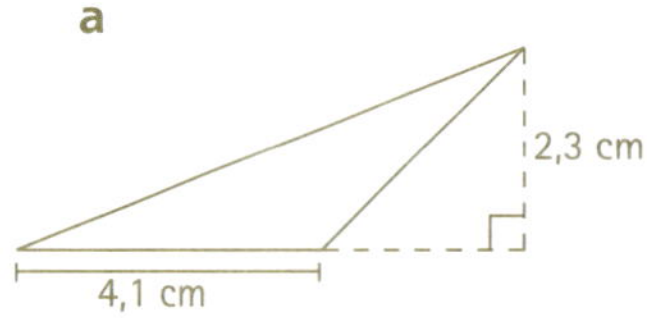

b

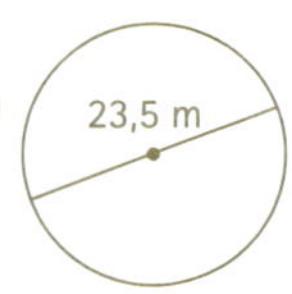

c

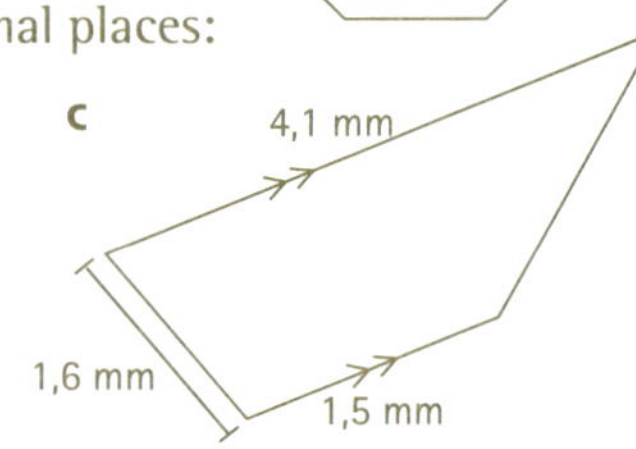

Use this grid to answer the following:

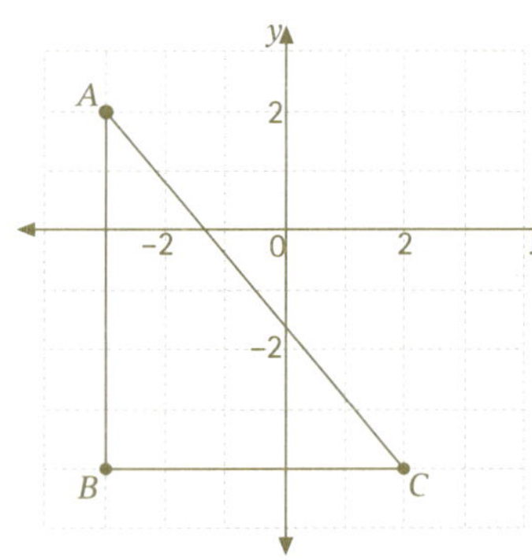

5 Find the gradient and equation of the line *BC*.

6 Find the gradient and equation of the line *AB*.

7 Calculate the length of the line segment *AC*.

8 Find the equation of the line *AC*.

9 Find the x-intercept and y-intercept of line *AC*.

10 Find the mid-point of *AC*.

11 $M(3; -2)$ is the mid-point of the line segment from $P(-3; 4)$ to T. Find the coordinates of T.

12 Find **a** the total surface area and **b** the volume of this solid to 2 decimal places.

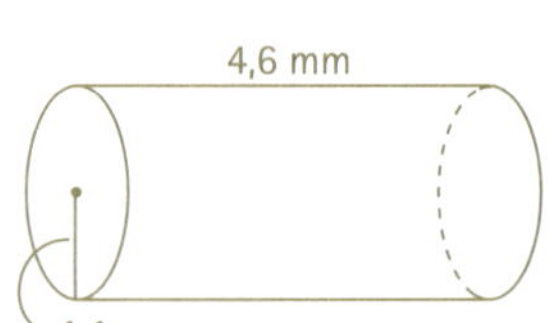

State which transformation could have been performed to obtain the image (in colour) from the original:

13

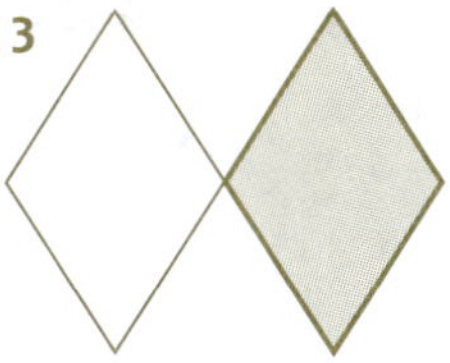

14

15

Copy each figure into your book, and show the image after performing the stated transformation:

16

A 180° rotation about the point indicated

17

A translation 3 units left

18

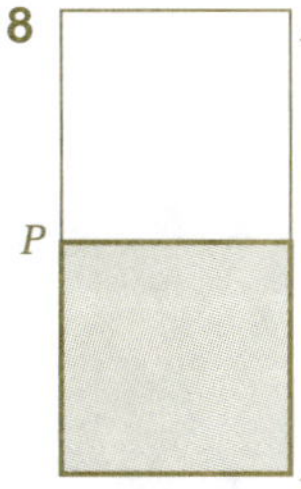

A reflection in the line PQ

19 Perform the following translations and write down the coordinates of the images A', B', C', D' and E':

a On point $A(-7; 4)$: a translation of +3 units horizontally and 4 units vertically.

b On point $B(-7; -2)$: a translation of −12 units horizontally and −7 units vertically.

c On point $C(-5; -2)$: a translation of +3 units horizontally and −6 units vertically.

d On point $D(-7; -3)$: a translation of −3 units horizontally and +3 units vertically.

e On point $E(-1; 2)$: a translation of −4 units horizontally and −6 units vertically.

20 The points $A(-2; 5)$, $B(4; -5)$ and $C(3; -11)$ are given. Perform the reflection in the x-axis on each point and write down the coordinates of the images A', B' and C'.

REVIEW

21 The points $A(2; 3)$, $B(4; -9)$ and $C(-3; -7)$ are given. Perform the reflection in the straight line $y = x$ on each point and write down the coordinates of the images A', B' and C'.

For the triangle shown, find:

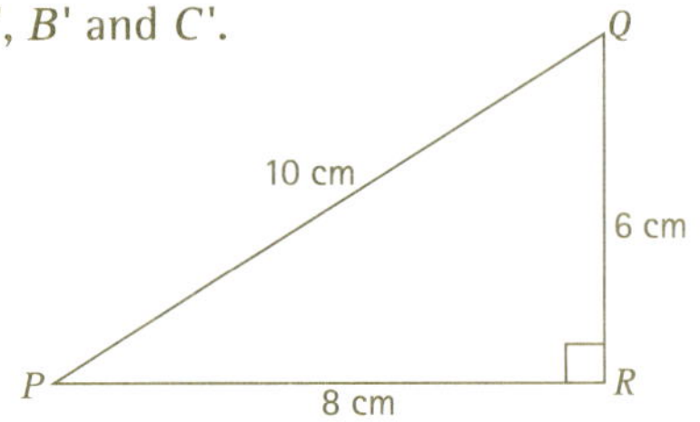

22 $\sin Q$ **23** $\tan P$

24 $\cos Q$

25 the size of $\hat{P}$, to the nearest degree.

Find the length of the side labelled d in each of the triangles, to 1 decimal place:

26

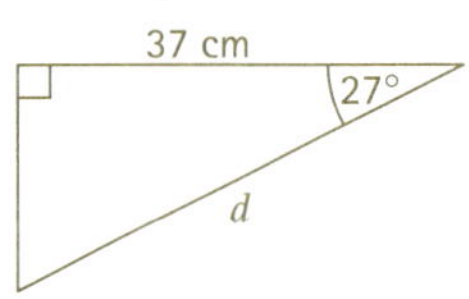

27

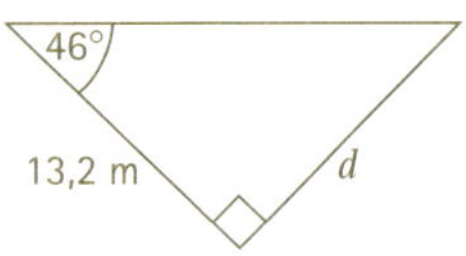

28

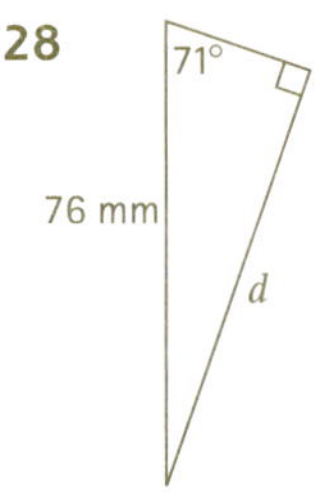

Find the size of angle θ in each triangle, to the nearest degree:

29

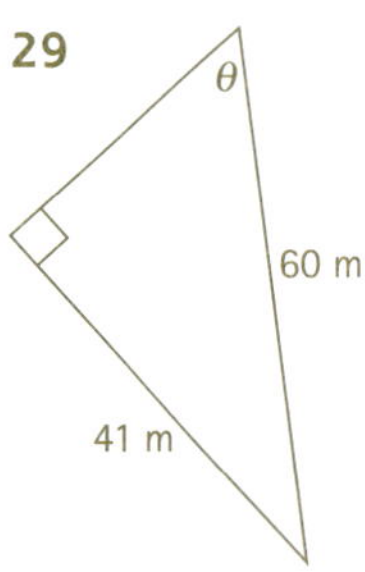

30

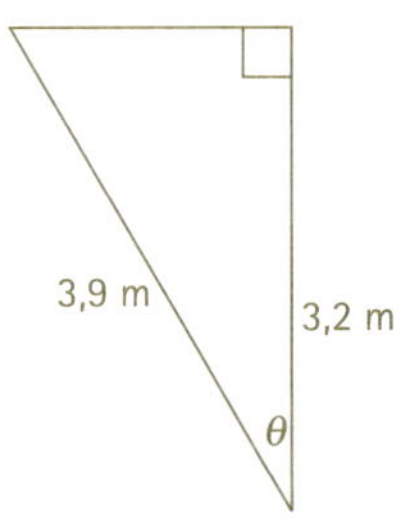

31

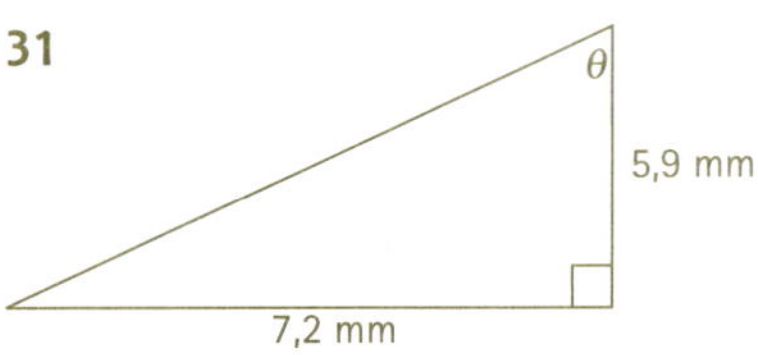

32 An owl perched on the top of a lamp post sees a mouse at an angle of depression of 26°. If the mouse is 9 m from the foot of the lamp post, how high is the lamp post, to 1 decimal place?

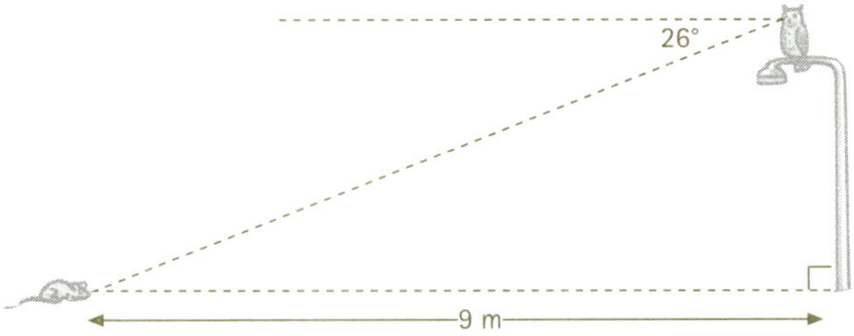

33 A 2 m long piece of wood is used to support the wall of a shed. If the wood makes an angle of 37° with the wall of the shed, how high up the wall is the wood attached, to the nearest centimetre?

34 Find x to 2 decimal places or θ to the nearest minute for these right-angled triangles.

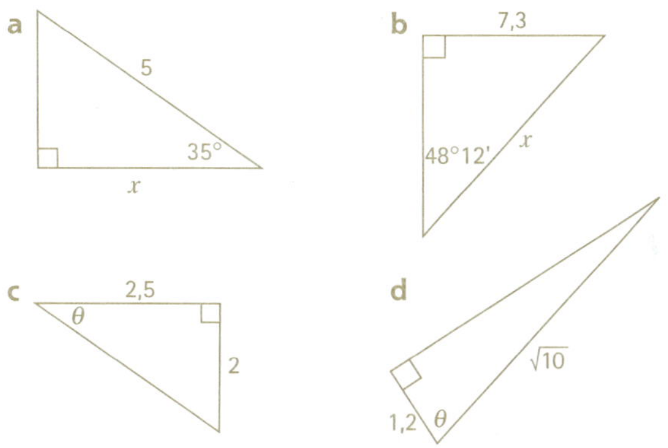

35 The angle of elevation from one flying aeroplane to another is 15° 36', while the horizontal distance between them is 540 m. Find the vertical height difference between the two planes.

36 Alan measures the angle of elevation of the top of an apartment block to be 35°. He is standing 42 m away from the block, and his eye-level is 1,6 m above the ground. Construct a scale drawing to show this information, and find the height of the apartment block. (Use a scale of 1 cm to 5 m.)

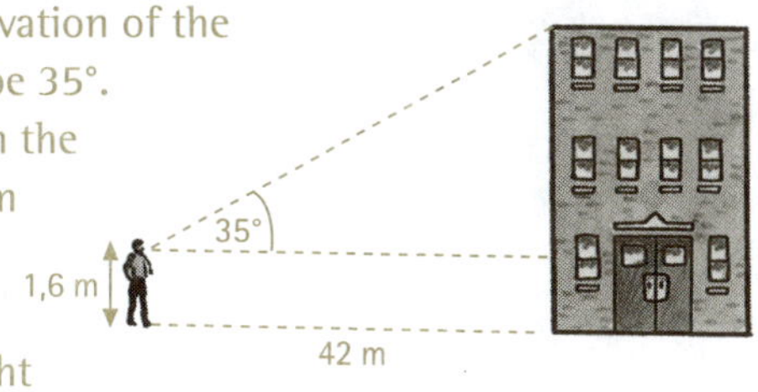

MODULE 2

PROJECTS

ASSESSMENT PORTFOLIO

See page 292 (rubric 8).

1 Research and describe the contributions of each of these cultures or periods to the development and use of geometry and trigonometry:

a Egyptian and Babylonian (3000 BC to AD 260)

b Greek (600 BC to AD 450)

c Hindu (200 BC to AD 1250)

d Arabian (AD 650 to 1200)

e Period of transmission (AD 950 to 1500)

f Modern (first half, AD 1450 to 1700)

g Modern (second half, AD 1700 to present).

(Framework from: Howard Eves, *An Introduction to the History of Mathematics*. New York, CBS College Publishing, 1983.)

ASSESSMENT PORTFOLIO

See page 292 (rubric 8).

2 **Designing milk cartons**

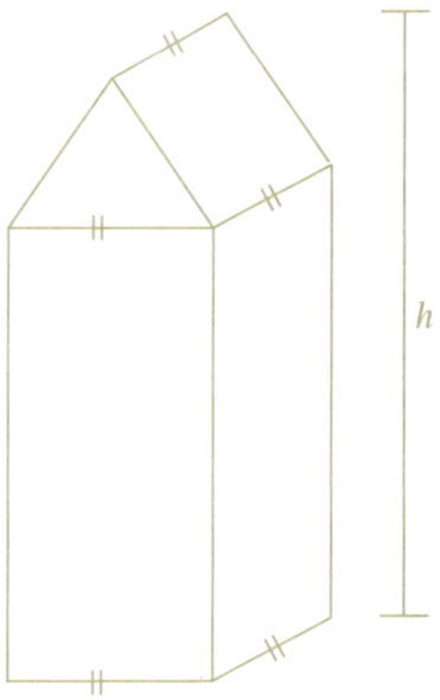

Milk cartons made at a cardboard factory are to be designed to hold 1 litre of milk. They are to be made into the shape of a square-based prism with a triangular top, as shown.

Between 10 cm^3 and 50 cm^3 of spare air space must be designed into each carton. The milk carton must be at least 22 cm high in total.

a Design the 1-litre milk carton and show your design by drawing a diagram marking all side measurements.

b Repeat **a** with the extra condition that the total surface area of the outside of the carton must be between 690 cm^2 and 700 cm^2.

SELF-ASSESSMENT

Copy the rubric into your workbook and put a tick (✓) in the column that describes how you feel about the work in this module.

Self-assessment rubric

How well do you understand these concepts from Module 2?	I understand this concept very well	I understand this concept	I need more help to understand this concept
Triangles and their properties			
Quadrilaterals and their properties			
Surface area of prisms and cylinders			
Volume of prisms and cylinders			
Effect of change in dimensions on surface area			
Effect of change in dimensions on volume			
Gradients of line segments			
Definitions of trigonometric functions			
Translations			
Reflection			
Scale drawing			

MODULE 3

FUNCTIONS AND ALGEBRA

MODULE CHAPTERS

1 Solving and graphing linear equations

2 Solving and graphing non-linear equations

3 More real-life graphs

KEY SKILLS

solving linear equations • graphing linear equations • sketching straight-line graphs • solving simultaneous equations using substitution • solving simultaneous equations using elimination • solving problems using simultaneous equations • solving linear inequalities • plotting quadratic graphs • investigating types of parabolas • multiplication and division • factorisation • grouping in pairs • factorising quadratic trinomials • solving quadratic equations algebraically • solving problems and modelling with quadratics • sketching parabolas • plotting hyperbolas • plotting and sketching exponential functions • solving exponential equations • plotting and sketching trigonometric functions • solving simultaneous equations graphically • interpreting graphs • modelling relationships

In this module, you will work through activities, projects and exercises that allow you to investigate, analyse, describe and represent a wide range of functions and solve related problems (Learning outcome 2).

By the end of this module you will:

- demonstrate the ability to work with various types of functions
- recognise relationships between variables in terms of numerical, graphical, verbal and symbolic representations and convert flexibly between them
- generate graphs of functions, initially by means of point-by-point plotting, supported by available technology, to make and test conjectures and to generalise the effects of the parameters (a and q) on the graphs for functions, including:
  $y = ax + q$
  $y = ax^2 + q$
  $y = \frac{a}{x} + q$
  $y = ab^x + q; b > 0$
  $y = a \sin(x); y = a \cos(x); y = a \tan(x);$
  $y = a \sin(x) + q; y = a \cos(x) + q; y = a \tan(x) + q$
- identify characteristics of the functions above and use them to sketch graphs of these functions; characteristics include domain and range; intercepts with the axes; turning points; minimum and maximum values; asymptotes; shape and symmetry; periodicity and amplitude; average gradient; intervals on which function increases/decreases; the discrete or continuous nature of the graph
- manipulate algebraic expressions: multiply a binomial by a binomial; factorise trinomials; factorise by grouping in pairs; simplify algebraic fractions with monomial denominators

MODULE 3 IN REAL LIFE

graphs clearly represent real-life trends and features and make them easier to understand, for example, population growth or decrease • you will need to calculate average rates of change in many real-life situations • you will learn to calculate areas of objects and the distances that projectiles (for example, cricket balls) travel • you will design bridges, building plans and parts of aeroplanes • you will find out how to calculate distances and angles in real life

- solve: linear equations; quadratic equations by factorisation; exponential equations of the form $ka^{x+p} = m$ (also by using trial and error); linear inequalities in one variable and illustrate the solution graphically; linear equations in two variables simultaneously (numerically, algebraically and graphically)
- use mathematical models to investigate real-life problems, including: making conjectures, demonstrating and explaining their validity; expressing and justifying mathematical generalisations of situations; using various representations to interpolate and extrapolate; describing a situation by interpreting graphs, or drawing graphs from a description of a situation, with special focus on trends and features
- investigate the average rate of change of a function between two values of the independent variable, demonstrating an intuitive understanding of average rate of change over different intervals.

Integration

In this module, you will also work with exponents, rounding and number patterns (Learning outcome 1) and interpret data (Learning outcome 4).

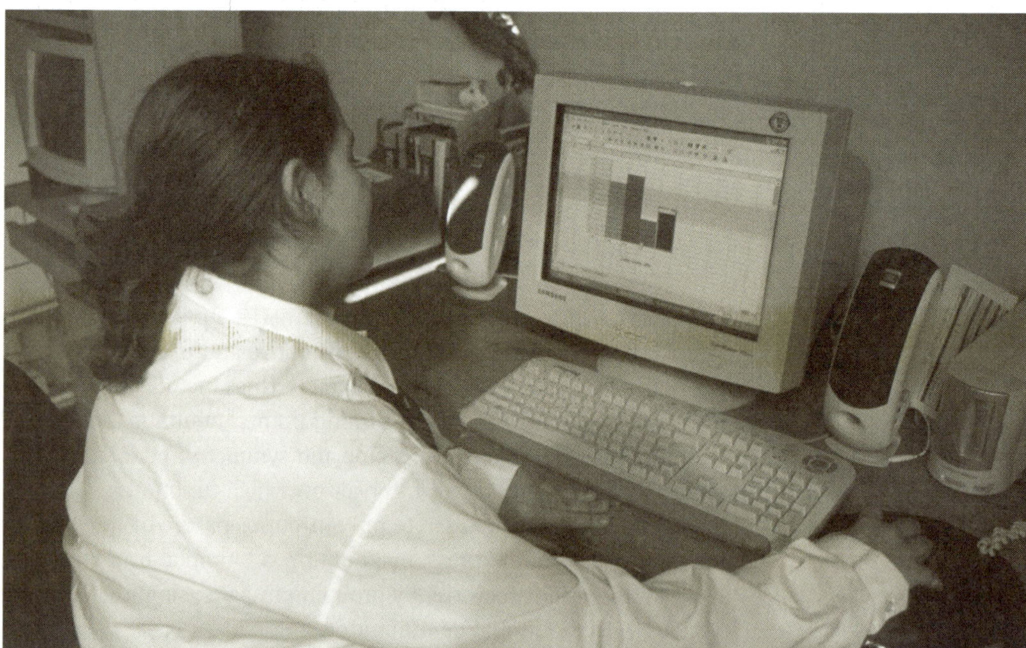

CHAPTER 1 SOLVING AND GRAPHING LINEAR EQUATIONS

KEY VOCABULARY

equation • linear • graph • straight line • gradient • positive gradient • constant • negative gradient • domain • range • y-intercept • x-intercept • coordinates • variable • simultaneous equations • substitution • elimination • inequality

Many real-life situations can be represented through linear equations. In this chapter you will learn to solve linear equations that describe real-life problems and find out how to represent real-life problems in the form of graphs.

Solving linear equations

Note: Linear equations are written in the form $ax + b = c$.

Often it is easier to solve a problem by using an equation than through other methods. In a linear equation, $ax + b = c$, a, b and c are constants.

Here are examples of linear equations of the form $ax + b = c$:

$3x + 6 = 12$ $-2x - 5 = 9$ $5x - 6 = 0$

For each equation, you can use three numbers for a, b and c. Once you have chosen the three numbers, they are "fixed". They are called **constants**, because they cannot change. But x is a placeholder for any number, so its value can vary. That's why x is called a **variable**. If you replace x with a number, you get a statement that is either true or false. If you get a true statement, you say the value of x you used is the **solution** of the equation.

Look at the equation $3x + 6 = 12$.

3, 6 and 12 are constants and x is a variable. For $x = 2$ you get the true statement $3(2) + 6 = 12$, but for any other value of x the statement is false. So the solution of the equation is $x = 2$.

Linear equations have only one solution.

EXAMPLES

1 A taxi company charges R8 up front plus R3 per kilometre travelled. If the fare is R83, what distance was travelled?

You can express this problem in a linear equation, for example:

$C = 3x + 8$, where C = cost of travel and
x = distance travelled.

Now, C = R83 $\therefore 83 = 3x + 8$

Using inverse operations you can solve for x:

$\therefore 75 = 3x$ Subtract 8 from both sides.

$\therefore 25 = x$ Divide both sides by 3.

$\therefore$ The distance travelled was 25 km.

2 Solve these linear equations:

a $4x + 5 = 17$ **b** $3(2x + 4) + 2(3x - 2) = 20$

a $4x + 5 = 17$

$\therefore 4x = 12$ Subtract 5 from both sides.

$\therefore x = 3$ Divide both sides by 4.

Check: Verify your answer either mentally or on paper, as required: $4(3) + 5 = 17$.

b $3(2x + 4) + 2(3x - 2) = 20$

$\therefore 6x + 12 + 6x - 4 = 20$ Expand brackets.

$\therefore 12x + 8 = 20$ Collect like terms.

$\therefore 12x = 12$ Subtract 8 from both sides.

$\therefore x = 1$ Divide both sides by 12.

Check: $\{3[2(1) + 4] + 2[3(1) - 2]\} = 20$

3 Solve these equations:

a $\frac{x + 3}{4} = 2$ **b** $\frac{x}{2} - 4 = 2$ **c** $\frac{4x - 2}{3} = \frac{3x - 1}{2}$

a $\frac{x + 3}{4} = 2$

$\therefore x + 3 = 8$ Multiply both sides by 4.

$\therefore x = 5$ Subtract 3 from both sides.

Check: $\frac{5 + 3}{4} = \frac{8}{4} = 2$

b $\frac{x}{2} - 4 = 2$

$\therefore \frac{x}{2} = 6$ Add 4 to both sides.

$\therefore x = 12$ Multiply both sides by 2.

Check: $\frac{12}{2} - 4 = 6 - 4 = 2$

c $\frac{4x - 2}{3} = \frac{3x - 1}{2}$

$\therefore \frac{{}^{2}\cancel{6}(4x - 2)}{\cancel{3}_1} = \frac{{}^{3}\cancel{6}(3x - 1)}{\cancel{2}_1}$ Multiply both sides by lowest common denominator.

$\therefore 2(4x - 2) = 3(3x - 1)$

$\therefore 8x - 4 = 9x - 3$ Expand both sides.

$\therefore -4 = x - 3$ Subtract $8x$ from both sides.

$\therefore -1 = x$ Add 3 to both sides.

Check: $\frac{4(-1) - 2}{3} = \frac{-6}{3} = -2$ and $\frac{3(-1) - 1}{2} = \frac{-4}{2} = -2$

Exercise

1 Solve these equations:

a $x + 4 = 5$ b $x - 2 = 4$ c $-x + 2 = 6$

d $-x - 3 = -7$ e $2x + 5 = 7$ f $3x + 5 = 11$

g $4x + 2 = -6$ h $5x + 4 = -11$ i $-2x + 7 = 9$

j $-3x + 12 = 6$ k $-5x - 4 = 11$ l $-4x + 7 = 5$

m $3 - x = 5$ n $6 - x = -3$ o $4 - 2x = 6$

p $6 - 7x = -14$

2 Now solve these equations:

a $3(x + 2) = 12$ b $2(x - 4) = 10$

c $2(2x + 3) = 14$ d $3(3x - 2) = 21$

e $4(2x + 3) = -4$ f $2(3x - 4) = -6$

g $3(2x + 3) + 2(x + 4) = 25$ h $2(2x - 3) + 3(4x - 1) = 23$

i $2(3x - 2) - 3(x + 1) = -7$ j $5(2x + 1) - 3(x - 3) = -21$

k $3(x - 4) = 6(x - 7)$ l $3(3x + 4) = 2(x + 13)$

m $4(x + 2) = 3(x - 5)$ n $4(4x - 1) = 5(2x + 4)$

3 Solve these equations:

a $\frac{x + 2}{3} = 5$ b $\frac{x - 4}{2} = 3$ c $\frac{x + 2}{5} = -4$

d $\frac{x}{3} - 1 = 4$ e $\frac{3x}{2} + 2 = 8$ f $\frac{2x}{5} - 6 = -10$

g $6 - \frac{3x}{2} = 0$ h $5 - \frac{2x}{3} = 1$ i $7 - \frac{4x}{5} = 3$

j $\frac{2x - 3}{5} = 5$ k $\frac{2x + 1}{3} = -3$ l $\frac{3x - 2}{4} = 4$

m $\frac{2x + 1}{7} = \frac{3x - 5}{4}$ n $\frac{5x - 4}{4} = \frac{x - 5}{5}$ o $\frac{3x - 5}{4} = \frac{2x - 8}{3}$

p $\frac{x + 2}{4} + 1 = 7$ q $\frac{x - 7}{2} - 3 = 2$ r $2 + \frac{x + 4}{3} = -2$

4 For each of these statements, write an equation and solve for the variable:

a If 4 is added to x the result is 6.

b If x is added to 12 the result is 8.

c If 5 is subtracted from x the result is 5.

d If x is subtracted from 13 the result is 20.

e Twice the value of x is added to 3 and the result is 9.

f 5 less than x doubled is -15.

g 3 times x plus 4 is 16.

h 5 less than twice x is 3 less than x.

5 The perimeter of a square is 52 cm. Determine the length of the side.

6 The perimeter of an isosceles triangle is 42 mm. If the similar sides are 10 mm, determine the length of the odd side.

7 The sum of two consecutive numbers is 23. What are the numbers?

8 I ride four times faster than I walk. If a trip took me 45 minutes and I spent 15 of these walking 3 km, how far did I ride?

9 A service technician charges R120 up front and R100 for each hour she works.

   a Find a linear equation for the total charge (C) of any job.

   b What will a 4-hour job cost?

   c If the technician works on a job for 3 days and averages 6 hours per day, what will be the overall cost?

10 A petrol tank holds 71 litres. If it contains 5 litres and a petrol pump fills it at 3 litres per 10 seconds, find:

   a a linear equation for the amount of fuel (x litres) in the tank at time t minutes

   b how long it will take to fill the tank to 23 litres

   c how long it will take to fill the tank.

Graphs of linear equations

In the previous section you saw that some real-life situations can be represented by linear equations. Now you will look at the graphical representation of these linear equations.

EXAMPLE

Look at the taxi example given in the last section. The initial charge was R8 and the charge per kilometre R3.

You can draw up a table of values for the information:

Kilometres travelled (x)	0	10	20	30	40	50
Cost (C)	8	38	68	98	128	158

Here is a graph for this information:

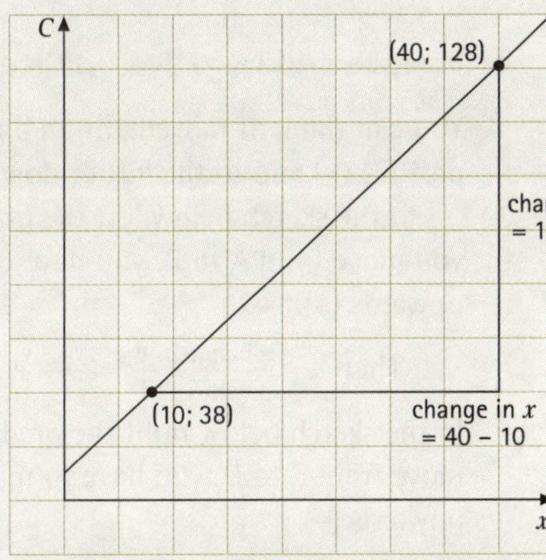

The slope of any line is constant and is called the **gradient**. To calculate the gradient in this case:

$$\text{Gradient} = \frac{\text{rise}}{\text{run}} = \frac{\text{change in } C}{\text{change in } x}$$

Any two points on the line may be used to calculate the gradient. Using (10; 38) and (40; 128):

$$\text{Gradient} = \frac{128 - 38}{40 - 10}$$
$$= 3$$

The algebraic relationship for the cost of the taxi is $C = 3x + 8$, where C = cost of travel (R) and x = the distance travelled (km).

The gradient is the same as the charge per kilometre and the C-intercept is 8.

In general, for a linear equation of the form $y = ax + q$, where a and q are constants:

$$a = \text{gradient} = \frac{\text{rise}}{\text{run}} = \frac{\text{change in } y}{\text{change in } x}$$

$$q = y\text{-intercept.}$$

The gradient of the line is **positive** if it **slopes up** from left to right, and **negative** if it **slopes down** from left to right.

The gradient of a line joining two points $(x_1; y_1)$ and $(x_2; y_2)$ is:

$$a = \frac{y_2 - y_1}{x_2 - x_1}$$

Positive and negative gradients

You can think of movement on a graph to the right or upwards as positive (+) and to the left or downwards as negative (–).

In the sketch below left the gradient of AB is **positive**, because if you move from A to B, you have to move to the right (+) and then upwards (+).

$$\text{Then } a = \frac{\text{change in } y}{\text{change in } x} = \frac{y_2 - y_1}{x_2 - x_1} = \frac{+}{+} = +$$

In the sketch below right the gradient of CD is negative, because if you move from C to D, you have to move to the left (–) and then upwards (+).

$$\text{Then } a = \frac{\text{change in } y}{\text{change in } x} = \frac{y_2 - y_1}{x_2 - x_1} = \frac{+}{-} = -$$

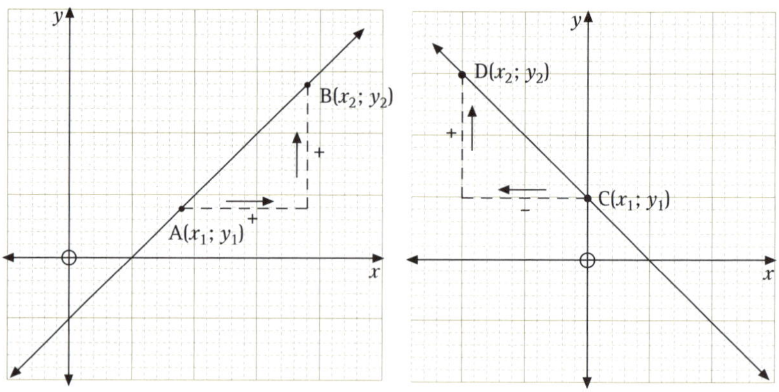

EXAMPLES

1 Work out the gradient for these straight lines:

a

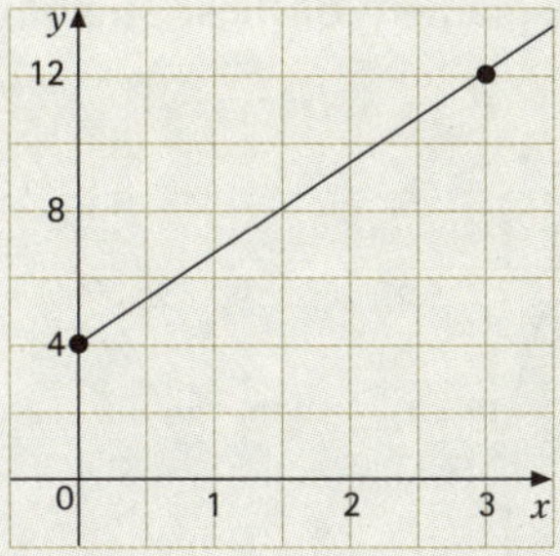

b

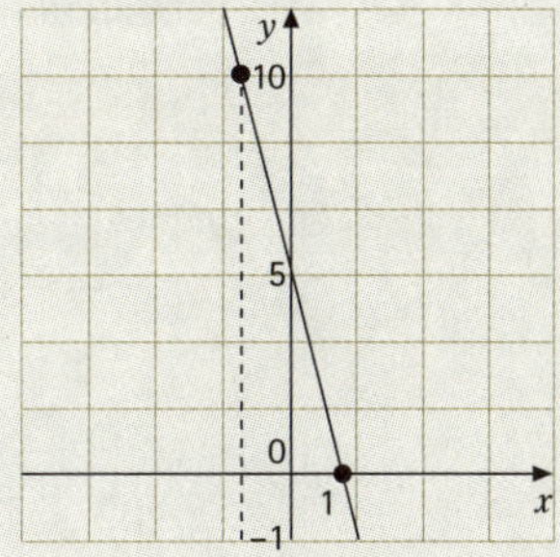

c

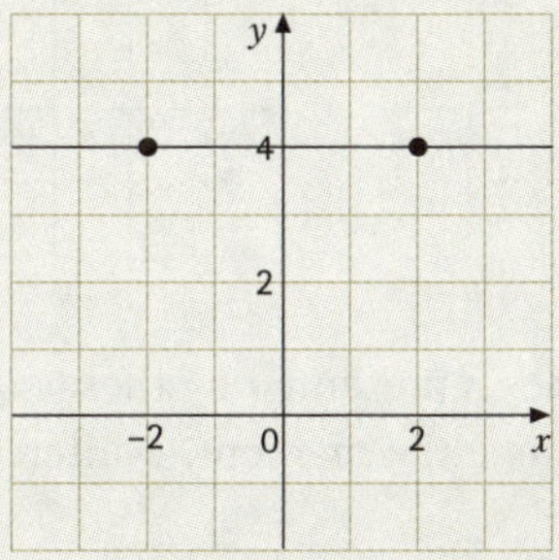

d

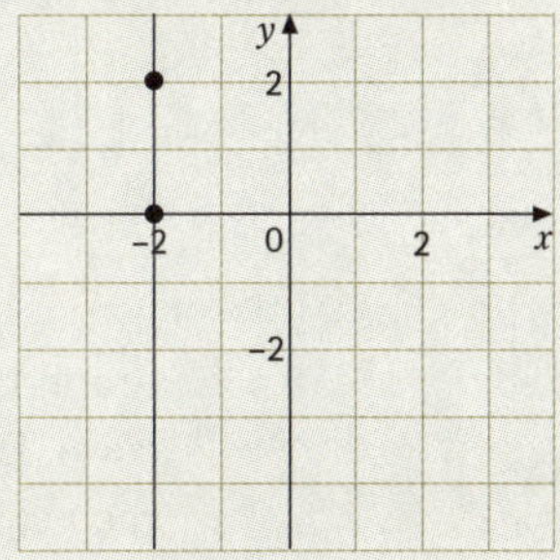

a $a = \frac{y_2 - y_1}{x_2 - x_1}$

$a = \frac{12 - 4}{3 - 0}$

$= \frac{8}{3}$

b $a = \frac{y_2 - y_1}{x_2 - x_1}$

$a = \frac{10 - 0}{-1 - 1}$

$= \frac{10}{-2}$

$= -5$

c $a = \frac{y_2 - y_1}{x_2 - x_1}$

$a = \frac{4 - 4}{2 - -2}$

$= \frac{0}{4}$

$= 0$

d $a = \frac{y_2 - y_1}{x_2 - x_1}$

$a = \frac{2 - 0}{-2 - -2}$

$= \frac{2}{0}$

= undefined

Remember: In the function $y = ax + q$, the **domain** is all the x-values for which the function is defined (true). The **range** is all the y-values of all the points that lie on the graph with a given domain.

Note: The **y-intercept** will always have an x-value of 0, because for each point on the y-axis the value of $x = 0$. The y-value varies, depending on the position of the y-intercept on the y-axis.

2 Graph $y = 4x - 3$ from $x = -1$ to $x = 3$ and determine its gradient, y-intercept, domain and range.

$a = \frac{y_2 - y_1}{x_2 - x_1}$

$a = \frac{9 - 1}{3 - 1}$

$= \frac{8}{2}$

$= 4$

y-intercept is (0; –3)

Domain: $-1 \leq x \leq 3$

Range: $-7 \leq y \leq 9$

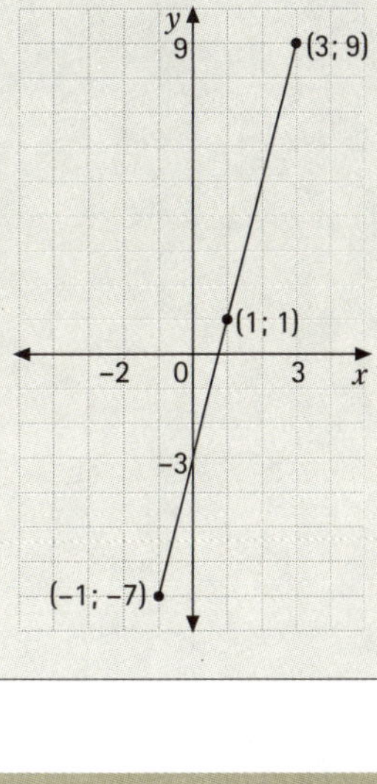

Exercise

1 Calculate the gradient of these straight lines:

a

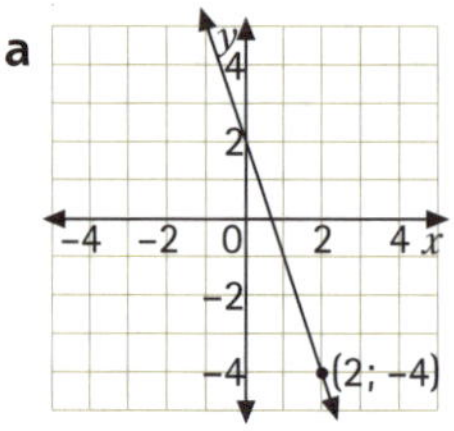

b

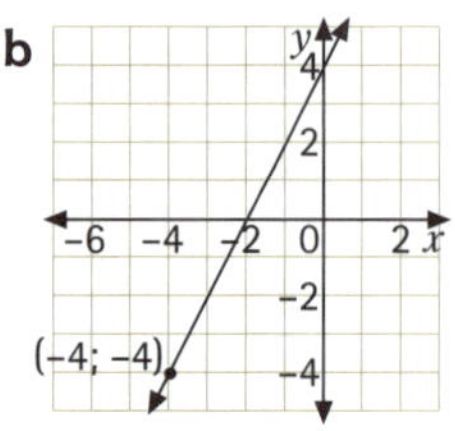

c

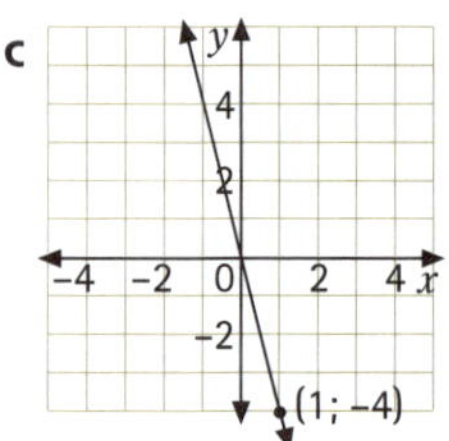

d

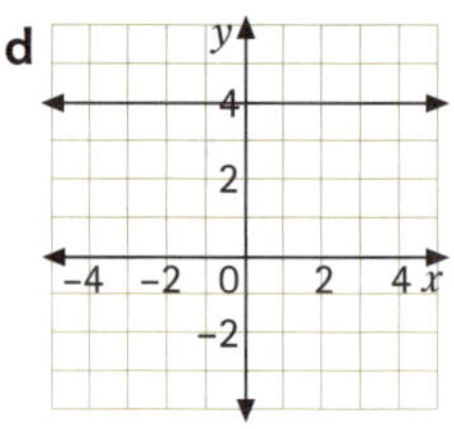

e

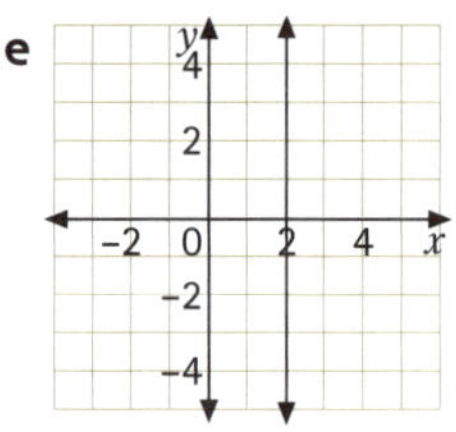

f

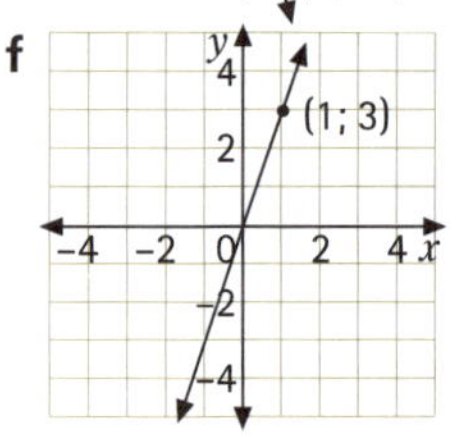

2 Graph the lines represented by these equations from $x = -3$ to $x = 3$. Determine the gradient, y-intercept, domain and range in each case.

a $y = 2x + 1$ **b** $y = 5x + 1$ **c** $y = 3x - 4$ **d** $y = x - 5$

e $y = -4x + 3$ **f** $y = -2x + 2$ **g** $y = -2x - 5$ **h** $y = -5x - 1$

i $y = 3$ **j** $y = -4$ **k** $x = 3$ **l** $x = -4$

3 Match each of these equations to one of the graphs:

a $y = 3x + 4$ **b** $y = -2x - 4$ **c** $y = 4$

d $y = x$ **e** $y = -2x + 4$ **f** $x = -3$

i

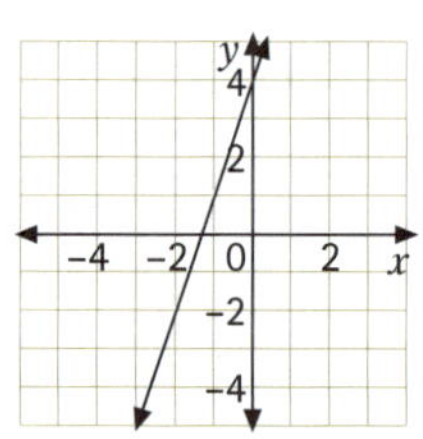

ii

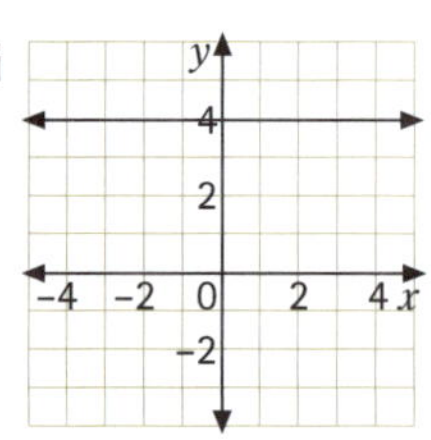

iii

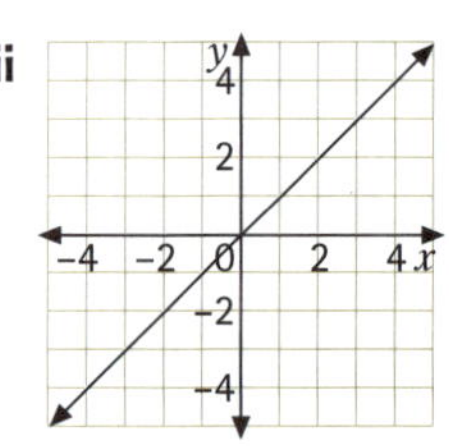

iv

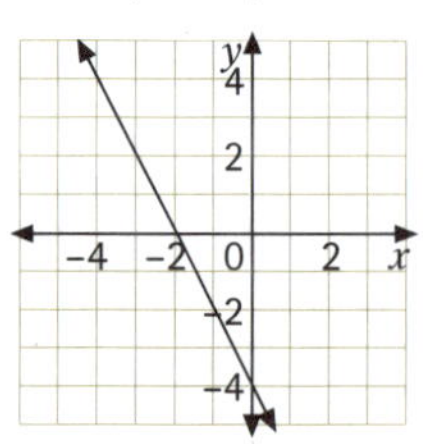

v

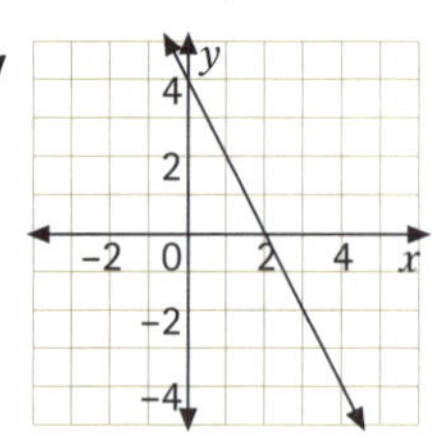

vi

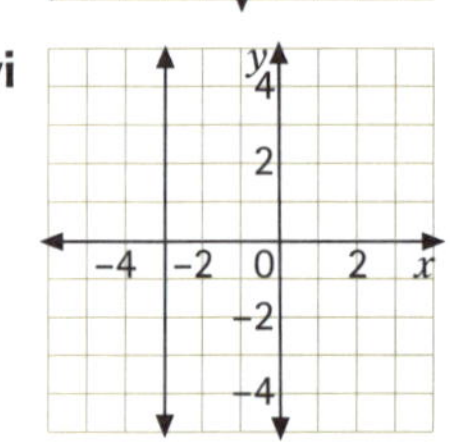

4 On the same set of axes, graph $y = 2x - 1$, $y = 2x$, $y = 2x + 1$, $y = 2x + 2$.

a Discuss what you see as the y-intercept increases while the gradient stays the same.

b What term describes the family of lines you have graphed?

5 On the same set of axes, graph $y = x - 1$, $y = 2x - 1$, $y = 3x - 1$, $y = 4x - 1$.

Discuss what you see as the gradient changes while the y-intercept stays the same.

Equations of straight-line graphs

As previously defined, a linear equation is of the form $y = ax + q$, where a and q are constants,

where $a = \text{gradient} = \dfrac{\text{change in } y}{\text{change in } x}$

$q = y\text{-intercept}$.

The gradient of a line joining two points $(x_1; y_1)$ and $(x_2; y_2)$ is given by

$$a = \frac{y_2 - y_1}{x_2 - x_1}$$

We use these definitions to determine the equation of a straight-line graph.

EXAMPLES

1 Find the equation of the straight line shown.

Remember: Decide on any 2 points on the line and substitute them.

$$a = \frac{y_2 - y_1}{x_2 - x_1} = \frac{13 - 4}{4 - 1} = \frac{9}{3} = 3$$

Remember: Substitute a and q into $y = ax + q$.

and $q = 1$

$\therefore y = 3x + 1$

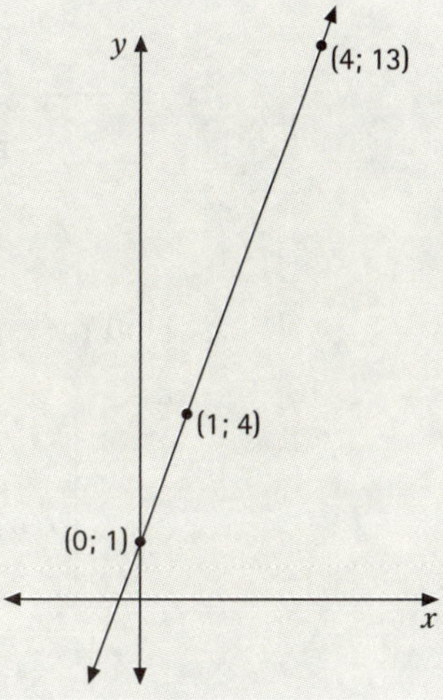

2 Find the equation of the straight line shown.

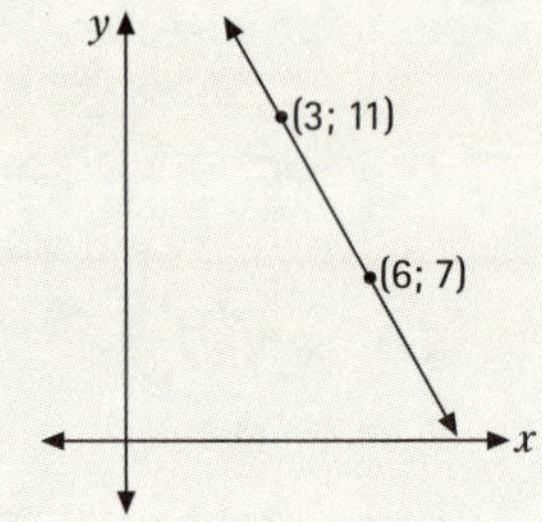

$a = \frac{y_2 - y_1}{x_2 - x_1}$

$= \frac{11 - 7}{3 - 6}$

$= -\frac{4}{3}$

So $y = -\frac{4}{3}x + q$, but (6; 7) is a point on the line

$\therefore 7 = -\frac{4}{3} \times (6) + q$

$\therefore 7 = -8 + q$

$\therefore 15 = q$

$\therefore y = -\frac{4}{3}x + 15$

Hint: No y-intercept is shown, so substitute a point and the gradient into $y = ax + q$.

Exercise

1 Find the equation of each straight line:

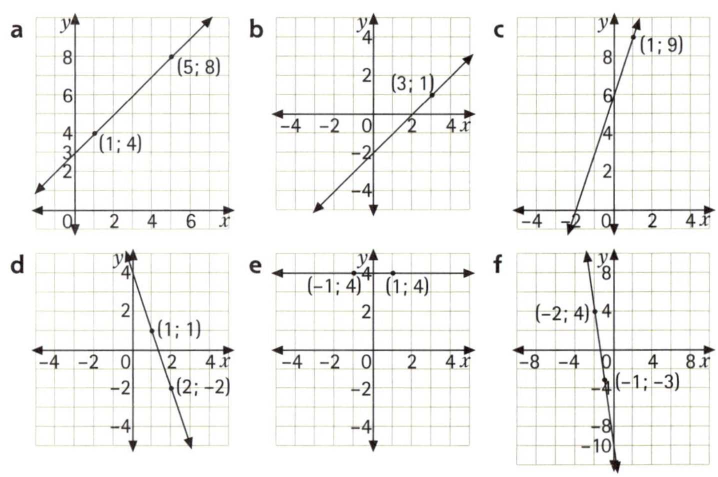

2 Find the equation of each straight line:

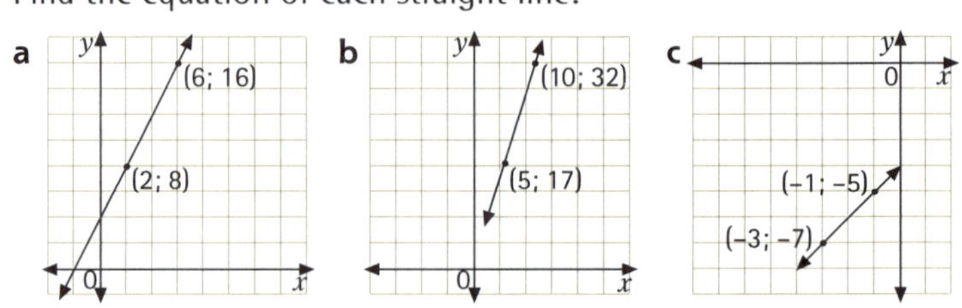

d
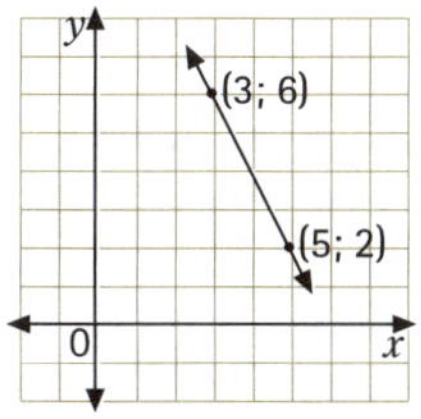

e
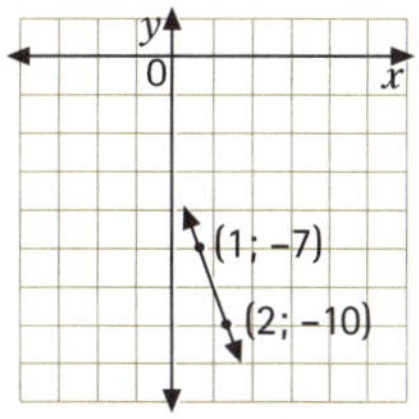

f
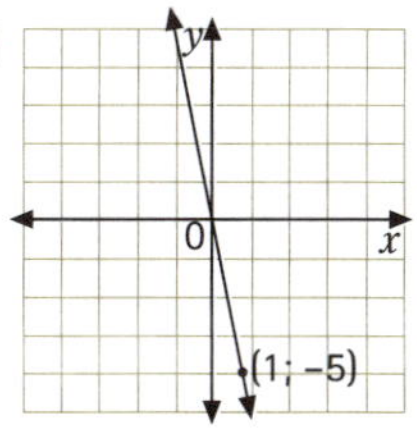

3 Given these tables of values, determine the equation relating x and y in each case. Also give the domain and range for each.

a

x	0	3	5
y	5	14	20

b

x	4	5	6
y	−4	−6	−8

Sketching straight-line graphs

To sketch a straight line, you need to show only two points of the graph. These are usually the x- and y-intercepts.

EXAMPLES

Sketch these lines:

1 $y = \frac{1}{2}x + 2$

2 $3x - 2y = 6$

Hint: Start at the y-intercept, move 2 down, then 4 to the left to find the x-intercept.

1 $y = \frac{1}{2}x + 2$

$\therefore$ gradient $= \frac{1}{2} = \frac{-2}{-4} = \frac{\text{2 down}}{\text{4 left}}$

$\therefore$ y-intercept $= 2$

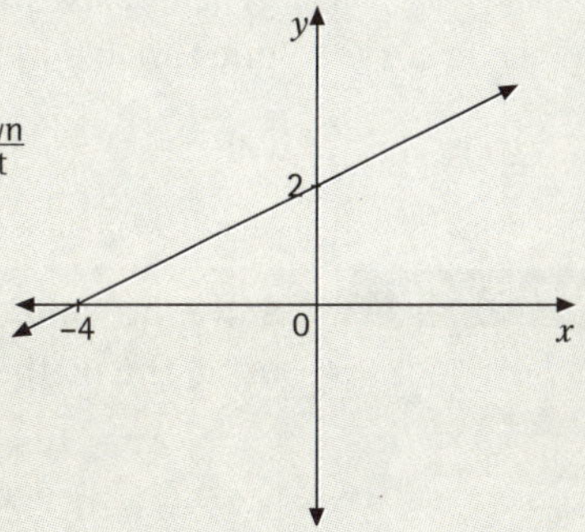

2 $3x - 2y = 6$

x-intercept at $y = 0$: $3x = 6$

$\therefore x = 2$

y-intercept at $x = 0$: $-2y = 6$

$\therefore y = -\frac{6}{2}$

$\therefore y = -3$

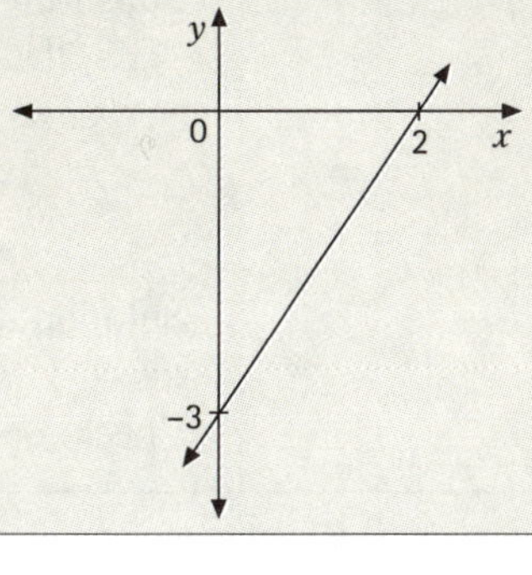

Exercise

1 Sketch these lines by considering the gradient and y-intercept:

a $y = 2x + 6$	**b** $y = 3x - 12$	**c** $y = -3x + 9$	**d** $y = -5x - 10$
e $y = 3$	**f** $y = -5$	**g** $x = 5$	**h** $x = -6$
i $y = 3x$	**j** $y = x$	**k** $y = -2x$	**l** $y = -4x$

2 Sketch these lines by finding the x- and y-intercepts:

a $3x + 2y = 6$	**b** $x + 3y = 9$	**c** $3x - 4y = 24$
d $-2x - 7y = 14$	**e** $3x + 6y = -6$	**f** $2x + 3y = 5$
g $-3x + 4y = 7$	**h** $3x = 12$	**i** $4y = 2$
j $-2x = 12$	**k** $-5y = 10$	**l** $-\frac{1}{2}x = 3$

Solving simultaneous equations using substitution

There are two methods for solving simultaneous equations algebraically: substitution and elimination.

In this section you will use the method of **substitution.** It is generally used when at least one of the equations is an expression of one variable in terms of the other,

e.g. $y = 3x + 4$, $3x + y = 2$ or $y = 2x - 6$, $y = -x - 1$

EXAMPLE

Hint: Number the equations you have been given.

Solve the simultaneous equations $y = x + 3$ and $2x + 3y = 19$ by using the method of substitution.

$2x + 3y = 19$ ....(1)
$y = x + 3$ ....(2)

Substitute (2) into (1):
$2x + 3(x + 3) = 19$
$\therefore 5x + 9 = 19$
$\therefore 5x = 10$
$\therefore x = 2$

Substitute $x = 2$ into (2):
$y = 2 + 3$
$= 5$

Solution: (2; 5)

Check: Substitute (2; 5) into equation (1), that is: $2(2) + 3(5) = 19$.

Exercise

Solve these pairs of simultaneous equations by using the method of substitution:

1 $y = x - 2$ and $3x - 2y = 7$
2 $y = x - 1$ and $3x + 2y = 8$
3 $y = x$ and $4x + 3y = 7$
4 $y = x$ and $7x + 3y = 10$
5 $y = x + 5$ and $3x + y = 13$
6 $y = x + 3$ and $6x + y = 17$
7 $y = 2x + 3$ and $11x - 5y = -14$
8 $y = 3x - 2$ and $7x - 2y = 8$
9 $y = 3x - 5$ and $3x + 5y = 11$
10 $y = 4x + 1$ and $2x - 3y = -23$

Solving simultaneous equations using elimination

Simultaneous equations are generally solved using elimination when both equations are in the form $ax + by = c$.

e.g. $2x - y = 6$, $3x + y = 10$ or $-5x + y = -2$, $6x + 3y = 5$

Adding or subtracting multiples of these two equations allows one of the variables to be eliminated.

EXAMPLES

Remember: Always check your solutions by substituting the coordinates into one of the original equations.

Solve the following pairs of simultaneous equations using the elimination method:

1 $x + y = 6$ and $3x - y = 10$

$x + y = 6$ ....(1)
$3x - y = 10$ ....(2)
(2) + (1): $4x = 16$
$\therefore x = 4$
Substitute $x = 4$ into (1):
$4 + y = 6$
$\therefore y = 2$
Solution: (4; 2)

Check: Substitute (4; 2) into equation (1):
(4) + (2) = 6.

2 $3x + 2y = 6$ and $5x + 3y = 11$

$3x + 2y = 6$ ....(1)
$5x + 3y = 11$ ....(2)
(1) × 3: $9x + 6y = 18$ ....(3)
(2) × 2: $10x + 6y = 22$ ....(4)
(3) − (4): $-x = -4$
$\therefore x = 4$
Substitute $x = 4$ into (1):
$3(4) + 2y = 6$
$\therefore y = -3$
Solution: (4; −3)

Check: Substitute (4; −3) into equation (1):
3(4) + 2(−3) = 6.

Exercise

1 Solve these pairs of simultaneous equations using the elimination method:

a $x + 7 = 7$ and $5x - y = 5$
b $x + y = 5$ and $3x - y = 3$
c $x - y = 2$ and $2x + y = 10$
d $x + y = 0$ and $4x + y = 9$
e $2x + 3y = 8$ and $4x - 3y = -2$
f $3x + 2y = 8$ and $x - 2y = 0$
g $-3x + 2y = -4$ and $5x - 2y = 8$
h $-2x + 3y = 8$ and $-4x - 3y = -2$

2 Solve these pairs of simultaneous equations using the elimination method:

a $3x + 4y = 7$ and $2x + 4y = 6$
b $x + 3y = 5$ and $4x + 3y = 11$
c $2x + 3y = 1$ and $2x + 5y = -1$
d $4x + y = 10$ and $4x + 4y = 16$
e $x + y = -5$ and $3x + y = -9$
f $-2x + y = -10$ and $3x + y = 10$
g $-2x - 4y = -4$ and $3x - 4y = 6$
h $-x - 6y = -19$ and $-2x - 6y = -20$

3 Solve these pairs of simultaneous equations using the elimination method:

a $3x + 5y = 8$ and $x - 2y = -1$
b $2x + y = 10$ and $3x - 2y = 8$
c $x + 2y = 4$ and $3x - y = 5$
d $3x - 4y = 24$ and $x - 2y = 10$
e $3x + 2y = 6$ and $5x + 3y = 11$
f $3x + 2y = 5$ and $2x + 3y = 5$
g $4x - 3y = 0$ and $3x + 4y = 25$
h $2x + y = 5$ and $3x - 2y = 4$
i $4x - 5y = -14$ and $7x + y = -5$
j $-2x + 3y = 13$ and $5x - 6y = -28$

4 First decide on an appropriate method and then solve these pairs of simultaneous equations:

a $5x + 3y = 18$ and $x = 3y$
b $3x - y = 13$ and $x + y = -9$
c $x - 2y = 2$ and $x + 2y = 22$
d $3x - 8y = 36$ and $x = -3y - 5$
e $2x + 7y = -25$ and $5x + 7y = -31$
f $4x - 2y = 6$ and $y = 2x - 5$
g $7x - 3y = 41$ and $3x - y = 17$
h $2x + 6y = 6$ and $3x - 2y = -2$

Simultaneous equations and problem solving

When you draw up equations from real situations, you need to define the unknown quantities with appropriate variables. When you solve simultaneous linear equations, there should be only two unknown quantities and two pieces of information given.

EXAMPLES

Hint: Define your variables with appropriate letters for convenience, form two equations from the information given, and solve the equations.

1 Neo and Sue have two children. The sum of their ages is 17 and the difference between their ages is 5. If Khanyi is the older sister of Bongani, determine their ages.

Let k be Khanyi's age and b be Bongani's age:

$k + b = 17$ ....(1)

$k - b = 5$...............(2)

(1) + (2): $2k = 22$

$k = 11$

Substitute $k = 11$ into (1):

$11 + b = 17$

$b = 6$

$\therefore$ Khanyi is 11 years old and Bongani is 6 years old.

2 Dumisani purchases 3 ericas and 5 gazania plants from the local nursery and the cost is R25. Mpho buys 4 ericas and 3 gazania plants at R26. Determine the cost of each type of plant.

Let e be the cost of an erica and g be the cost of a gazania.

$3e + 5g = 25$.................(1)

$4e + 3g = 26$.................(2)

(1) × 4: $12e + 20g = 100$...............(3)

(2) × 3: $12e + 9g = 78$.................(4)

(3) – (4) $11g = 22$

$\therefore g = 2$

Substitute $g = 2$ into (1):

$3e + 5(2) = 25$

$\therefore 3e + 10 = 25$

$\therefore 3e = 15$

$\therefore e = 5$

$\therefore$ Ericas cost R5 each and gazanias R2 each.

Exercise

1 Let x and y be two numbers that satisfy statements **a** to **d**. Set up two linear equations according to the information and solve them simultaneously to determine the numbers in each case.

a Their sum is 16 but their difference is 2.

b Their sum is 30 but their difference is 10.

c Twice the larger number plus the smaller is 12 and their sum is 7.

d The sum of twice the first plus three times the second is 10 and the difference between four times the first and three times the second is 16.

2 The perimeter of a rectangle is 48 cm. If the length of the rectangle is three times the width, determine its dimensions.

3 Paul is 3 years older than Nolwezi. If their combined age is 63, determine their ages.

4 A vanilla milkshake is R2 more than a fruit shake. If 3 vanilla milkshakes and 5 fruit shakes cost R30, determine their individual prices.

5 Find the values of the variables in each case:

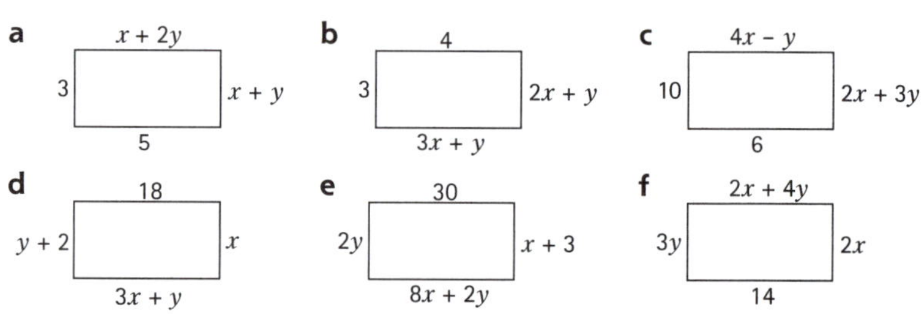

6 Tickets for a soccer match cost R30 for children and R70 for adults. If 5 000 people attended the game and the total takings at the gate were R250 000, determine the number of children and adults who attended the match.

7 Four tyres and a four-wheel balance cost R1 840. If a tyre costs 22 times more than a single-wheel balance, determine the price of each tyre and the price to balance its wheel.

8 Con the fruiterer sells two fruit packs.

Pack 1: 10 apples and 5 mangos (R28,50)

Pack 2: 15 apples and 4 mangos (R38,20)

a Define two variables and set up a pair of linear equations.

b Solve the two simultaneous equations to determine the individual prices of each piece of fruit.

c Determine the cost of 1 apple and 5 mangos.

9 A maths test contains multiple-choice questions worth 2 marks each and short questions worth 3 marks each. The test is out of 50 marks and there are 22 questions.

a Define two variables.

b Set up two linear equations.

c Solve the two equations simultaneously to determine the number of multiple-choice questions.

d If the test was 1 hour long, how much of your time would you allocate to answering the short questions?

10 The graph represents the rental cost of a small car from two car rental firms called Magalies Motor Mart and Speedy Car Rental. Determine:

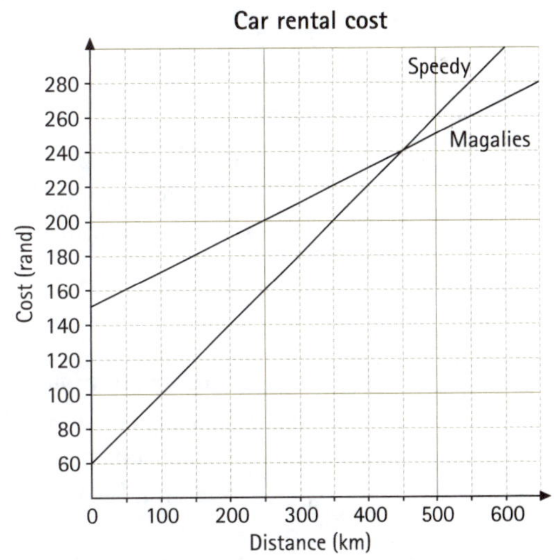

a the initial rental cost from each company

b the cost per kilometre when renting from each company

c the linear equations for the total rental cost from each company

d the number of kilometres at which the cost is the same from both rental firms.

e If you had to travel 300 km, which company would you choose?

f If you had R240 to spend on travel, who would give you the most kilometres?

Solving linear inequalities

Number lines are particularly useful when representing either single numbers or sets of numbers.

For example, the set of x from $x = 3$ to $x = -3$ is represented as

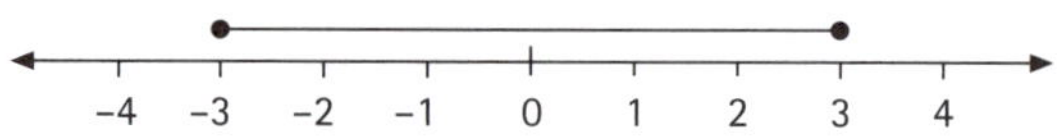

or as an inequality: $-3 \leq x \leq 3$

Do you remember these statements and their corresponding number lines?

$x > a$ (x is greater than a)

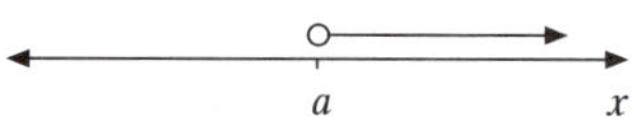

$x \geq a$ (x is greater than or equal to a)

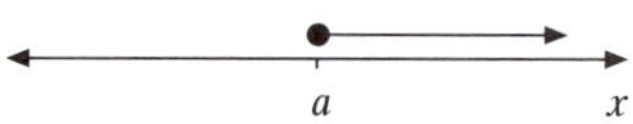

$x < a$ (x is less than a)

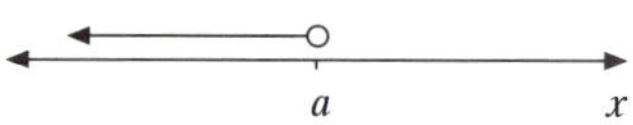

$x \leq a$ (x is less than or equal to a)

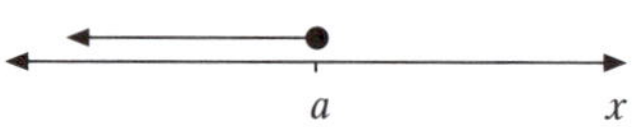

Note: When you solve linear inequalities, you follow the same rules as for solving linear equations, but you reverse the inequality if you multiply or divide by a negative number.

Consider the statement $6 > 4$, which is known to be true.

- Interchanging signs: $6 > 4$ or $4 < 6$
- Under addition: $6 + 2 > 4 + 2$
  $8 > 6$ (true statement)
- Under subtraction: $6 - 2 > 4 - 2$
  $4 > 2$ (true statement)
- Under multiplication and division by a positive number: $12 > 8$, $3 > 2$ (true statements)
- Under multiplication and division by a negative number: $-12 > -8$, $-3 > -2$ (not true)

If you reverse the signs, then the statement will be true: $-12 < -8$, $-3 < -2$ (true statements)

EXAMPLES

Solve these inequalities and graph their solutions on a number line:

1 $3x + 4 > 13$

$3x > 9$

$\therefore\ x > 3$

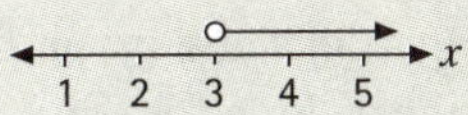

2 $4 - \frac{x}{3} \le 6$

$-\frac{x}{3} \le 2$

$\therefore\ x \ge -6$

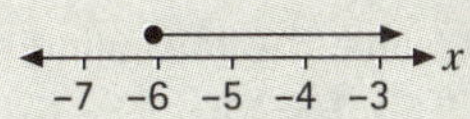

3 $3x + 2 > 6x - 4$

$2 > 3x - 4$

$6 > 3x$

$2 > x$

$\therefore\ x < 2$

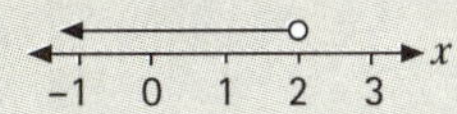

Exercise

1 Represent these inequalities as graphs on a number line:

a $x > 3$ **b** $x \le 4$ **c** $x < -2$ **d** $x \ge 7$

e $x < 5$ **f** $x \ge 5$ **g** $x \le 6$ **h** $x \ge -2$

i $x > 7$ **j** $x < -3$ **k** $x > -6$ **l** $x \le 20$

2 Write as inequalities:

a 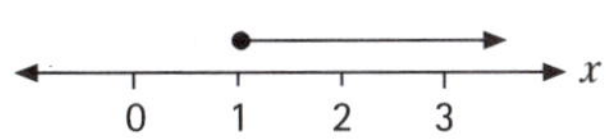

b 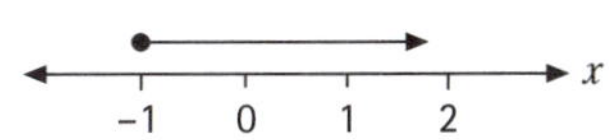

c 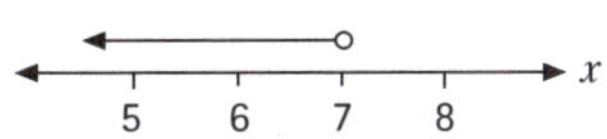

d 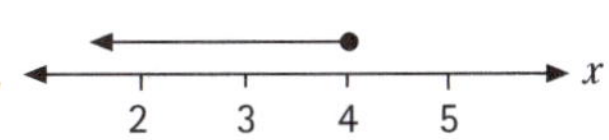

e 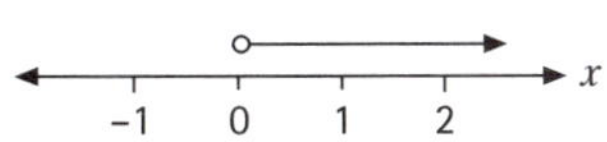

f 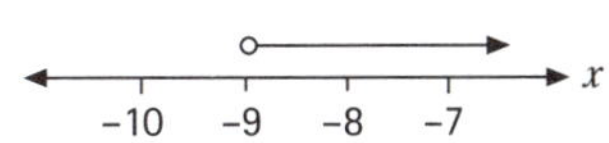

g 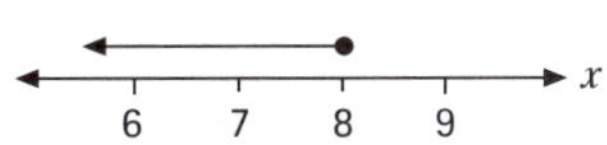

h 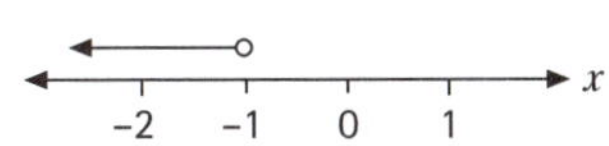

3 Solve these inequalities and graph their solutions on a number line:

a $x + 3 > 5$ **b** $x + 7 \le 9$ **c** $x - 4 \ge 7$ **d** $x - 4 > -6$

e $3x \ge 9$ **f** $4x < -8$ **g** $-2x > 10$ **h** $-5x \le 10$

i $\frac{x}{2} \le 4$ **j** $\frac{x}{5} > -6$ **k** $-\frac{x}{5} > 0$ **l** $-\frac{x}{3} \le -7$

4 Solve the inequalities and graph their solutions on a number line:

a $2x + 4 < 10$ **b** $3x + 1 \ge 10$ **c** $3x - 6 \ge 9$

d $4x - 5 < 7$ **e** $-3x + 2 \le -4$ **f** $-5x - 7 \ge 13$

g $2(3x + 7) > 8$ **h** $3(2x - 3) \le 9$ **i** $4 - 3x \le -2$

5 Solve these inequalities:

a $\frac{x}{3} + 4 \leq 6$ **b** $\frac{x}{7} - 2 > 4$ **c** $\frac{x}{9} + 6 < 4$ **d** $-\frac{x}{3} - 4 > 2$

e $3 - \frac{x}{4} \leq 6$ **f** $-2 + \frac{x}{5} > 6$ **g** $\frac{x+3}{8} \leq 2$ **h** $\frac{5x-3}{2} < 6$

i $\frac{2x+3}{5} > 3$ **j** $\frac{4-3x}{2} \geq 5$ **k** $\frac{7-5x}{4} > 8$ **l** $\frac{4-6x}{5} \leq -4$

Challenge:

6 Solve the following inequations and graph their solutions on a number line. The first one is done for you.

a $-2 \leq x - 3 \leq 6$ $\qquad -2 \leq x - 3 \leq 6$

$1 \leq x \leq 9 \qquad (+3)$

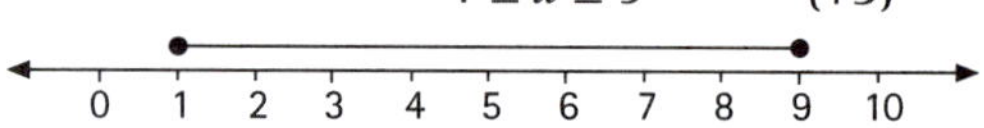

SELF-ASSESSMENT

See page 286 (rubric 1).

b $1 \leq x - 2 \leq 7$ **c** $-4 \leq x + 3 \leq 6$ **d** $-2 \leq x + 7 < 0$

e $0 \leq 2x + 3 \leq 7$ **f** $-5 \leq 3x + 4 \leq 11$ **g** $-16 \leq 3x - 4 \leq -10$

h $7 \leq 7x - 70 \leq 14$

CHAPTER 2 SOLVING AND GRAPHING NON-LINEAR EQUATIONS

KEY VOCABULARY

plot • quadratic • parabola • turning point • axis • symmetry • dilation • factorisation • grouping • trinomials • modelling • curve • hyperbolas • reciprocal function • constant product • inverse proportion • exponential function • quadrant • asymptote • trigonometric function • sine • cosine • tan • amplitude • frequency period • maximum • minimum

Many real-life problems cannot be represented by a linear equation. In this chapter you will look at real-life problems that can be represented through non-linear or quadratic equations. Problems involving area can be solved using quadratic equations. Many engineering problems involve quadratic equations. If you want to work out the trajectory (path) of a projectile (for example, a cricket ball or a spacecraft), you also work with quadratic equations.

Remember:
gradient = $\frac{y_2 - y_1}{x_2 - x_1}$

Plotting quadratic graphs

When you dealt with graphing straight lines, the equations were in the form $y = mx + c$, where m and c were constants.

The key features of straight-line graphs are:

① gradient
② x-intercept
③ y-intercept.

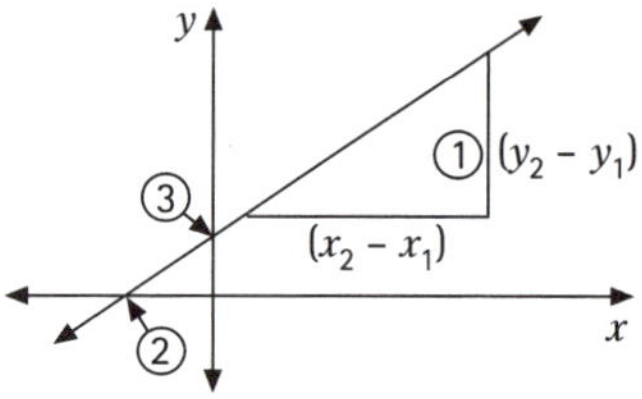

In this section you will focus on quadratic equations. Quadratic equations are written as $y = ax^2 + bx + c$, where a, b and c are constants and $a \neq 0$.

The graphs of these equations are called **parabolas**. Parabolas have some important features.

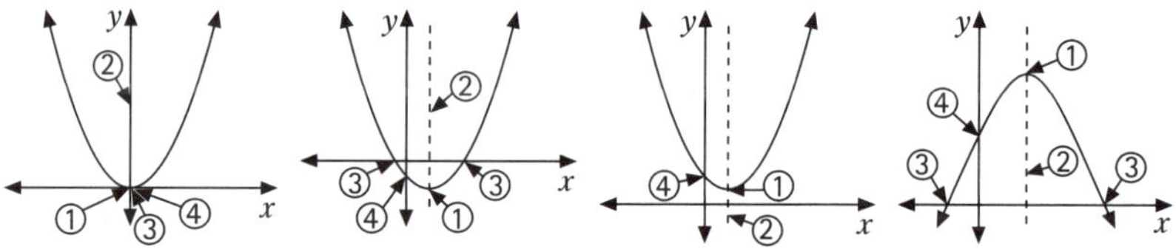

The key features of a parabola are:

① turning point: a maximum or a minimum
② axis of symmetry
③ x-intercepts (there can be 0, 1 or 2 intercepts)
④ y-intercept.

EXAMPLE

Determine the key features of parabolas **a** and **b**:

i turning point **ii** axis of symmetry
iii x-intercepts **iv** y-intercept.

Also write down the domain and range.

a

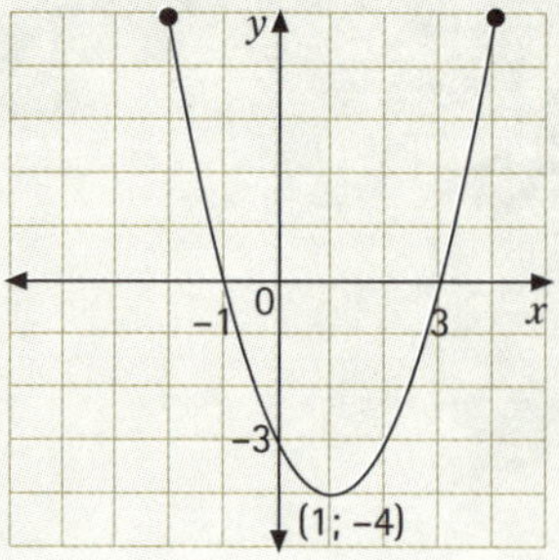

b

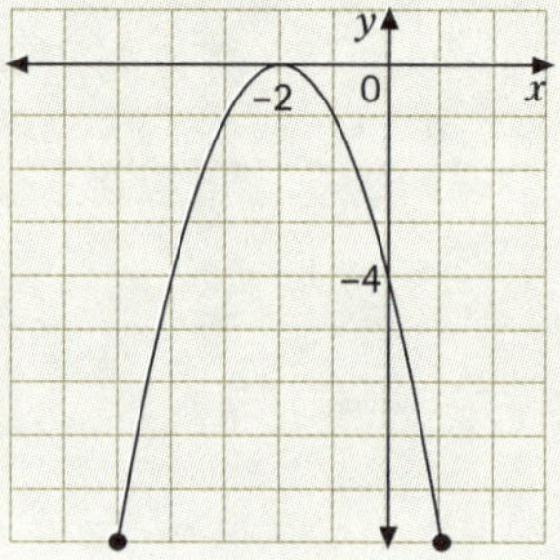

a Turning point is at its minimum at (1; –4)   Domain: $-2 \leq x \leq 4$
Axis of symmetry is $x = 1$
x-intercepts at (–1; 0) and (3; 0)   Range: $-4 \leq y \leq 5$
y-intercept at (0; –3)

b Turning point is at its maximum at (–2; 0)   Domain: $-5 \leq x \leq 1$
Axis of symmetry is $x = -2$ Range: $-9 \leq y \leq 0$
x-intercept at (–2; 0)
y-intercept at (0; –4)

Exercise

1 Determine the key features of these parabolas:

i turning point   **ii** axis of symmetry

iii x-intercepts   **iv** y-intercept.

Also write down the domain and range.

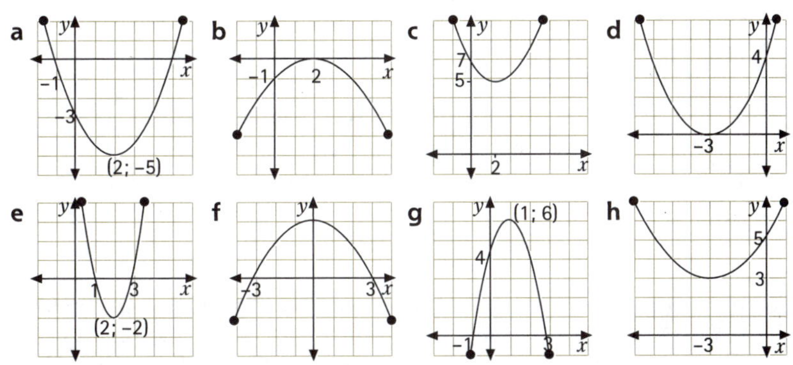

2 Use graph paper to plot the following graphs and determine the:

i turning point   **ii** axis of symmetry

iii x-intercepts   **iv** y-intercept.

a $y = x^2$ for $-4 \leq x \leq -4$ **b** $y = x^2 + 3$ for $-5 \leq x \leq 2$

Investigating types of parabolas

The simplest parabolic shape is generated by the quadratic equation $y = x^2$.

x	−3	−2	−1	0	1	2	3
y	9	4	1	0	1	4	9

As you can see, the parabola is symmetrical about the y-axis and cup-shaped.

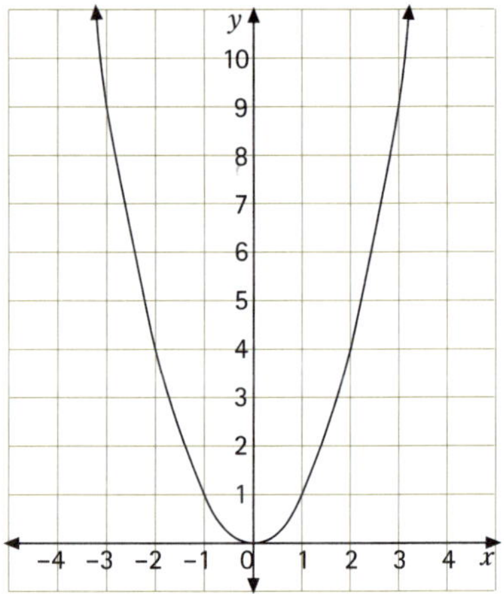

Like all basic graphs, it can be transformed by:

- dilation – making it wider or narrower
- translation in the x-direction – moving it left or right
- translation in the y-direction – moving it up or down
- inverting – reflection in the x-axis.

In the next exercise you will explore how you can make these transformations take place.

Exercise

1 Copy and complete the table on page 167 for graphs **a** to **f**:

a

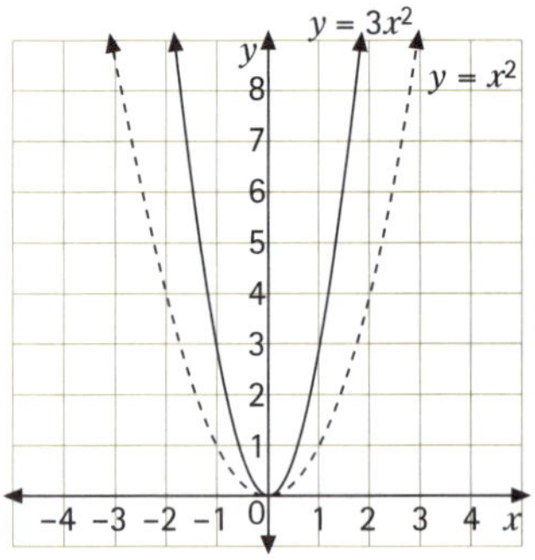

b

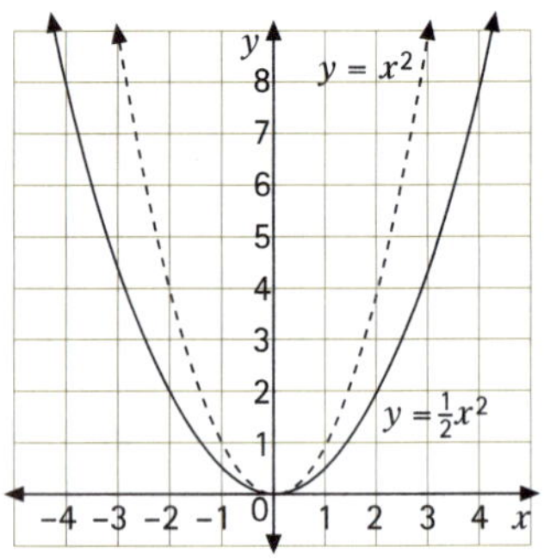

c

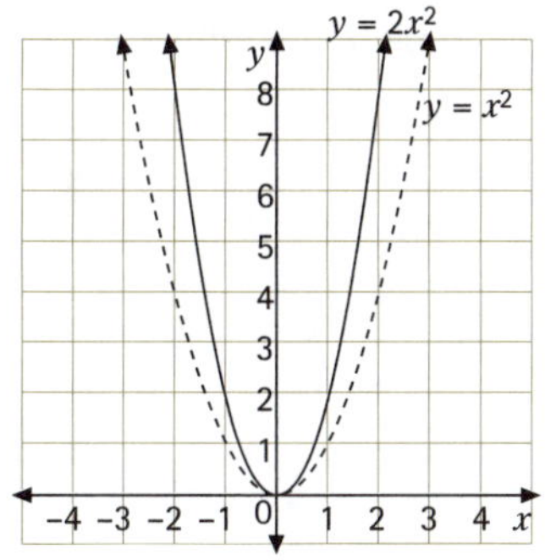

d

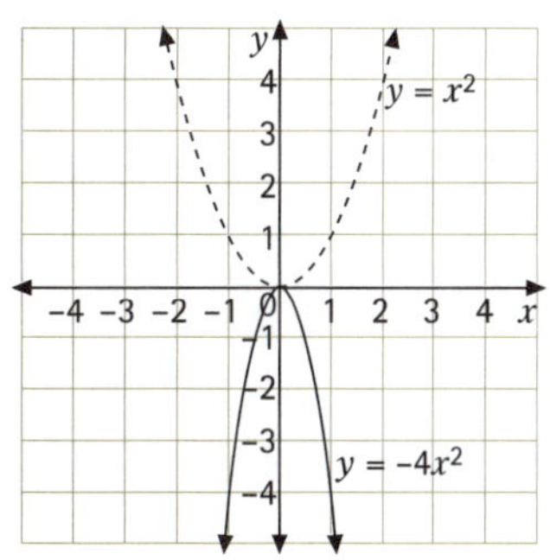

e

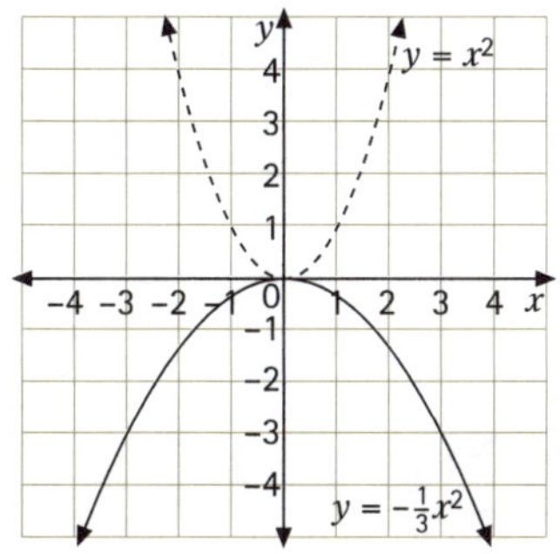

f

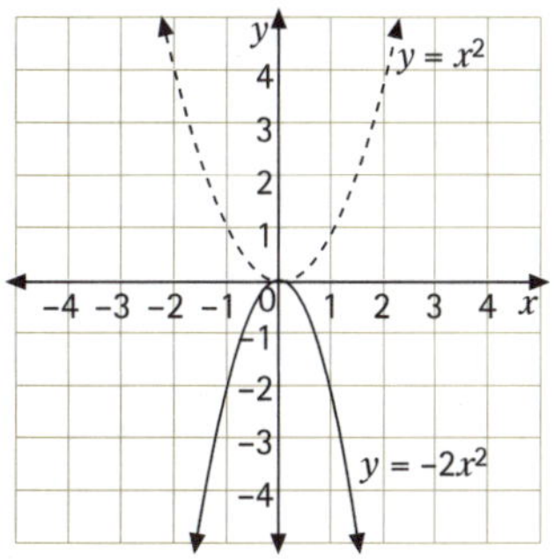

	Formula	Maximum or minimum	Inverted yes/no	Turning point	y-value when $x = 1$	Wider/narrower than $y = x^2$
a	$y = 3x^2$					
b	$y = \frac{1}{2}x^2$					
c	$y = 2x^2$					
d	$y = -4x^2$					
e	$y = -\frac{1}{3}x^2$					
f	$y = -2x^2$					

2 Plot these pairs of graphs on the same axis for x-values between -5 and 5 and compare their tables of values and their shapes.

a $y = x^2$ and $y = 4x^2$

b $y = x^2$ and $y = \frac{1}{3}x^2$

c $y = x^2$ and $y = 6x^2$

d $y = x^2$ and $y = \frac{1}{4}x^2$

e $y = x^2$ and $y = 7x^2$

f $y = x^2$ and $y = \frac{2}{5}x^2$

Suggest how the constant a in $y = ax^2$ transforms the graph of $y = x^2$.

3 Copy and complete the table for these graphs:

a

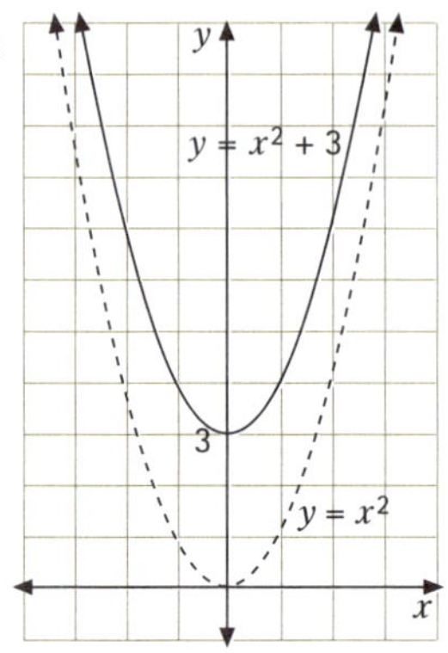

b

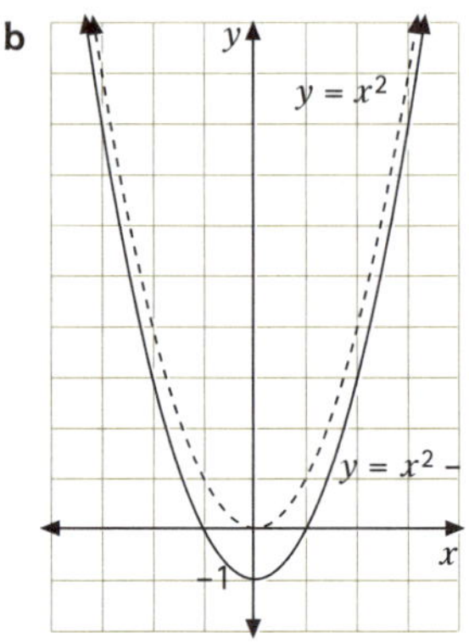

c

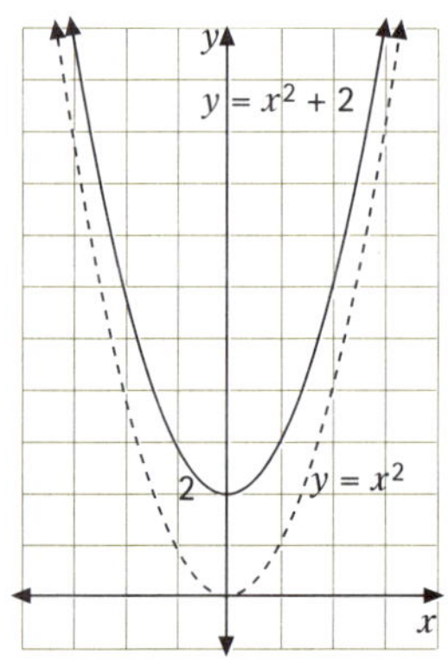

d

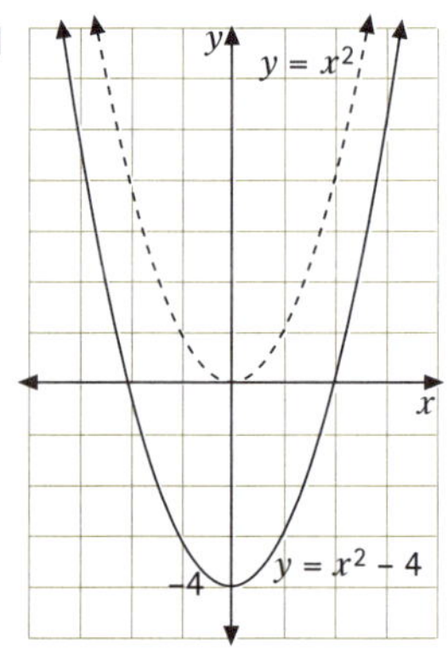

	Formula	Turning point	y-value when $x = 1$
a	$y = x^2 + 3$		
b	$y = x^2 - 1$		
c	$y = x^2 + 2$		
d	$y = x^2 - 4$		

4 Plot these sets of graphs on the same axes from –5 to 5 and compare the turning point of each:

a $y = x^2; y = x^2 + 1; y = x^2 + 2; y = x^2 + 3$

b $y = x^2; y = x^2 - 1; y = x^2 - 3; y = x^2 - 5$

Explain how the constant c in $y = x^2 + c$ transforms the graph of $y = x^2$.

Revision of multiplication and division

EXAMPLES

1 Expand:
$5(x + 3)$
$= 5x + 15$

2 Simplify:
$\frac{15x + 3}{3}$
$= \frac{15x}{3} + \frac{3}{3}$
$= 5x + 1$

3 Expand $(a + 4)(b + 3)$:

$(a + 4)(b + 3)$
$= (a + 4)b + (a + 4)3$
$= ab + 4b + 3a + 12$

Think of $(a + 4)$ as a single quantity, X.
$(a + 4)(b + 3) = X(b + 3) = Xb + X3$
$= (a + 4)b + (a + 4)3$
Check for like terms to simplify.

4 Expand $(a + 2)(a + 5)$:

$(a + 2)(a + 5)$
$= (a + 2)a + (a + 2)5$
$= a^2 + 5a + 2a + 10$
$= a^2 + 7a + 10$ Collect like terms to simplify: $2a + 5a = 7a$.

5 Expand $(2x - 1)(x + 6)$:

$(2x + -1)(x + 6)$ Rewrite subtractions as additions of negatives.
$= (2x + -1)x + (2x + -1)6$
$= 2x^2 + -1x + 12x + -6$
$= 2x^2 + 11x - 6$ Collect like terms: $-1x + 12x = 11x$.

Exercise

Simplify:

1 $\frac{3x - 6}{3}$ **2** $\frac{5x - 4}{2}$ **3** $\frac{7x + 3}{2}$

Expand and simplify where possible:

4 $(b + 2)(b + 4)$ **5** $(d + 4)(d + 1)$ **6** $(a + 3)(b + 5)$

7 $(x + 3)(x + 3)$ **8** $(a + 4)(a + 4)$ **9** $(c + 6)(c - 2)$

10 $(x - 1)(x + 7)$	**11** $(a + 5)(a - 5)$	**12** $(x + 4)(x - 4)$
13 $(c - 6)(c + 6)$	**14** $(d - 4)(d - 2)$	**15** $(k - 7)(k - 7)$
16 $(a + 3)(b - 4)$	**17** $(2x + 7)(x - 4)$	**18** $(x - 6)(3x + 5)$
19 $(8t - 6)(2t - 3)$	**20** $(2k - 5)(4k - 3)$	**21** $(4n - 3)^2$

Expansion by inspection

Example 4 on page 169 showed you that:
$(a + 2)(a + 5) = a^2 + 5a + 2a + 10$

By multiplying **each term** in the second bracket by **each term** in the first, you obtain all four terms of the expansion:

Note that **collecting like terms** is necessary.

$$(a + 2)(a + 5) = \overset{①}{a^2} + \overset{②}{5a} + \overset{③}{2a} + \overset{④}{10}$$
$$= a^2 + 7a + 10$$

Exercise

Expand by inspection:

1 $(a + 2)(a + 6)$	**2** $(b + 4)(b + 3)$	**3** $(x + 7)(x + 7)$
4 $(c + 1)(c + 1)$	**5** $(d + 8)(d + 4)$	**6** $(a + 4)(a - 2)$
7 $(y + 3)(y - 1)$	**8** $(g - 4)(g - 3)$	**9** $(7 + a)(3 - a)$
10 $(2x + 1)(x + 5)$	**11** $(k + 6)(2k + 3)$	**12** $(3m + 2)(m - 4)$
13 $(1 - 3n)(2 + n)$	**14** $(3x + 4)(3x - 4)$	**15** $(2t + 4)^2$
16 $(2t - 4)^2$	**17** $(2x - 5)(3x - 1)$	**18** $(5a + 3)(4a - 5)$
19 $(8 - 2p)(3 + 5p)$	**20** $(7 - 3y)(3 - 5y)$	**21** $(-3x - 8)(7 - 4x)$

Special expansion patterns

Hint: Square each term, and double the product of the two terms.

You will come across two special expansion patterns: perfect squares and the difference of two squares (see page 171).

Perfect squares

$(a + b)^2 = (a + b)(a + b) = a^2 + 2ab + b^2$
$(a - b)^2 = (a - b)(a - b) = a^2 - 2ab + b^2$

EXAMPLES

1 Expand $(x-5)^2$.

$(x-5)^2$

$= x^2 - 2 \times x \times 5 + 5^2$ $\quad (a-b)^2 = a^2 - 2ab + b^2$ Match the pattern.

$= x^2 - 10x + 25$

$(x-5)^2 = x^2 - 2 \times x \times 5 + 5^2$.

Simplify.

2 Expand $(3d+4)^2$.

Note: Use brackets around $3d$, since $a \rightarrow 3d$.

$(3d+4)^2$ $\quad (a+b)^2 = a^2 + 2ab + b^2$ Match the pattern.

$= (3d)^2 + 2 \times 3d \times 4 + 4^2$

$= 9d^2 + 24d + 16$ $\quad (3d+4)^2 = (3d)^2 + 2 \times 3d \times 4 + 4^2$

$(3d)^2 = 3d \times 3d = 9d^2$

Exercise

Expand:

1 $(a+2)^2$	**2** $(a-2)^2$	**3** $(x+3)^2$	**4** $(x-3)^2$
5 $(c+8)^2$	**6** $(d+10)^2$	**7** $(b-4)^2$	**8** $(c-5)^2$
9 $(d+1)^2$	**10** $(x-1)^2$	**11** $(2k+1)^2$	**12** $(2k-1)^2$
13 $(2n+3)^2$	**14** $(3x+4)^2$	**15** $(5a+7)^2$	**16** $(5+3f)^2$
17 $(5x-4)^2$	**18** $(2a+\frac{1}{2})^2$	**19** $(8-4p)^2$	**20** $(-6x-\frac{3}{4})^2$

The difference of two squares

Remember: Opposites add up to zero.

Another special expansion pattern involves two terms that contain the same variables, but which have different signs.

same variable as terms, but opposite signs

$(a+b)(a-b) = a^2 - b^2$

first term squared   second term squared

$(a+2)(a-2)$

$= a^2 - 2a + 2a - 2^2$

$= a^2 - 2^2$

EXAMPLES

1 Expand $(x + 3)(x - 3)$:

$(x + 3)(x - 3)$ $\quad (a + b)(a - b) = a^2 - b^2$ $\quad$ Match the pattern.

$= x^2 - 3^2$

$= x^2 - 9$ $\quad (x + 3)(x - 3) = x^2 - 3^2$

Simplify where possible.

2 Expand $(2m - 7)(2m + 7)$:

$(2m - 7)(2m + 7)$ $\quad (a + b)(a - b) = a^2 - b^2$ $\quad$ Match the pattern.

$= (2m)^2 - 7^2$

$= 4m^2 - 49$ $\quad (2m + 7m)(2m - 7m) = (2m)^2 - 7^2$

Simplify.

Note: Use brackets around $2m$.

Exercise

Expand:

1 $(a + 4)(a - 4)$

2 $(b + 6)(b - 6)$

3 $(x + 5)(x - 5)$

4 $(c - 7)(c + 7)$

5 $(d - 8)(d + 8)$

6 $(10 + h)(10 - h)$

7 $(9 - t)(9 + t)$

8 $(x + y)(x - y)$

9 $(h - k)(h + k)$

10 $(2a + 1)(2a - 1)$

11 $(3x + 5)(3x - 5)$

12 $(5p - 3)(5p + 3)$

13 $(9n - 8)(9n + 8)$

14 $(11 - 6k)(11 + 6k)$

15 $(2x + y)(2x - y)$

16 $(3a + 2b)(3a - 2b)$

17 $(4n - 5p)(4n + 5p)$

18 $(\frac{1}{2}x - \frac{3}{4}y)(\frac{1}{2}x + \frac{3}{4}y)$

Evaluate using the difference of two squares:

19 $8^2 - 3^2$

20 $11^2 - 9^2$

21 $17^2 - 13^2$

22 $65^2 - 35^2$

23 $5{,}7^2 - 4{,}3^2$

24 $(\frac{3}{4})^2 - (\frac{1}{4})^2$

Verify the following using replacements $a = 3$ and $b = 2$:

25 $a^2 + b^2 \neq (a + b)^2$

26 $a^2 + b^2 \neq (a + b)(a - b)$

Revision of factorisation

A mathematical statement can be written as a product of its factors. The process of writing statements in this form is called factorisation. To solve quadratic equations later on, you must be able to factorise expressions.

In the past you learned to look for **common factors**. This is always the first thing to do when you simplify algebraic expressions.

EXAMPLES

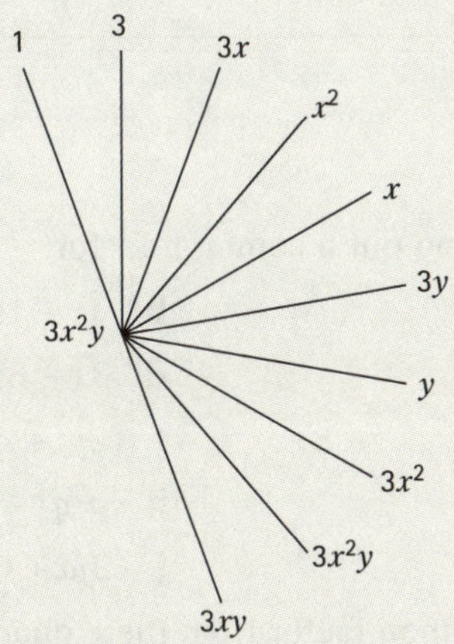

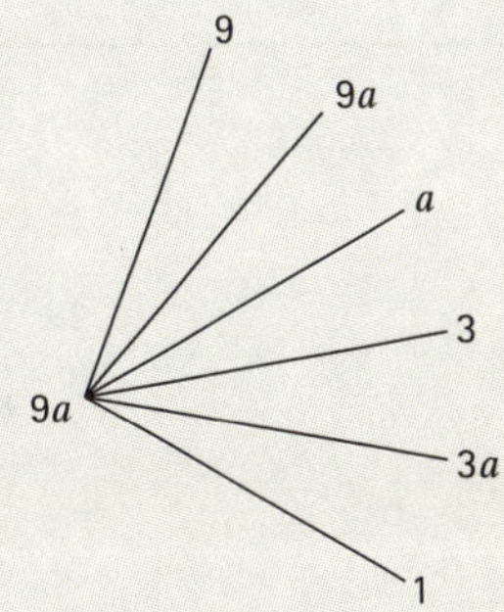

$4x + 2 = 2(x + 2)$

$9ax^2 - 12a^2x = 3ax(3x - 4a)$

$3x^2 - 6x + 18 = 3(x^2 - 2x + 6)$

Since $(a - b)(a + b) = a^2 - b^2$, there is a rule that can help you to factorise certain quadratic expressions.

For the difference of perfect squares:

$a^2 - b^2 = (a - b)(a + b)$

1 Factorise by determining the common factor:

a $5x - 20$
$= 5(x - 4)$

b $2f^2g^2 + 10f^4g$
$= 2f^2g(g + 5f^2)$

c $6a^2 + 3a - 12$
$= 3(2a^2 + a - 4)$

2 Use the rule above to factorise:

a $x^2 - 16$
$= (x)^2 - (4)^2$
$= (x - 4)(x + 4)$

b $9a^2 - 4b^2$
$= (3a)^2 - (2b)^2$
$= (3a - 2b)(3a + 2b)$

3 Factorise:

a $12b^2 - 1\,452$
$= 12(b^2 - 121)$ (common factor)
$= 12(b - 11)(b + 11)$

b $(x + 3)^2 - 4$
$= (x + 3)^2 - (2)^2$
$= (x + 3 - 2)(x + 3 + 2)$
$= (x + 1)(x + 5)$

Exercise

1 Factorise by taking out a common factor:

a $4x - 12$ **b** $3x - 15$
c $2x + 2y$ **d** $3x - 6y$
e $xy + xz$ **f** $2pj + 4pf$
g $ac^2 - a^2c$ **h** $p^3q^4 - p^2q^4$
i $2d^3h + 6dh^2$ **j** $3jd + 6j^2$

2 Remove the common factor from these quadratic expressions:

a $3x^2 - 3x + 6$ **b** $4a^2 + 2a - 6$ **c** $2c^2 - 4c + 6$
d $8d^2 - 4d + 6$ **e** $5h^2 + 10h - 15$ **f** $4k^2 + 8k - 10$
g $2e^2 - 8e$ **h** $3r^2 + 6r$ **i** $4f^2 - 10f$
j $15k^2 + 10$ **k** $6f^2 - 15$ **l** $4t^2 + 14$

3 Factorise:

a $x^2 - 4$ **b** $d^2 - 25$ **c** $f^2 - 16$
d $4c^2 - 9$ **e** $9e^3 - 16$ **f** $25w^2 - 100$
g $16y^2 - 25x^2$ **h** $9s^2 - 100t^2$ **i** $81g^2 - 4h^2$
j $49a^2 - 81b^2$ **k** $100r^2 - 9c^2$ **l** $25v^2 - 16k^2$

4 Factorise by first removing the common factor:

a $3e^2 - 12$ **b** $5d^2 - 45$ **c** $2t^2 - 32$
d $4f^2 - 36$ **e** $3j^2 - 27$ **f** $6p^2 - 24$
g $2d^2 - 18s^2$ **h** $3f^2 - 48j^2$ **i** $5t^2 - 500s^2$
j $8a^2 - 200b^2$ **k** $12w^2 - 108q^2$ **l** $27s^2 - 243t^2$

5 Factorise:

a $(x + 4)^2 - 9$ **b** $(d - 6)^2 - 16$ **c** $(h - 5)^2 - 81$

d $(s - 9)^2 - 100$ **e** $(f + 8)^2 - 16$ **f** $(k - 4)^2 - 25$

More factorisation

Factorising by grouping in pairs

In some expressions, there is no common factor in all terms. In such cases, you can regroup or pair the terms to factorise sections of an expression. Sometimes you will then find a common factor.

EXAMPLES

Hint: Always check by expansion to see if the original expression can be recreated.

1 Factorise, if possible, $ab + ac + bx + cx$.

$(ab + ac) + (bx + cx)$ **or**	$(ab + bx) + (ac + cx)$	Group terms into pairs with common factor(s). Factorise each pair, then again using the bracketed common factor.
$= a(b + c) + x(b + c)$	$= b(a + x) + c(a + x)$	
$= (b + c)(a + x)$	$= (a + x)(b + c)$	

Check:

$$(b + c)(a + x) = ba + bx + ca + cx$$

2 Factorise, if possible, $3t - ax^2 + tx^2 - 3a$.

$(3t - 3a) + (tx^2 - ax^2)$ Group the first and fourth terms, and the second and third.

$= 3(t - a) + x^2(t - a)$

$= (t - a)(3 + x^2)$

Challenge: Is there an alternative grouping that works?

Check:

$$(t - a)(3 + x^2) = 3t + tx^2 - 3a - ax^2$$

3 Factorise, if possible, $x^2 + 5x - 3x - 15$.

$(x^2 + 5x) + (-3x + -15)$	Group terms into pairs.
$= x(x + 5) + -3(x + 5)$	Note that the **negative** common factor
$= (x + 5)(x + -3)$	(−3) creates the bracketed factor to match. $3(-x - 5)$ is a correct set of factors, but does not create a common factor.

Check:

$(x + 5)(x + -3) = x^2 - 3x + 5x - 15.$

Exercise

1 True or false: $x(a - b) = -x(-a + b)$? Where could this pattern be useful?

Complete each pattern:

2 $3x - 6 = 3(. . . - . . .)$

3 $3x - 6 = -3(. . . + . . .)$

4 $mn - mp = m(. . .)$

5 $mn - mp = -m(. . .)$

6 $-10ab + 15b = 5b(. . .)$

7 $-10ab + 15b = -5b(. . .)$

Factorise:

8 $(ax + ay) + (bx + by)$

9 $(mp + pt) + (xm + tx)$

10 $(x^2 + 4x) + (3x + 12)$

11 $(c^2 + 2c) + (5c + 10)$

12 $(tp - ts) + (ap - as)$

13 $(10x + 20y) - (xc + 2yc)$

14 $(ax + bx) - (ay + by)$

15 $x^2 + ab + ax + ab$

16 $ab + b + 4a + 4$

17 $x^3 + x^2 + x + 1$

18 $xy + 5a + 5x + ay$

19 $a^2 + nt + ta + na$

20 $ab - 2b - 5ax + 10x$

21 $ab^2 - ab - b + 1$

22 $11x + 22 - xy - 2y$

23 $x^2 + 2x + 3x + 6$

24 $y^2 + 2y + 5y + 10$

25 $a^2 + 3a + 4a + 12$

26 $a^2 - 2a + 3a - 6$

27 $n^2 + 3n - 5n - 15$

28 $t^2 + 3t + t + 3$

29 $x^2 - 3x - 4x + 12$

30 $y^2 - 4y + 7y - 28$

31 $d^2 - 3d - 8d + 24$

32 $f^2 - 5f + 2f - 10$

33 $h^2 + 8h - 10h - 80$

34 $k^2 - 7k - 5k + 35$

Factorise:

35 $2n^2 + 2n + 3n + 3$

36 $3x^2 + 2x + 12x + 8$

37 $2p^2 - 5p + 10p - 25$

38 $4a^2 - 4a + 3a - 3$

39 $c^2 + nc + kc + kn$

40 $d^2 - 2d + pd - 2p$

Factorising quadratic trinomials

Consider the following expansions:

$$(x + 3)(x + 4) = x^2 + 3x + 4x + 12 = x^2 + 7x + 12$$

$7 = 3 + 4 \quad 12 = 3 \times 4$

$$(x + 2)(x - 1) = x^2 + 2x - 1x - 2 = x^2 + 1x - 2$$

$1 = 2 - 1 \quad -2 = 2 \times -1$

So: The **middle term** is determined by **adding** the last term in each bracket.
The **end term** is determined by **multiplying** the last term in each bracket.

In general, $(x + a)(x + b) = x^2 + ax + bx + ab$
$= x^2 + (a + b)\,x + ab$

You will use these facts to help you factorise quadratic trinomials of this type.

For example, to factorise $x^2 + 5x + 6$ you must find factors of 6 that add up to 5.

$1 \times 6 = 6$ and $1 + 6 = 7 \neq 5$

$2 \times 3 = 6$ and $2 + 3 = 5$, therefore the numbers you need are 2 and 3,

so $x^2 + 5x + 6 = (x + 2)(x + 3)$.

In general, for $x^2 + bx + c$, you ask what factors of c add up to b.

EXAMPLES

Factorise:

1 $3x^2 - 6x$ **2** $x^2 + 7x + 12$

3 $x^2 + x - 6$ **4** $x^2 - 5x + 6$

Note: A common factor is all that's needed in example 1.

1 $3x^2 - 6x = 3x(x - 2)$

2 $x^2 + 7x + 12$

$1 \times 12 = 12 \qquad 1 + 12 = 13$

$2 \times 6 = 12 \qquad 2 + 6 = 8$

$3 \times 4 = 12 \qquad 3 + 4 = 7$

$\therefore x^2 + 7x + 12 = (x + 3)(x + 4)$

3 $x^2 + x - 6$

$1 \times -6 = -6 \qquad 1 + -6 = -5$

$-1 \times 6 = -6 \qquad -1 + 6 = 5$

$-2 \times 3 = -6 \qquad -2 + 3 = 1$

$\therefore x^2 + x - 6 = (x - 2)(x + 3)$

4 $x^2 - 5x - 6$

$-1 \times -6 = 6 \qquad -1 + 6 = -7$

$-3 \times -2 = 6 \qquad -3 + -2 = -5$

$\therefore x^2 -5x + 6 = (x - 3)(x - 2)$

Exercise

1 Simplify by removing the common factor:

a $x^2 - 3x$ **b** $x^2 + 4x$ **c** $x^2 - 6x$ **d** $x^2 - 10x$

e $4x^2 + 2x$ **f** $5x^2 - 15x$ **g** $8x^2 - 2x$ **h** $20x^2 - 8x$

2 Factorise these quadratic trinomials:

a $x^2 + 3x + 2$ **b** $x^2 + 6x + 5$ **c** $x^2 + 8x + 7$

d $x^2 + 7x + 6$ **e** $x^2 + 5x + 6$ **f** $x^2 + 6x + 9$

g $x^2 + 7x + 10$ **h** $x^2 + 7x + 12$ **i** $x^2 + 11x + 18$

j $x^2 + 11x + 30$ **k** $x^2 + 5x + 4$ **l** $x^2 + 13x + 12$

3 Factorise these quadratic trinomials:

a $x^2 + x - 2$ **b** $x^2 + 4x - 5$ **c** $x^2 + 6x - 7$

d $x^2 + 5x - 6$ **e** $x^2 + x - 6$ **f** $x^2 + 2x - 8$

g $x^2 - x - 2$ **h** $x^2 - 4x - 5$ **i** $x^2 - 3x - 10$

j $x^2 - 4x - 21$ **k** $x^2 - 6x - 16$ **l** $x^2 - x - 12$

4 Factorise the quadratic trinomials:

a $x^2 - 3x + 2$ **b** $x^2 - 4x + 3$ **c** $x^2 - 6x + 5$

d $x^2 - 7x + 10$ **e** $x^2 - 6x + 8$ **f** $x^2 - 7x + 12$

g $x^2 - 2x + 1$ **h** $x^2 - 9x + 18$ **i** $x^2 - 11x + 18$

j $x^2 - 9x + 20$ **k** $x^2 - 5x + 6$ **l** $x^2 - 13x + 12$

5 Factorise the quadratic trinomials:

a $x^2 - x - 2$	**b** $x^2 + 4x + 3$	**c** $x^2 - 7x + 6$
d $x^2 + 7x + 12$	**e** $x^2 - 3x + 2$	**f** $x^2 - 5x - 14$
g $x^2 + 4x + 4$	**h** $x^2 - 7x - 30$	**i** $x^2 - 9x + 20$
j $x^2 - 9x - 36$	**k** $x^2 - 5x - 36$	**l** $x^2 + 5x - 24$

6 Factorise completely by removing the common factor first:

a $2x^2 + 14x + 20$	**b** $3x^2 + 21x + 36$	**c** $2x^2 + 22x + 36$
d $5x^2 - 5x - 10$	**e** $4x^2 - 16x - 20$	**f** $3x^2 - 9x - 30$
g $3x^2 - 6x + 3$	**h** $2x^2 - 14x - 36$	**i** $5x^2 - 45x + 90$

Solving quadratic equations algebraically

When you factorise a quadratic equation, you can solve it algebraically by using the zero factor law. The **zero factor law** is based on the fact that if you multiply any number by zero the result is zero.

Zero factor law
If $ab = 0$, then $a = 0$ or $b = 0$ or both a and $b = 0$.

EXAMPLES

Remember: Use the zero factor law after you have factorised.

1 Solve these equations:

a $x(x - 2) = 0$
$\therefore x = 0$ or $x - 2 = 0$
$\therefore x = 0$ or $x = 2$

b $(2x - 5)(x - 3) = 0$
$\therefore (2x - 5) = 0$ or $(x - 3) = 0$
$\therefore 2x = 5$ or $x = 3$
$\therefore x = \frac{5}{2}$ or $x = 3$

2 Solve these equations and verify your solutions:

a $x^2 - 3x = 0$ **b** $x^2 - 25 = 0$
c $x^2 + 3x + 2 = 0$ **d** $x^2 - 5x = 6$

a $x^2 - 3x = 0$
$\therefore x(x - 3) = 0$ Factorise.
$\therefore x = 0$ or $x - 3 = 0$ Zero factor law.
$\therefore x = 0$ or $x = 3$ Simplify.

$(0)^2 - 3(0) = 0$
$(3)^2 - 3(3) = 0$ Verify.

b $x^2 - 25 = 0$

$\therefore (x-5)(x+5) = 0$ — Factorise.

$\therefore x - 5 = 0$ or $x + 5 = 0$ — Zero factor law.

$\therefore x = 5$ or $x = -5$ — Simplify.

$(5)^2 - 25 = 0$

$(-5)^2 - 25 = 0$ — Verify.

c $x^2 + 3x + 2 = 0$

$\therefore (x+2)(x+1) = 0$

$\therefore x + 2 = 0$ or $x + 1 = 0$

$\therefore x = -2$ or $x = -1$

$(-2)^2 + 3(-2) + 2 = 0$

$(-1)^2 + 3(-1) + 2 = 0$

d $x^2 - 5x = 6$

$\therefore x^2 - 5x - 6 = 0$

$\therefore (x-6)(x+1) = 0$

$\therefore x - 6 = 0$ or $x + 1 = 0$

$\therefore x = 6$ or $x = -1$

$(6)^2 - 5(6) = 6$

$(-1)^2 - 5(-1) = 6$

3 Solve these pairs of simultaneous equations by using substitution:

a $y = 3 - 4x$
$y = 4x^2$

$y = 3 - 4x$.......(1)
$y = 4x^2$...........(2)

Substitute (1) into (2):

$\therefore 3 - 4x = 4x^2$

$\therefore 0 = 4x^2 + 4x - 3$

$\therefore 0 = (2x-1)(2x+3)$

$\therefore 2x - 1 = 0$ or $2x + 3 = 0$

$\therefore 2x = 1$ or $2x = -3$

$\therefore x = \frac{1}{2}$ $\quad \therefore x = -\frac{3}{2}$

Substitute into (2):

So $y = 4 \times (\frac{1}{2})^2$

$\therefore y = 1$

or $y = 4(-\frac{3}{2})^2$

$\therefore y = 9$

Solutions: (0,5; 1) or (−1,5; 9)

b $y = x^2 + 6$
$-x + y = 18$

$y = x^2 + 6$.......(1)
$-x + y = 18$
$\therefore y = 18 + x$.......(2)

Substitute (2) into (1):

$\therefore 18 + x = x^2 + 6$

$\therefore 0 = x^2 - x - 12$

$\therefore 0 = (x-4)(x+3)$

$\therefore x - 4 = 0$ or $x + 3 = 0$

$\therefore x = 4$ $\quad \therefore x = -3$

Substitute into (2):

So $y = 18 + 4$

$\therefore y = 22$

or $y = 18 - 3$

$\therefore y = 15$

Solutions: (4; 22) or (−3; 15)

Exercise

1 Solve these equations:

a $x(x - 2) = 0$
b $x(x - 5) = 0$
c $x(2x - 1) = 0$
d $x(3x + 2) = 0$
e $(x - 2)(x - 3) = 0$
f $(x - 5)(x - 3) = 0$
g $(x - 2)(x + 1) = 0$
h $(x - 3)(x + 4) = 0$
i $(x + 4)(x + 2) = 0$
j $(x + 6)(x + 7) = 0$
k $(2x - 1)(x + 3) = 0$
l $(x - 5)(2x - 3) = 0$
m $(3x - 4)(2x + 3) = 0$
n $(4x + 5)(5x - 3) = 0$

2 Solve the equations and check your solutions:

a $x^2 - 2x = 0$
b $x^2 - 5x = 0$
c $x^2 + 4x = 0$
d $x^2 + 5x = 0$
e $x^2 - 8x = 0$
f $x^2 + 10x = 0$
g $x^2 + 11x = 0$
h $x^2 + 3x = 0$

3 Solve the equations and check your solutions:

a $(x - 2)^2 = 0$
b $(x - 3)^2 = 0$
c $(2x - 1)^2 = 0$
d $(3x + 2)^2 = 0$
e $(x - 7)^2 = 0$
f $(4x - 3)^2 = 0$
g $(2x - 5)^2 = 0$
h $(6x + 5)^2 = 0$

4 Solve the equations and check your solutions:

a $x^2 - 4 = 0$
b $x^2 - 25 = 0$
c $x^2 - 16 = 0$
d $x^2 - 81 = 0$
e $x^2 - 100 = 0$
f $x^2 - 36 = 0$
g $x^2 = 1$
h $x^2 = 144$
i $x^2 = 9$

5 Solve the equations and check your solutions:

a $x^2 - x - 2 = 0$
b $x^2 + 4x + 3 = 0$
c $x^2 - 7x + 6 = 0$
d $x^2 + 7x + 12 = 0$
e $x^2 - 3x + 2 = 0$
f $x^2 - 5x - 14 = 0$
g $x^2 + 4x + 3 = 0$
h $x^2 - 7x - 30 = 0$
i $x^2 - 9x + 20 = 0$
j $x^2 - 9x = 36$
k $x^2 - 5x = 36$
l $x^2 + 5x = 24$

6 Solve these equations using factorisation and the zero factor law. Remember to check all your solutions.

a $x^2 - 4x = 0$
b $x^2 + 6x + 5 = 0$
c $x^2 - 9 = 0$
d $x^2 + 15x = 0$
e $x^2 - 3x = 18$
f $x^2 + 4x - 12 = 0$
g $5x^2 = 100$
h $4x^2 - 2x - 6 = 0$
i $3x^2 - 6x = 0$
j $5x^2 - 45 = 0$
k $8x^2 + 16x + 8 = 0$
l $4x^2 - 24x = 64$

7 Solve these pairs of simultaneous equations by substitution:

a $y = x^2$, $y = 8x - 15$ **b** $y = x^2$, $y = 5x - 6$ **c** $y = -x^2$, $y + 2x = -8$ **d** $y = 3 + 7x$, $y = 2x^2 - 12$

Problem solving and modelling with quadratics

You can represent many practical problems by quadratic equations. Here are some examples of common problems that can be represented (modelled) using quadratics.

Area

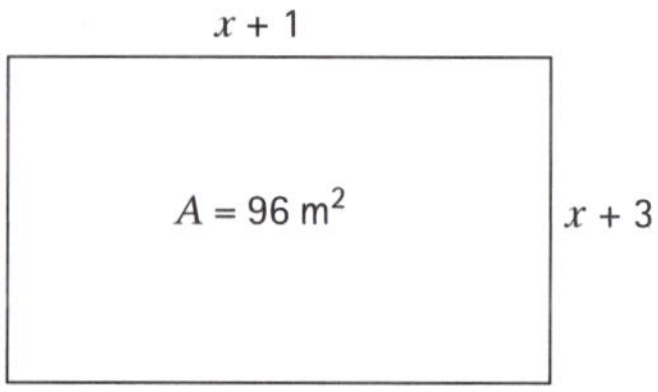

Projectiles, for example, a soccer ball:

Engineering, for example, bridges:

EXAMPLES

1 A rectangle has an area of 6 cm² and its breadth is 5 cm less than its length. Find the dimensions of the rectangle.

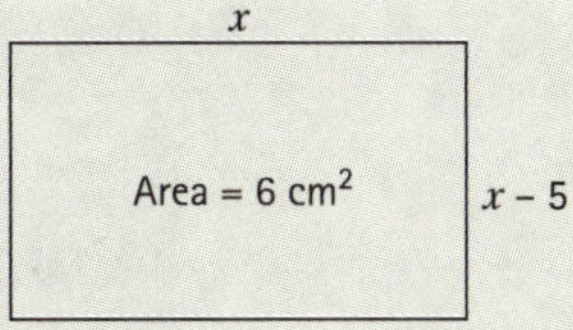

Let x = length of the rectangle. (Decide on a variable if you are not given one.)

Area = $l \times b$

$\therefore x(x - 5) = 6$ Form an equation.
$\therefore x^2 - 5x = 6$
$\therefore x^2 - 5x - 6 = 0$
$\therefore (x - 6)(x + 1) = 0$
$\therefore x - 6 = 0$ or $x + 1 = 0$
$\therefore x = 6$ or $x = -1$ Solve the equation.
$\therefore x = 6$

Hint: Check that your solution is reasonable. $x = -1$ is not a realistic solution, as you can't have a negative length.

The dimensions of the rectangle are 6 cm and 1 cm.

2 A boat travels under a bridge that is 8 m wide and 4 m high and has a parabolic shape.

a Sketch a graph of the bridge's arch, showing the information you know.

b Check that the arch can be modelled by the equation $h = -\frac{1}{4}(d - 4)^2 + 4$.

c Plot the equation and determine how close to the edge a 3 m high boat may travel.

a

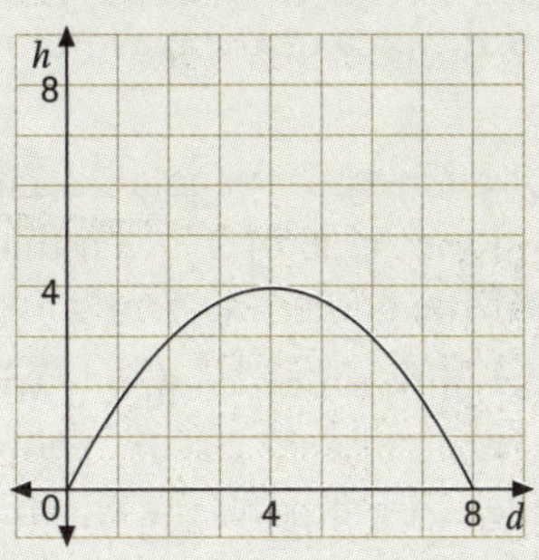

b $h = -\frac{1}{4}(d - 4)^2 + 4$

The turning point is (4; 4).

$d = 0$: $h = -\frac{1}{4}(0 - 4)^2 + 4$
$= -\frac{1}{4}(-4)^2 + 4$
$= -4 + 4$
$= 0$

$d = 8$: $h = -\frac{1}{4}(8 - 4)^2 + 4$
$= -\frac{1}{4}(4)^2 + 4$
$= -4 + 4$
$= 0$

∴ All conditions are satisfied and the equation is a good model.

c Find the d-values that give $h = 3$.

$h = 3$: $3 = -\frac{1}{4}(d - 4)^2 + 4$

$-1 = -\frac{1}{4}(d - 4)^2$ Subtract 4 from both sides.

$\therefore 4 = (d - 4)^2$ Multiply both sides by –4.

$\therefore 4 = d^2 - 8d + 16$

$\therefore 0 = d^2 - 8d + 12$

$\therefore 0 = (d - 2)(d - 6)$

$\therefore d - 2 = 0$ or $d - 6 = 0$

$\therefore d = 2$ or $d = 6$

The boat must travel a minimum of 2 m away from the edge.

Exercise

1 The dimensions of a rectangular plot are represented by $(x + 2)$ m and $(x - 1)$ m. Find the value of x if the area of the plot is 28 m^2.

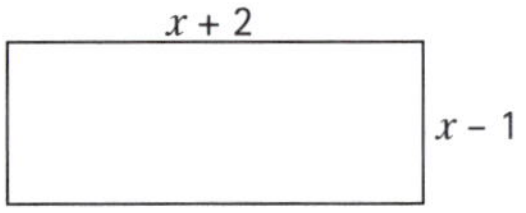

2 The length of a rectangular children's playground is 3 m more than its width. If the area of the playground is 70 m^2, determine its length.

3 The area of an irregular vegetable garden is shown. Find the value of x, given that the total area of the garden is 47 m^2.

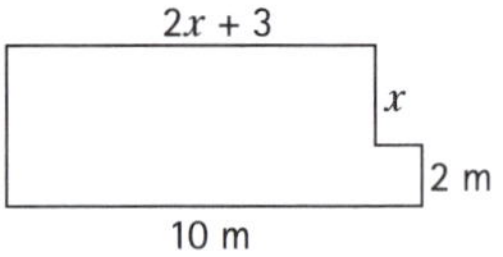

4 Mashome's construction company wants to build a house of area 234 m^2 shown in the diagram. Determine the width of land left around the house.

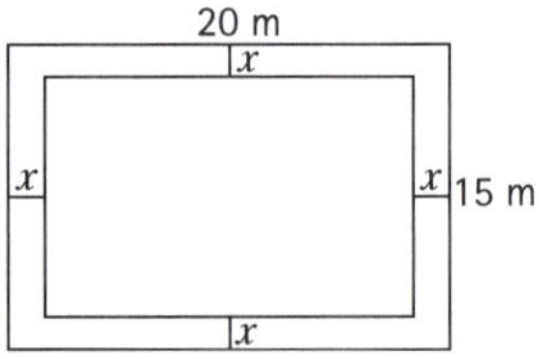

5 A plot is in the shape of a right-angled triangle. The longest side of the plot is 4 times longer than the shortest side. The third side is 3 m longer than the shortest side. Find the length of each side.

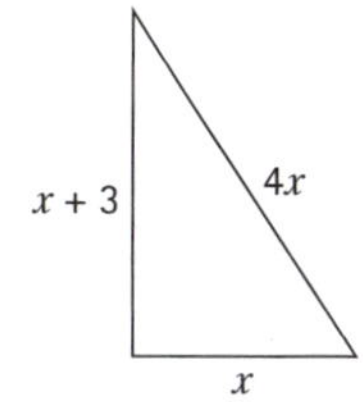

6 A ball is thrown up from the top of a 120 m tall building. The position of the ball above the ground, in metres, after a certain time, is given by:

$$H = -\frac{1}{5}(t - 10)^2 + 130$$

where t is the time in seconds.

a How long will it take for the ball to be 50 m above the ground?

b When will the ball hit the ground?

7 Stanley is d years old.

a What is Stanley's age in 1 year's time?

b What was Stanley's age 4 years ago?

c Write an expression for Stanley's age 4 years ago times his age in 1 year's time.

d If the expression in **c** equals 50, determine Stanley's age.

Sketching parabolas

Parabolas, like other graphs, are a visual representation of a real-life situation. When you sketch a graph you need to show all or some of its key features. These are the key features you should show when sketching a parabola:

- Its shape shows a maximum or minimum.
- The turning point shows the translation of $y = x^2$.
- The x-intercepts (if any).
- Any other point (usually the y-intercept or the point where $x = 1$ if the y-intercept is at the turning point) to show the dilation of $y = x^2$.

On pages 164 to 168 you looked at parabolas and their general features. On the next page you will see two general examples of parabolas for $a > 0$. The two graphs show which points you should show when sketching a parabola:

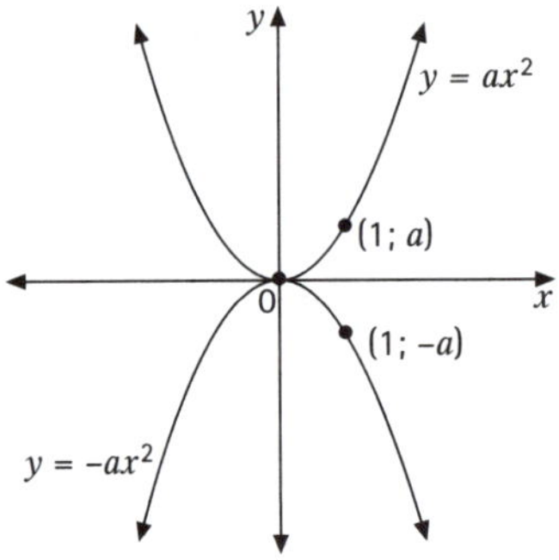

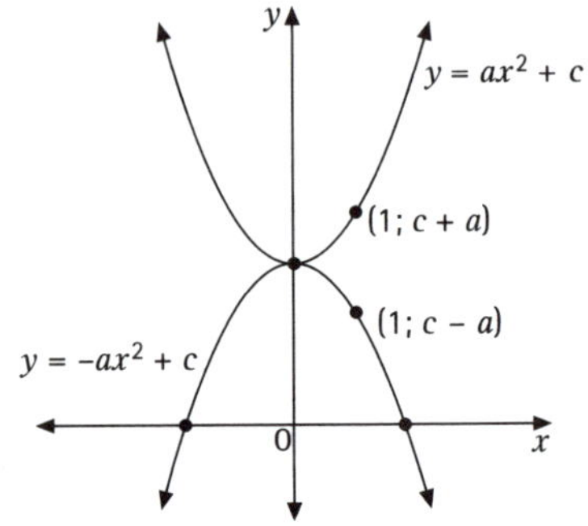

EXAMPLES

Sketch these graphs:

1 $y = 3x^2$ **2** $y = -x^2 + 4$ **3** $y = 2x^2 + 2$

1 Shape of graph: ∪
Turning point: (0; 0)
$x = 1$: $y = 3(1)^2$
$= 3$

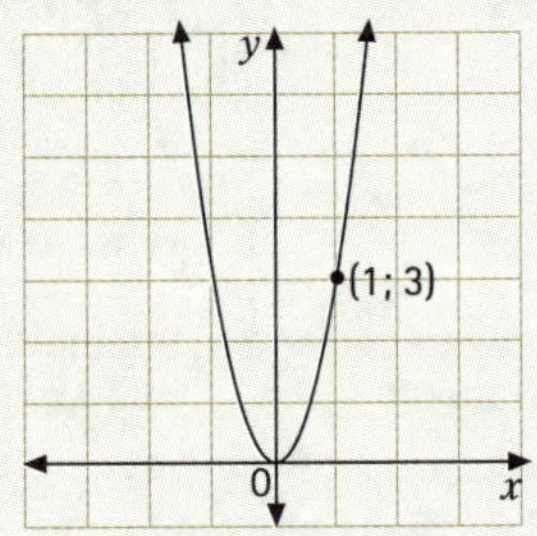

2 Shape of graph: ∩
Turning point: (0; 4)
$x = 1$: $y = -(1)^2 + 4$
$= 3$
x-intercepts at $y = 0$:
$0 = -x^2 + 4$
$\therefore x^2 - 4 = 0$
$\therefore (x + 2)(x - 2) = 0$
$\therefore x + 2 = 0$ or $x - 2 = 0$
$\therefore x = -2$ or $x = 2$
x-intercepts:
(–2; 0); (2; 0)

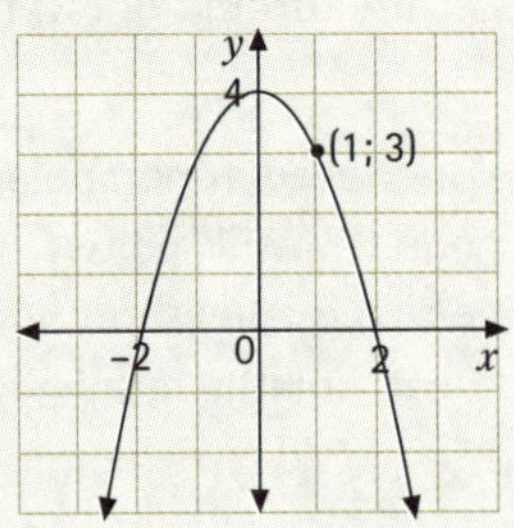

3 Shape of graph: ∪
Turning point: (0; 2)
$x = 1$: $y = 2(1)^2 + 2$
$= 4$
x-intercepts at $y = 0$:
$\therefore 0 = 2x^2 + 2$
$\therefore 0 = x^2 + 1$
$\therefore -1 = x^2$
No solutions, because a square cannot be negative. So there are no x-intercepts.

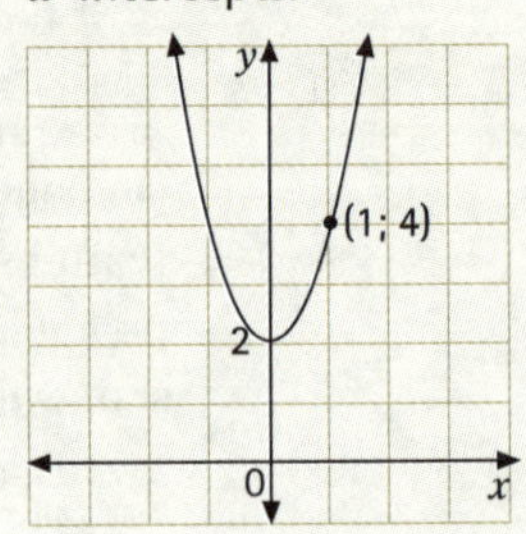

Exercise

1 Sketch the graphs of:

a $y = 2x^2$ **b** $y = -3x^2$ **c** $y = \frac{1}{2}x^2$

d $y = -\frac{1}{3}x^2$ **e** $y = x^2 + 2$ **f** $y = x^2 - 4$

g $y = -x^2 + 1$ **h** $y = -x^2 - 3$ **i** $y = 2x^2 - 2$

j $y = \frac{1}{2}x^2 - 2$ **k** $y = -\frac{1}{4}x^2 + 4$ **l** $y = -3x^2 - 2$

2 Compare these graphs with the general graphs on page 186 and determine the equation for each.

a
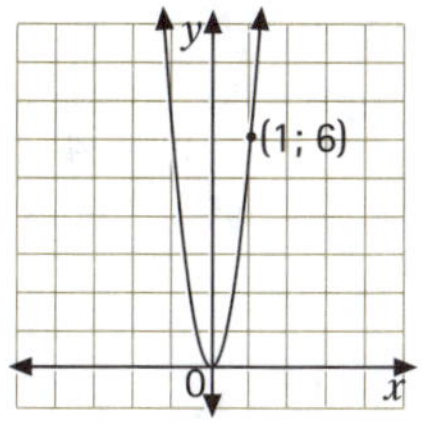

b
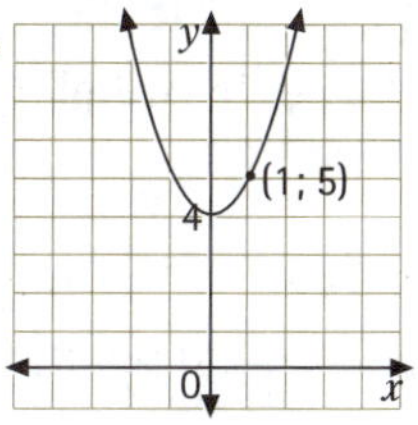

c
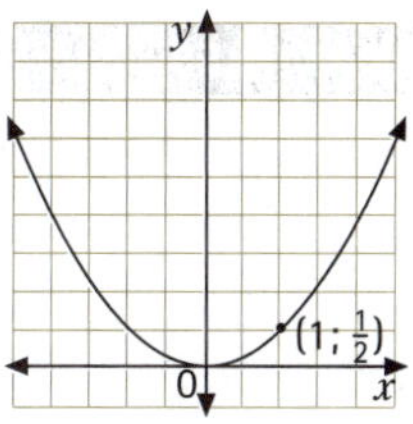

d
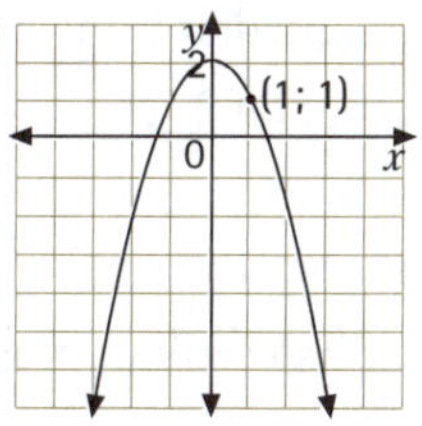

e
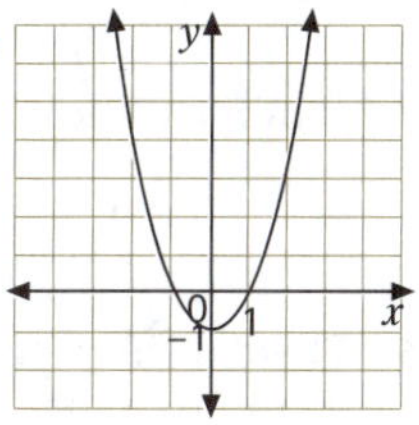

f
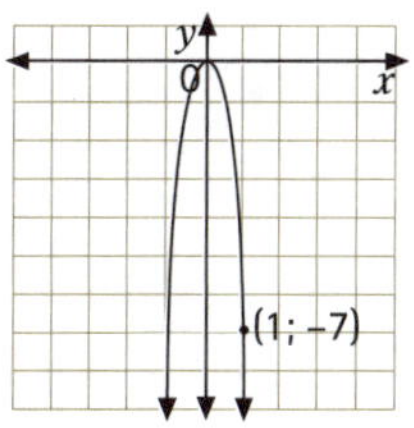

Plotting hyperbolas

The graphs of hyperbolas have a characteristic shape, with distinct mirror images in opposite quadrants. Hyperbolic functions are also called **reciprocal functions**. Reciprocal functions have a constant product. In the next investigation you will look at a real-life problem that can be solved using a hyperbolic function.

GROUP ASSESSMENT

See page 288 (rubric 3).

Investigation: Designing a townhouse

Work in a group.

You will need: pencil, eraser, graph paper or graphics calculator.

You are the architects planning the possible courtyard shapes for townhouses. Space is tight, and the developer allows for only 20 m^2. What possible dimensions could a rectangular shape have, with this limited area?

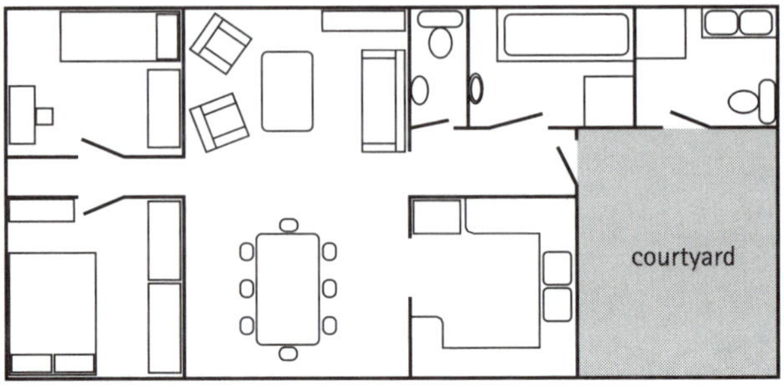

1 Complete the table of values:

Area = 20 m^2 ($A = lb$)

l	1	1,25	2	2,5	4					
b						4	2,5	2	1,25	1

2 Scale the x-axis (length) and y-axis (breadth) so 1 cm represents 2 m, showing the first quadrant only.

3 Plot all 10 points as accurately as possible.

4 Join all points with a smooth curve. Discuss the shape of the curve you have drawn in your group.

5 Write a short paragraph describing the shape and features of the curve.

6 Would it be correct to write the equation for this graph as $xy = 20$? Why must the product of the variables equal 20?

Rewrite the equation in the form $y = \dots$.

7 Write a recommendation on the best dimensions for the courtyard.

EXAMPLE

Graph the function $y = \frac{1}{x}$ for $-2 \leq x \leq 2$.

a Using appropriate values for x, draw up a table of ordered pairs:

$xy = 1$ or $y = \frac{1}{x}$.

x	−2	−1,25	−1	−0,8	−0,5	0	0,5	0,8		1,25	2
y	−0,5	−0,8	−1	−1,25	−2		2	1,25	1	0,8	0,5

b The x-values range from −2 to 2, perfectly balanced about the y-axis, so locate the y-axis in the centre.

c The *y*-values also range from –2 to 2, so locate the *x*-axis in the centre.

d Use a scale of 2 cm for 1 unit to allow for fractional values.

e Sketch a smooth curve through each set of five points.

f Check the shape using a graphics calculator, if possible.

Note: The curve does not exist at $x = 0$. This would mean that $y = \frac{1}{x}$ becomes $y = \frac{1}{0}$. (Division by 0 is not possible.)

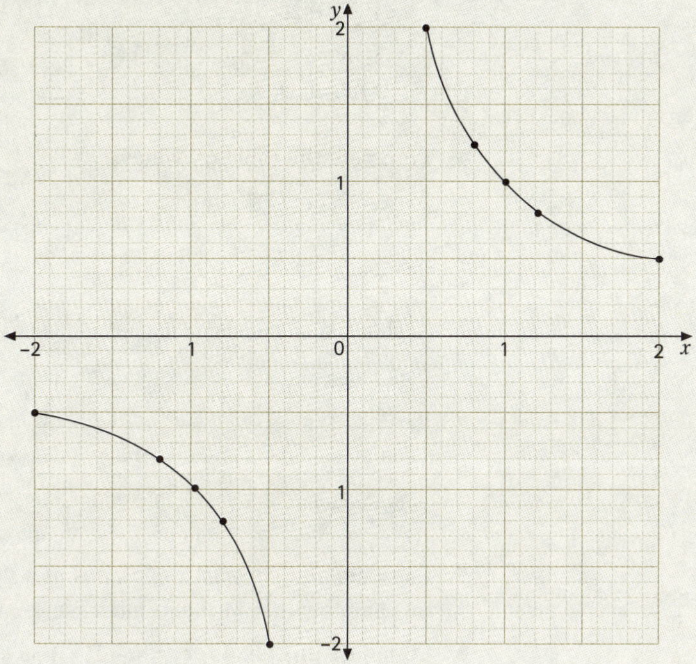

The graph shows you that:

- The hyperbola has two distinct parts, each a mirror image of the other. Can you give the equation of the mirror line?
- The "arms" of the curve are in opposite quadrants.
- The curve moves ever closer to the axes but never touches them. Test this using large values for x, for example, 1 000, –10 000. Also, use very small values for x, for example, $\frac{1}{1\,000}$.
  We say that the axes are the **asymptotes** of the curve.

Exercise

Complete the following tables, and then graph all four hyperbolas on the one number plane. Place the x-axis and y-axis in the centre.

1 $y = \frac{1}{x}$

x	0,25	0,5	0,8	1	1,25	2	4
y							

2 $y = \frac{2}{x}$

x	0,5	0,8	1	1,25	2	2,5	4
y							

3 $y = \frac{4}{x}$

x	1	1,3	2	3	4
y					

4 $y = \frac{8}{x}$

x	2	2,4	2,5	3	4
y					

5 Compare and contrast the graphs in a short paragraph. Comment on their location in the number plane and on their shape.

6 Complete this table for $xy = -2$:

x	2	4	8	10	20	40	100	200	2 000	2 000 000
y										

a Complete this sentence: As the x-values become very large, the y-values ... .

b Complete this sentence: y can never equal zero, because ..., so the curve will move ever closer to, but never ... .

DISCUSS

Will negatives have any meaning in this problem? Discuss this in your groups.

7 A certain journey between two towns is exactly 120 km. Use a graph model to show how variation in speed (x) over that journey causes variation in time taken (y) to complete the trip. Use speeds of 10, 20, 30, 40, 60, 80 and 120 km/h in your table of values.

From your graph, find:

a the time taken to do the trip at 100 km/h

b the speed necessary to cover the distance in $2\frac{1}{2}$ hours.

c Complete this sentence: The faster the vehicle travels, the ….

d Is it possible that a vehicle could travel so fast that the trip could be done in **no time** at all? How does the trend of the graph relate to this question?

8 The current needed to work your car's headlights is inversely proportional to the voltage. That is, the current times voltage product is always 144. Use a graph to show the variation in current (measured in ampere) for variation in voltage (measured in volt) from 0 to 360 volt. Scale the y-axis (ampere) up to 5 and the x-axis (volt) in multiples of 60 to 360 volt.

From your graph, find:

a the current that would be needed at 110 volts

b the voltage when the current is half an ampere.

The general form of reciprocal functions

Note: k is a constant product.

The general equation for a reciprocal function can be summarised like this:

$y = \frac{a}{x}$ or $xy = a$ A reciprocal function has a **constant product**. It represents **inverse** proportion.

As x increases, y decreases correspondingly, and vice versa. The greater the value of a, the flatter the shape of the curve, and the further from the origin it is located (depending on the scales used). Hyperbolas have a characteristic shape, with distinct mirror images in opposite quadrants.

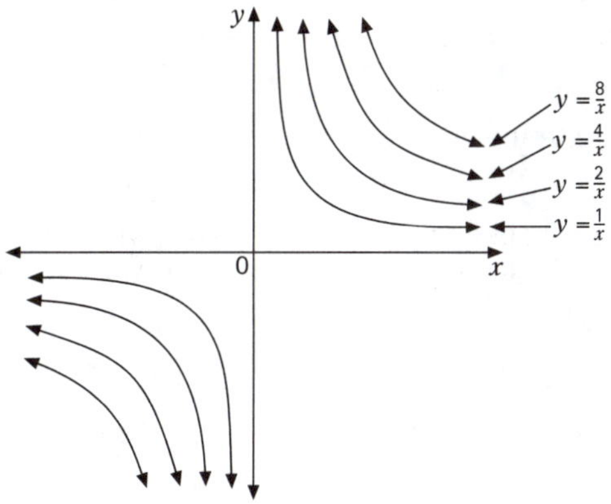

In this module, you investigated only those reciprocal functions in which the constant (a) is a positive integer.

What would you expect for graphs of functions for which a is a negative integer?

Exercise

1 Plot these graphs on the same axis for x-values between –5 and +5 and compare their tables of values.

$y = \frac{2}{x}$; $y = \frac{2}{x} + 1$; $y = \frac{2}{x} + 2$.

Discuss how the constant c in $y = \frac{2}{x} + c$ transforms the graph of $y = \frac{2}{x}$.

Match each graph with the appropriate equation:

2 $y = x^2$ **3** $y = 2 - x$ **4** $y = 2$ **5** $xy = 2$

6 $y = \frac{2}{x} + 1$ **7** $y = 2x$ **8** $y = x$ **9** $y = -\frac{1}{x}$

a

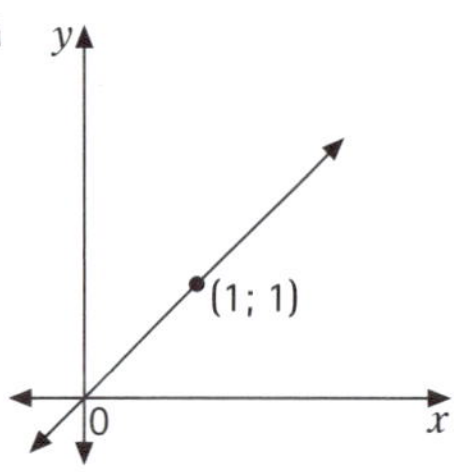

b

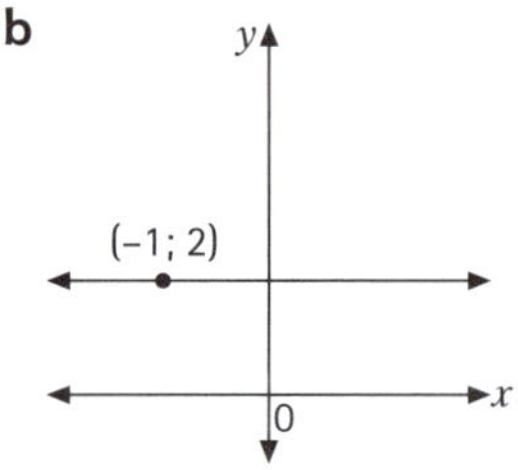

c

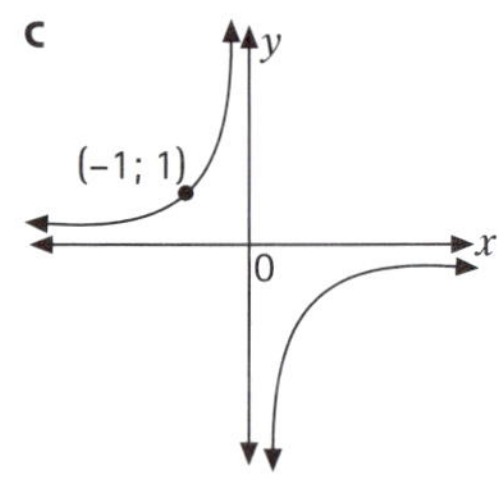

d

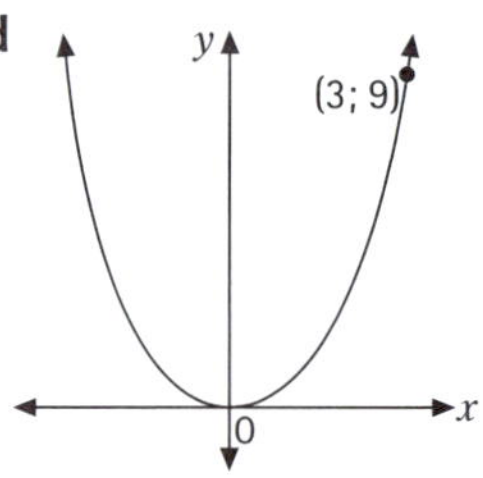

e

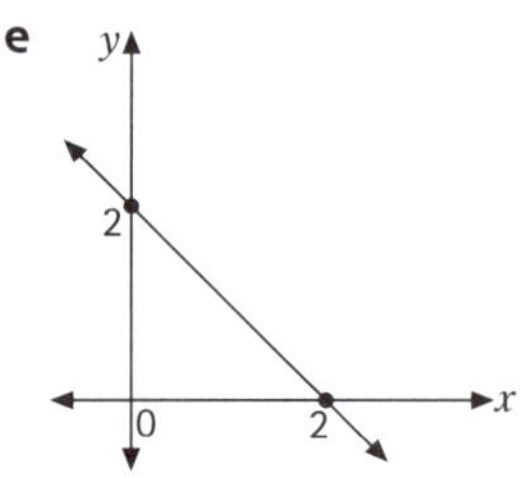

f

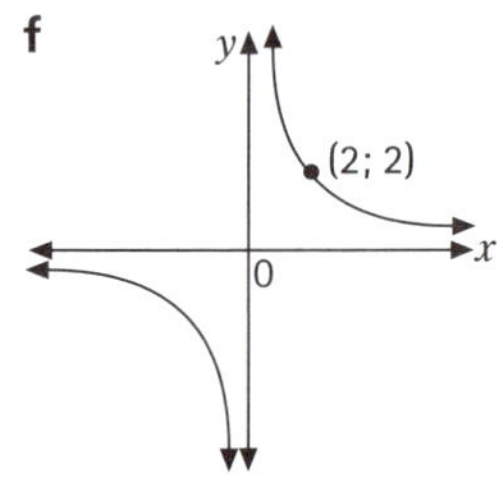

g

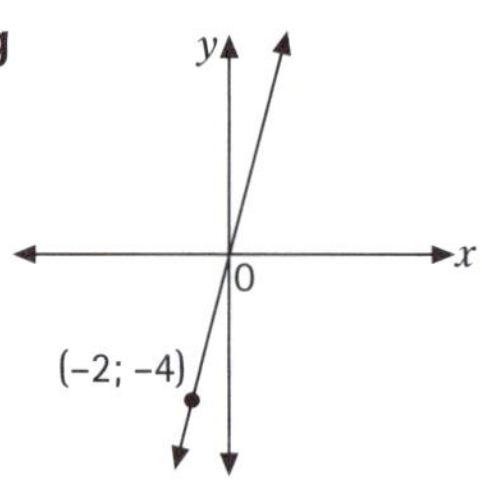

h

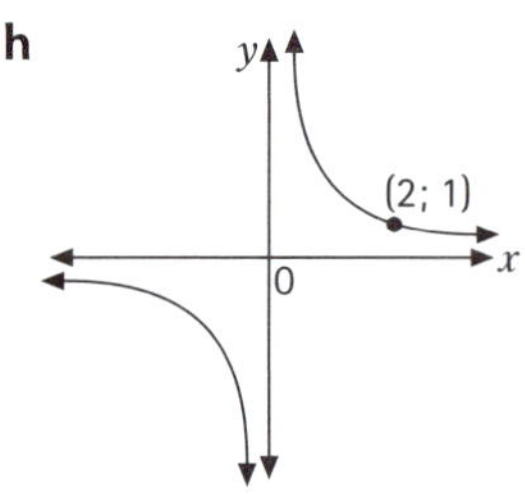

Match each equation with the appropriate description of its graph:

10 $y = -2x - 4$ **11** $x = -4$ **12** $y = \frac{4}{x} - 2$ **13** $y = 3x^2 - 4$

a a straight line parallel to the y-axis, with x-intercept –4

b a steep-sided upright parabola with y-intercept –4

c a linear function of negative gradient passing through (0; –4)

d a hyperbola passing through (–2; –4).

Are these statements true or false? (If false, write a correct statement.)

14 $y = -2x^2$ is an inverted parabola with the y-axis as axis of symmetry.

15 $y = -2 - 2x^2$ is identical in shape to the graph in question **14**, but moved 2 units down the y-axis.

16 $y = -2$ is a line of zero gradient.

17 $y = \frac{x}{2}$ is a hyperbola.

18 $yx = 5$ is the same function as $y = \frac{5}{x}$.

19 $2x^2 - 3y + 1 = 0$ is a straight line of gradient $\frac{2}{3}$.

20 $2x - 5y + 3 = 0$ is parallel to $3y - 1{,}2x = 0$.

21 $y = -x$ is an axis of symmetry of $xy = 2$.

Plotting and sketching exponential functions

DO YOU REMEMBER?

An asymptote is a line that the graph approaches but never meets, shown as a dotted line – in this case the x-axis.

Exponential functions are written in the form

$y = a^x$, where a is constant and $a > 0$.

The figure shows the basic shape of a graph of an exponential function.

This graph differs from a parabola. It is not symmetrical and has no turning points.

Exponential functions have an asymptote at $y = 0$.

You will now plot and sketch graphs of some exponential functions.

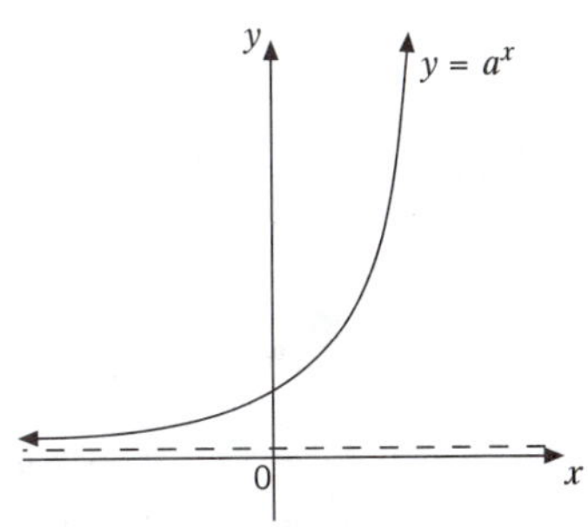

EXAMPLES

Graph the function $y = 2^x$ for $-2 \leq x \leq 2$.

1 Use appropriate values for x to draw up a table of ordered pairs:

x	−2	−1	0	1	2
y	0,25	0,5	1	2	4

2 The x-values range from −2 to +2, perfectly balanced about the y-axis, so you can locate the y-axis in the centre of the graph.

3 The y-values range from 0,25 to 4, so you can locate the x-axis at the bottom of the graph.

4 Use a scale of 2 cm for 1 unit to allow for decimal values.
5 Sketch a smooth curve through the five points.
6 The curve approaches ever closer to the x-axis, but will never touch it. Test this by using a very small value for x, for example, −8. The x-axis is called the asymptote of the curve.

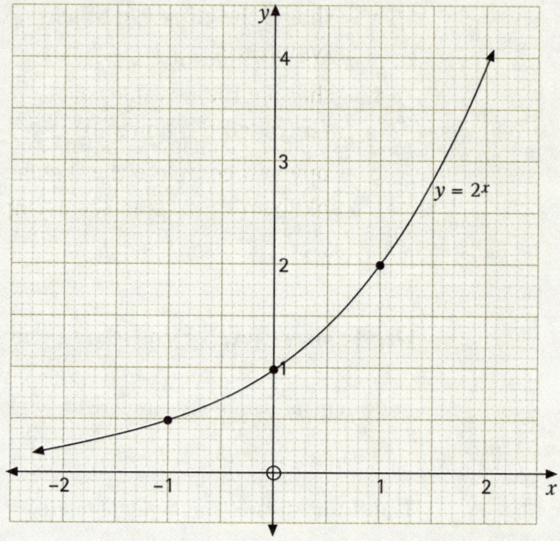

GROUP ASSESSMENT

See page 288 (rubric 3).

Investigation: Exponential functions

Work in a group.

You will need: a pencil, graph paper.

1 Plot $y = 2^x$, $y = 3^x$ and $y = 4^x$ on the same axes for $-3 \leq x \leq 5$.
   - **a** Determine approximate values of the following sets of numbers from the graphs:
     - **i** $2^0, 3^0, 4^0$ **ii** $2^1, 3^1, 4^1$ **iii** $2^{0,5}, 3^{0,5}, 4^{0,5}$
     - **iv** $2^{1,5}, 3^{1,5}, 4^{1,5}$ **v** $2^2, 3^2, 4^2$ **vi** $2^{-1}, 3^{-1}, 4^{-1}$
   - **b** Discuss how the graph is affected as a increases in $y = a^x$.

2 The graphs you investigated have some key features that are common to all exponential graphs. Discuss these questions in your group:
   - **a** Do all the graphs you investigated have the same basic shape?
   - **b** Does $a^0 = 1$ for all the graphs you have investigated?

c Does $a^1 = a$ for all the graphs you have investigated?

d What are the possible values of y (called the range)?

e What is the equation of the asymptote?

f What are the possible values of x (called the domain)?

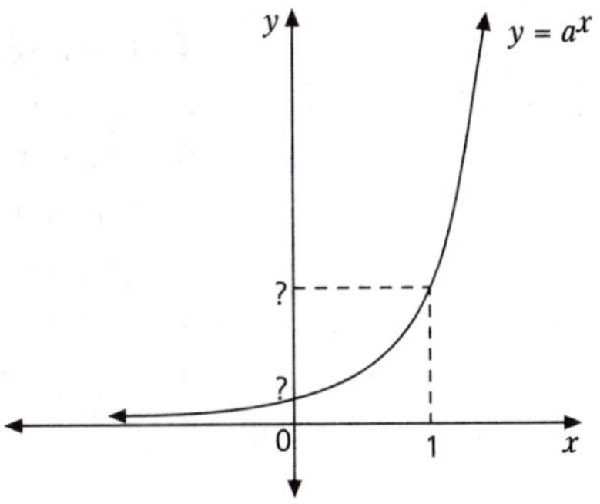

g Copy and complete the sketch shown by finding the two unknown values and the asymptote.

Exercise

1 Plot the graphs of this set of equations on the same pair of axes:

$y = 2^x$, $y = 2^x + 1$, $y = 2^x + 2$, $y = 2^x + 3$ for $-3 \leq x \leq 3$.

Use it to determine, for all graphs:

a the y-intercepts

b the values of y when $x = 1$

c the range of all possible y-values and the equation of the asymptote.

d Repeat the process for $y = 3^x$, $y = 3^x - 1$, $y = 3^x - 2$, $y = 3^x - 3$.

2 Explain in words how the graph of $y = a^x$ is transformed by c in $y = a^x + c$.

3 Make neat sketches of the graphs of these equations. In each case show the value of the y-intercept, the value of y when $x = 1$, and draw a dotted line to represent the asymptote.

a $y = 5^x$

b $y = 9^x$

c $y = 5^x + 3$

d $y = 7^x - 2$

Solving exponential equations

Solutions to exponential equations can be found in different ways.

Here you will find some methods of solving x:

- graphically (using graph paper)
- by trial and error (using a table of values)
- algebraically.

$3 = 2^x \quad x = ?$

Most solutions will have decimal answers. You will need to give your answers to a certain number of decimal places.

EXAMPLES

1 Use a calculator to evaluate to 3 decimal places where necessary:

a $y = 4^5$
$\approx 1\,024$

b $y = 2^{0,7}$
$\approx 0,616$

2 By plotting an appropriate graph over an appropriate domain, find an approximate solution to $10 = 3,5^x$ to 1 decimal place.

Consider $y = 3,5^x$.

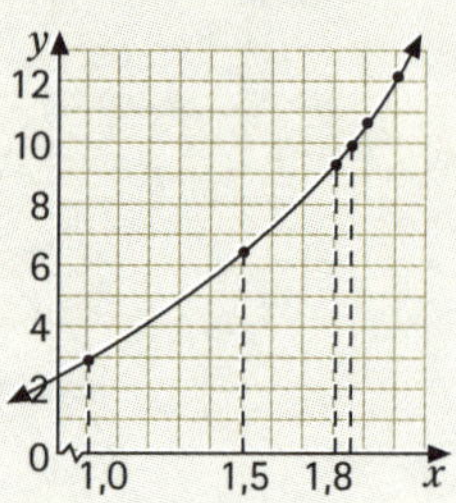

$y = 3,5^1 = 3,5$
$y = 3,5^2$ has a value less than 4^2 or 16.
So, since $y = 10$, x must be between 1 and 2.

Remember: The larger the scale, the more accurate your answer.

x	1	1,5	1,8	1,9	2,0
y	3,5	6,55	9,54	10,80	12,25

$x \approx 1,8$

3 Use trial and error to solve $8 = 2,4^x$ to two decimal places.

x	$2,4^x$	
2	5,76	too small
3	13,82	too big
2,5	8,92	too big
2,4	8,18	too big
2,3	7,49	too small
2,38	8,033	closest answer
2,37	7,963	
2,39	8,104	

$x \approx 2,38$

Remember: For the power x^7, x is called the base and 7 is called the exponent.

4 Some simple exponential equations can be solved by writing the powers on either side of the equal sign to the same base, and then equalising the exponents.

Solve these equations:

a $3^x = 9$

$\therefore 3^x = 3^2$

$\therefore x = 2$ (powers are equal, bases are equal, so exponents must be equal)

b $16^x = 32$

$\therefore (2^4)^x = 2^5$

$\therefore 2^{4x} = 2^5$

$\therefore 4x = 5$

$\therefore x = \frac{5}{4}$

$\therefore x = 1{,}25$

5 Solve $3 \times 5^x = 75$.

$3 \times 5^x = 75$

$\therefore 5^x = 25$

$\therefore 5^x = 5^2$

$\therefore x = 2$

6 Solve $2 \times 3^{x+2} = 486$.

$2 \times 3^{x+2} = 486$

$\therefore 3^{x+2} = 243$

$\therefore 3^{x+2} = 3^5$

$\therefore x + 2 = 5$

$\therefore x = 3$

Exercise

1 Use a calculator to evaluate y in these equations to 3 decimal places:

a $y = 4^5$ **b** $y = 2^4$ **c** $y = 4^{-3}$ **d** $y = 2^{-2}$

e $y = 3^{1,4}$ **f** $y = 6^{1,2}$ **g** $y = 4{,}2^{3,1}$ **h** $y = 2{,}6^{2,3}$

i $y = 4{,}7^{-0,7}$ **j** $y = 2{,}6^{-0,6}$ **k** $y = 2^{3,2}$ **l** $y = (-5{,}6)^{-1}$

2 By plotting an appropriate graph over an appropriate domain, find an approximate solution to these equations to 1 decimal place:

a $12 = 2^x$ **b** $2 = 3^x$ **c** $3 = 1{,}6^x$ **d** $11 = 1{,}7^x$

e $4 = 17{,}2^x$ **f** $2 = 2{,}6^x$ **g** $2{,}6 = 6^x$ **h** $17{,}2 = 5^x$

3 Use trial and error to solve these equations to 2 decimal places:

a $6 = 3{,}4^x$ **b** $8 = 2{,}4^x$ **c** $2 = 1{,}6^x$ **d** $11 = 3{,}9^x$

e $0{,}2 = 10^x$ **f** $0{,}3 = 9^x$ **g** $2{,}6 = 3{,}2^x$ **h** $3{,}7 = 1{,}06^x$

4 Solve for x by writing the powers to the same base:

a $2^x = 8$ **b** $3^x = 27$ **c** $5^x = 125$

d $6^x = 1\ 296$ **e** $9^x = 27$ **f** $27^x = 81$

g $64^x = 16$ **h** $49^x = 2\ 401$ **i** $9 \times 9^x = 243$

j $3 \times 27^x = 243$ **k** $2^{x+2} = 64$ **l** $7 \times 5^{x-3} = 875$

5 The population of a small country town with 3 000 people is increasing at 10% per year.

a Write the percentage as a decimal.

b What will the population be in

i 1 year's time? **ii** 2 years' time?

c Copy and complete the table:

No. of years	0	1	2	3	4
Population	3 000				

d Plot a graph of number of years vs. population.

e Using the graph, determine (to 2 decimal places):

i the population after 2,4 years

ii when the population will be 4 000.

Plotting and sketching trigonometric functions

In Module 2 you looked at the definitions for the three trigonometric functions sin θ, cos θ and tan θ for angles less than 90°. These trigonometric functions can be defined for all positive angles.

First quadrant

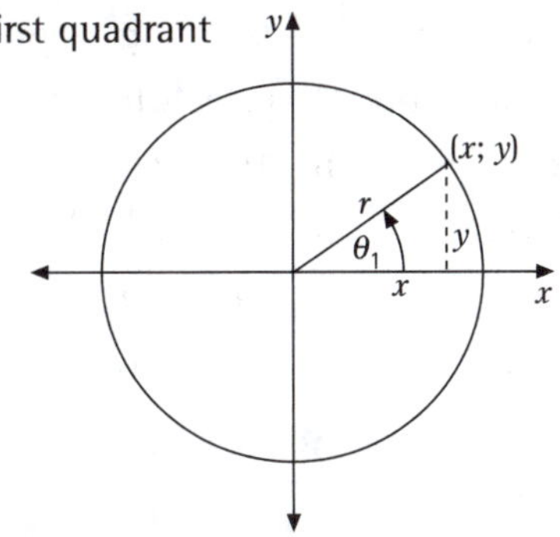

$$\sin \theta_1 = \frac{y\text{-coordinate}}{\text{radius}} = \frac{y}{r}$$

$$\cos \theta_1 = \frac{x\text{-coordinate}}{\text{radius}} = \frac{x}{r}$$

$$\tan \theta_1 = \frac{y\text{-coordinate}}{x\text{-coordinate}} = \frac{y}{x}$$

Second quadrant

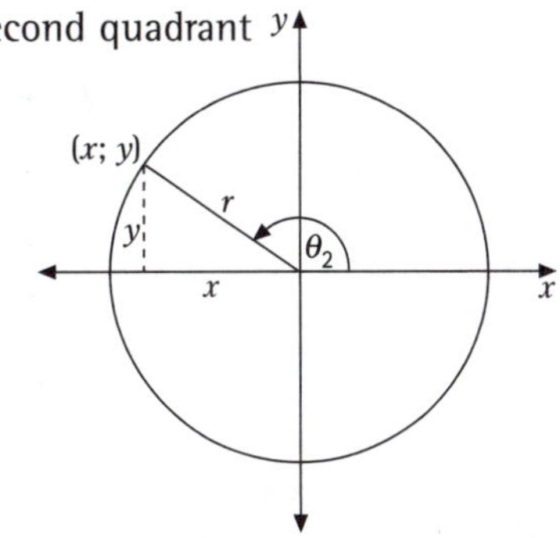

$$\sin \theta_2 = \frac{y\text{-coordinate}}{\text{radius}} = \frac{y}{r}$$

$$\cos \theta_2 = \frac{x\text{-coordinate}}{\text{radius}} = \frac{x}{r}$$

$$\tan \theta_2 = \frac{y\text{-coordinate}}{x\text{-coordinate}} = \frac{y}{x}$$

Third quadrant

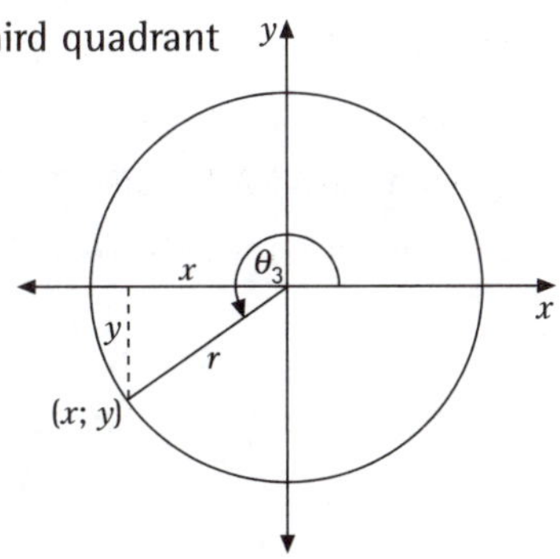

$$\sin \theta_3 = \frac{y\text{-coordinate}}{\text{radius}} = \frac{y}{r}$$

$$\cos \theta_3 = \frac{x\text{-coordinate}}{\text{radius}} = \frac{x}{r}$$

$$\tan \theta_3 = \frac{y\text{-coordinate}}{x\text{-coordinate}} = \frac{y}{x}$$

Fourth quadrant

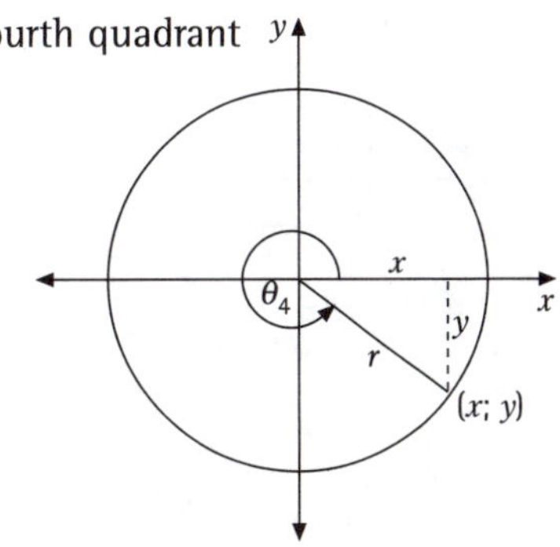

$$\sin \theta_4 = \frac{y\text{-coordinate}}{\text{radius}} = \frac{y}{r}$$

$$\cos \theta_4 = \frac{x\text{-coordinate}}{\text{radius}} = \frac{x}{r}$$

$$\tan \theta_4 = \frac{y\text{-coordinate}}{x\text{-coordinate}} = \frac{y}{x}$$

The values of sin, cos and tan for angles larger than 90° can be found by constructing and measuring drawings similar to those above, or by using a calculator.

Make sure that you know how to use your calculator to find the values (to 3 decimals) for:

sin 100° = 0,985 cos 250° = −0,342 tan 350° = −0,176

sin 460° = 0,985 cos 300° = 0,500 tan 10° = 0,176

Did you notice that sin 100° = sin 460°? The sin, cos and tan functions repeat the same values at regular intervals when you work with angles larger than 360°. Sin, cos and tan are called periodic functions.

Plotting the graphs of sin θ, cos θ and tan θ

Investigation: Trigonometric graphs

PEER ASSESSMENT

See page 288 (rubric 3).

1 Draw the graph of $y = \sin \theta$ for θ from 0° to 360°.

θ	0°	30°	60°	90°	120°	150°	180°
$\sin \theta$	0	0,5	0,9	1	0,9	0,5	0

θ	210°	240°	270°	300°	330°	360°
$\sin \theta$	−0,5	−0,9	−1	−0,9	−0,5	0

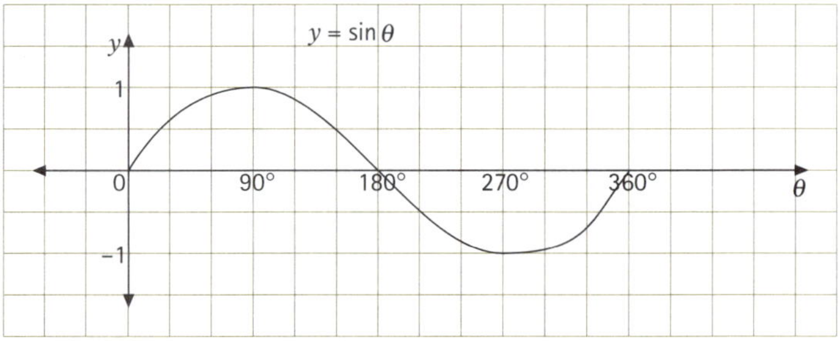

- The **amplitude** is 1 (no point on the graph has a perpendicular distance from the x-axis that is larger than one unit).
- The period is 360° because it has one complete S-shape in one revolution.
- The function values (y-values) **increase** for the intervals 0° to 90° and 270° to 360°.
- The function values (y-values) **decrease** for the interval 90° to 270°.
- The graph is **continuous** for the interval 0° to 360°.

Remember:
- The **amplitude** of a trigonometric function is half the vertical distance between the minimum and the maximum points on its graph.
- The **period** of a trigonometric function is the number of degrees between two consecutive identical points on its graph.
- The **frequency** of a trigonometric function is the number of complete cycles that will fit on the degree axis in a normal period of its graph.

2 Draw the graph of $y = \cos\theta$ for $0° \leq \theta \leq 360°$.

θ	0°	30°	60°	90°	120°	150°	180°
$\cos\theta$	1	0,9	0,5	0	–0,5	–0,9	–1

θ	210°	240°	270°	300°	330°	360°
$\cos\theta$	–0,9	–0,5	0	0,5	0,9	1

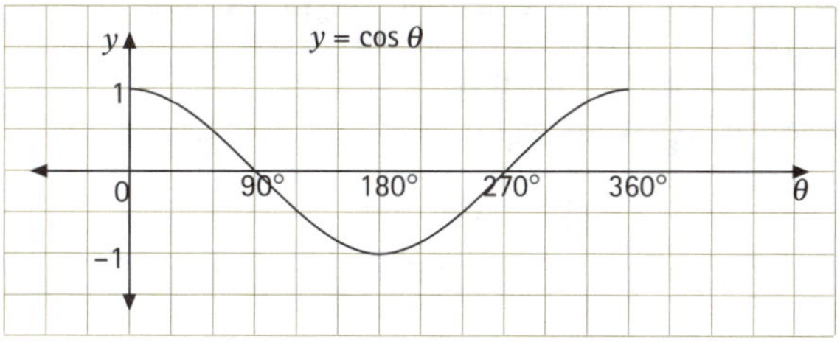

- The **amplitude** is 1 (no point on the graph has a perpendicular distance from the x-axis that is larger than one unit).
- The period is 360° because it has one complete C-Shape in one revolution.
- The function values (y-values) **decrease** for the interval 0° to 180° and **increase** for the interval 180° to 360°.
- The graph is **continuous** for the interval 0° to 360°.

3 Draw the graph of $y = \tan\theta$ for $0° \leq \theta \leq 360°$.

θ	0°	30°	60°	90°	120°	150°	180°
$\tan\theta$	0	0,6	1,7	–	–1,7	–0,6	0

θ	210°	240°	270°	300°	330°	360°
$\tan\theta$	0,6	1,7	–	–1,7	–0,6	0

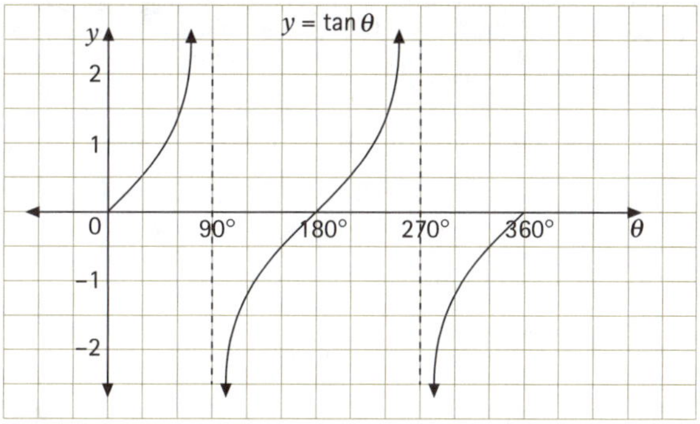

DO YOU REMEMBER?
An asymptote is a straight line that a curve approaches but never touches (meets).

- The **period** is 180° because there are two complete tan-shapes in 360° (one revolution).
- The function values (y-values) **increase** for the intervals 0° to 90°, 90° to 270° and 270° to 360°.
- The graph is **discontinuous** for $\theta = 90°$ and $\theta = 270°$.
- We call the vertical lines through $\theta = 90°$ and $\theta = 270°$ asymptotes.

GROUP ASSESSMENT

See page 288 (rubric 3).

Investigation: Amplitudes of trigonometric functions

Work in groups.

You will need: graph paper, pencils.

1 Complete this table:

θ	0°	30°	60°	90°	120°	150°	180°
$\sin\theta$	0	0,5	0,9	1	0,9	0,5	0
$2\sin\theta$							

θ	210°	240°	270°	300°	330°	360°
$\sin\theta$	–0,5	–0,9	–1	–0,9	–0,5	0
$2\sin\theta$						

Draw the graphs of $y = \sin\theta$ and $y = 2\sin\theta$ on the same set of axes. Then compare the amplitudes of the graphs. What do you notice? Discuss your findings in your group and write down your conclusion.

2 Complete this table:

θ	0°	30°	60°	90°	120°	150°	180°
$\cos\theta$	0	0,9	0,5	0	–0,5	–0,9	–1
$3\cos\theta$							

θ	210°	240°	270°	300°	330°	360°
$\cos\theta$	–0,9	–0,5	0	0,5	0,9	1
$3\cos\theta$						

Draw the graphs of $y = \cos\theta$ and $y = 3\cos\theta$ on the same set of axes. Then compare the amplitudes of the graphs. Discuss and write down your findings.

3 Complete this table:

θ	0°	30°	60°	90°	120°	150°	180°
$\tan\theta$	0	0,6	1,7	–	–1,7	–0,6	0
$\tan\theta + 1$							

θ	210°	240°	270°	300°	330°	360°
tan θ	0,6	1,7	–	–1,7	–0,6	0
tan θ + 1						

Draw the graphs of $y = \tan\theta$ and $y = \tan\theta + 1$ on the same set of axes. Then compare the graphs. What do you notice? Discuss and write down your findings.

4 Complete this table:

θ	0°	30°	60°	90°	120°	150°	180°
cos θ	1	0,9	0,5	0	–0,5	–0,9	–1
cos θ + 2							

θ	210°	240°	270°	300°	330°	360°
cos θ	–0,9	–0,5	0	0,5	0,9	1
cos θ + 2						

Draw the graphs of $y = \cos\theta$ and $y = \cos\theta + 2$ on the same set of axes. Then compare the graphs. What do you notice? Discuss and write down your findings.

5 Complete this table:

θ	0°	30°	60°	90°	120°	150°	180°
sin θ	0	0,5	0,9	1	0,9	0,5	0
2 sin θ + 3							

θ	210°	240°	270°	300°	330°	360°
sin θ	–0,5	–0,9	–1	–0,9	–0,5	0
2 sin θ + 3						

Draw the graphs of $y = \sin\theta$ and $y = 2\sin\theta + 3$ on the same set of axes. Then compare the graphs. What do you notice? Discuss and write down your findings.

6 Look at your results for questions 1 to 5 and complete these statements:

a If a is positive, the graph of $y = a\sin\theta$ looks like the graph of $y = \sin\theta$ that has been stretched to have an amplitude of ... units. Now write down a reason for your statement.

b If q is negative, the graph of $y = \tan\theta + q$ looks like the graph of $y = \tan\theta$ that has been moved (translated) $-q$ units .... Write down a reason for your statement.

c If a and q are both positive, the graph of $y = a \sin \theta + q$ looks like the graph of $y = \sin \theta$ that has been stretched to have an ... of a units and then moved (translated) ... units upwards. Write down a reason for your statement.

7 Swap your group's answers to question 6 with another group. Check each other's answers.

Exercise

1 Make neat sketches of these graphs, using the statements you completed in question 6 of the investigation above.

a $y = 4 \cos \theta$ **b** $y = 3 \sin \theta$

c $y = \tan \theta + 3$ **d** $y = 3 \sin \theta + 2$

e $y = 2 \cos \theta - 3$

2 Write down the equations of these graphs:

a

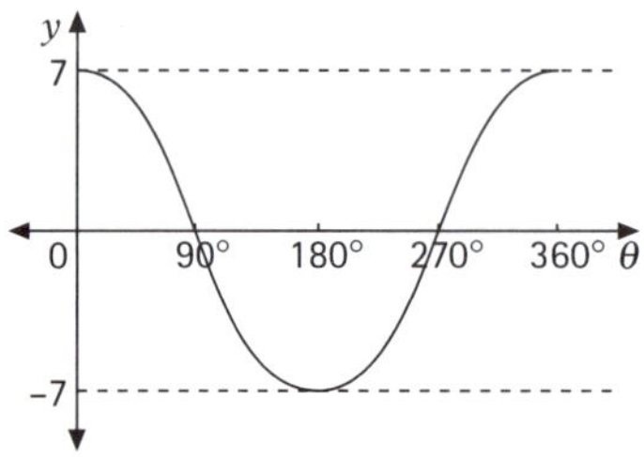

b

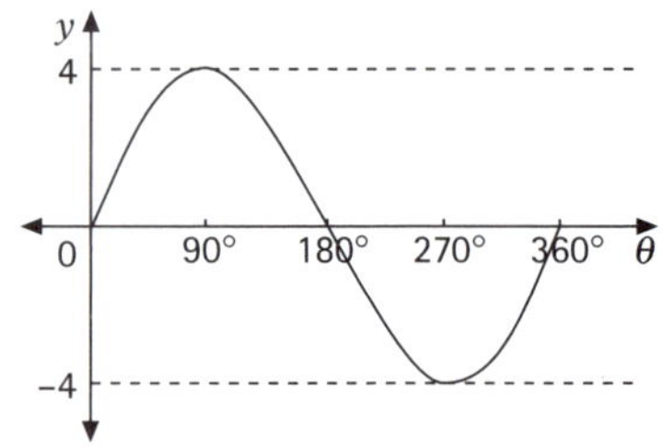

c

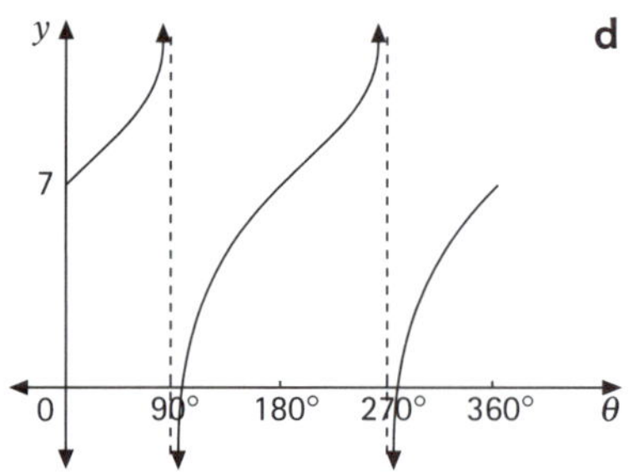

d

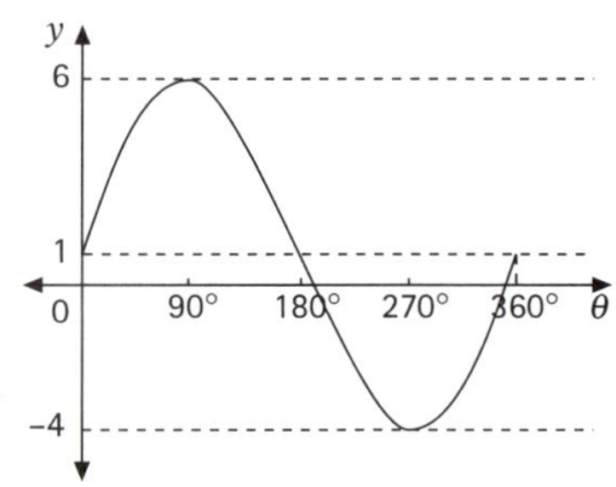

e

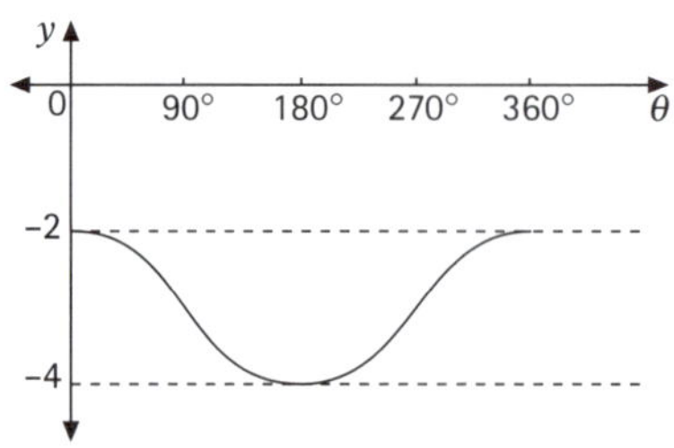

SELF-ASSESSMENT

See page 286 (rubric 1).

KEY VOCABULARY

intersect • parallel • point of intersection • best fit • modelling

Any problem that can be represented and solved algebraically, as you did in Chapter 2, can also be solved graphically. In this chapter you will represent real-life problems through graphs.

Solving simultaneous equations graphically

Remember: If you try to find the point of intersection, you are solving the equations simultaneously.

Two straight lines will always intersect unless they are parallel. At the point of intersection the two lines will have the same coordinates.

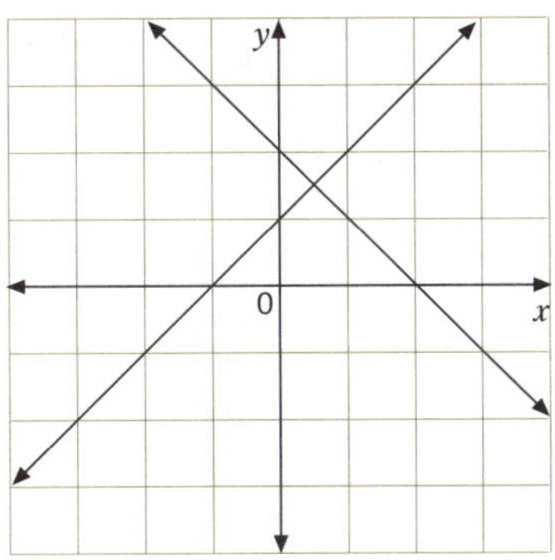

1 point of intersection

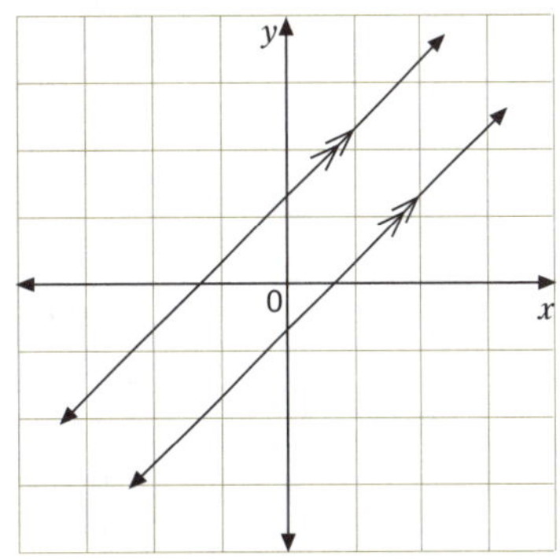

0 points of intersection

DISCUSS

The graphs in examples 1 and 2 represent real-life situations. Make up a real-life situation to which these graphs apply. Check each other's suggestions.

EXAMPLES

1 Determine the point of intersection of $y = 2x + 4$ and $3x + y = 9$ by drawing an accurate graph.

$y = 2x + 4$

x-intercept at $y = 0$: $0 = 2x + 4$

$-4 = 2x$

$\therefore -2 = x$

y-intercept at $x = 0$: $y = 4$

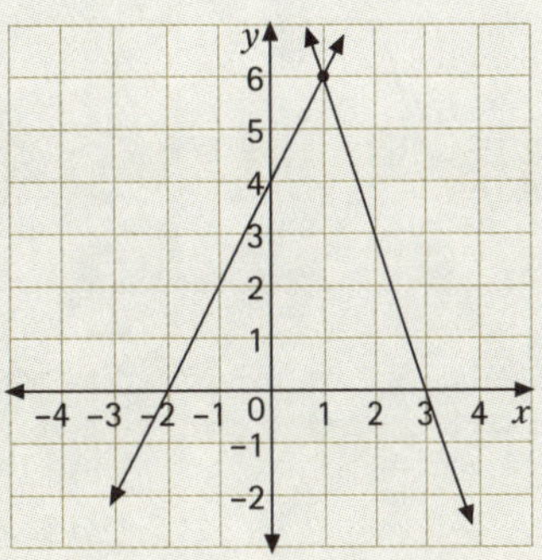

$3x + y = 9$

x-intercept at $y = 0$: $3x = 9$

$\therefore x = 3$

y-intercept at $x = 0$: $y = 9$

The point of intersection is (1; 6).

2 Find the points of intersection of $y = x^2$ and $x + y = 2$ ($-3 \leq x \leq 3$). Start by drawing up tables of ordered pairs using the given set of values for x.

- $y = -x + 2$

x	-3	0	3
y	5	2	-1

- $y = x^2$

x	-3	-2	-1	0	1	2	3
y	9	4	1	0	1	4	9

Now use the information in the tables to graph each equation. Choose a suitable scale for the x-values and y-values from the tables.

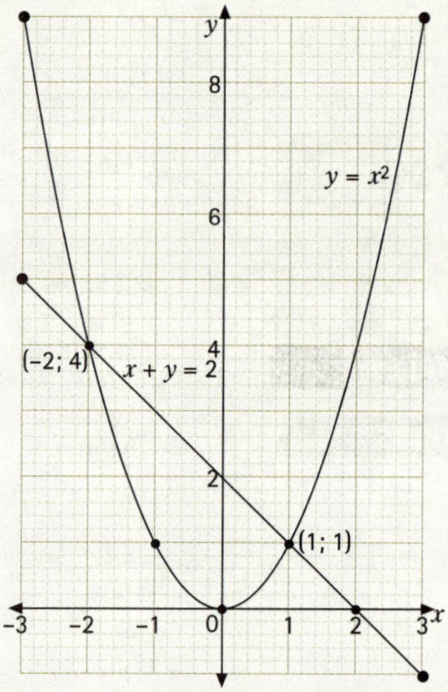

Exercise

Find the points of intersection (if any) for each of these graphs ($-3 \leq x \leq 3$):

1 $y = x^2$
$y = x + 2$

2 $y = x^2$
$y - x = 6$

3 $y + 5 = 0$
$y = -x^2 - 5$

4 $y = 2x^2 + 3$
$y = 3$

5 $y = \frac{1}{x}$
$y = x$

6 $y = \frac{x}{2}$
$xy = 2$

7 $xy = 4$
$x + 2y - 6 = 0$

8 $y = \frac{2}{x}$
$y = x^2 - 3$
$y = x - 1$

Interpreting graphs

Graphs displaying real-life information can tell you a great deal. To unfold any story they contain, always pay attention to their significant features.

These include:

- **slope** – whether positive (increasing), or negative (decreasing) or zero
- **intercepts** – both x and y
- **sign of the function** – whether its y-values are positive or negative
- **maximum and minimum points** – the largest and smallest value of y.

If the graph seems to show a trend, you can make educated statements about parts of the graph not shown, by using:

- **interpolation** – where you look within the graph produced from the given data
- **extrapolation** – where you extend a trend line on the graph produced from the original data to help you make predictions.

EXAMPLES

1 Which of the following graphs might best fit the rise and fall of the sea level at a jetty over a period of time? Explain your choice.

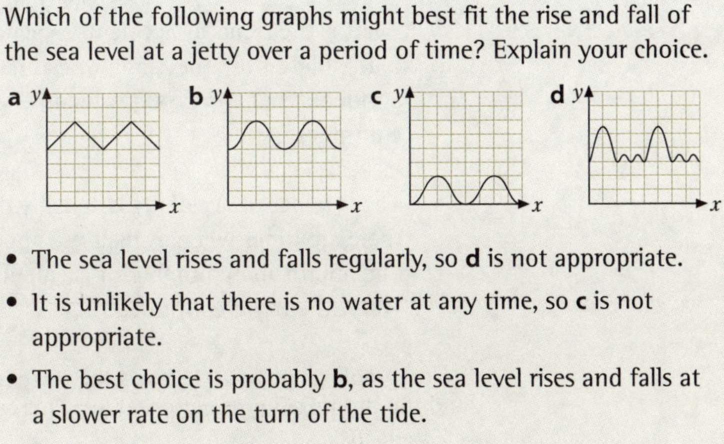

- The sea level rises and falls regularly, so **d** is not appropriate.
- It is unlikely that there is no water at any time, so **c** is not appropriate.
- The best choice is probably **b**, as the sea level rises and falls at a slower rate on the turn of the tide.

2 A car runs out of petrol near the top of a hill and rolls down onto a flat road where it finally stops. Sketch a graph of the car's speed (s) versus time (t). Label and explain the key features of your graph.

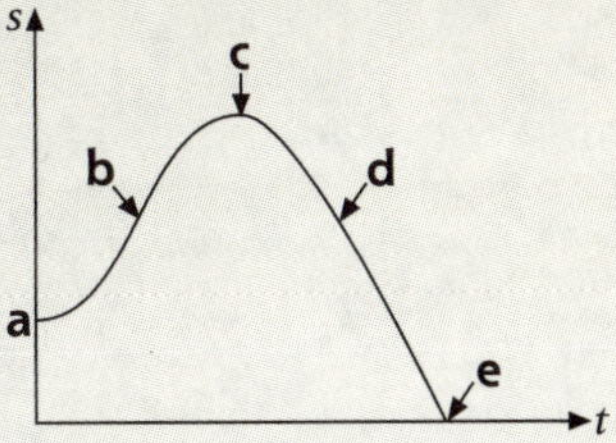

The labels **a** to **e** show:

a The car is moving when it runs out of petrol.

b The car gains speed going down the hill.

c It reaches a maximum speed at the bottom of the hill.

d It decelerates on the flat.

e It stops.

3 This graph shows the number of accidents in a factory over 4 years.

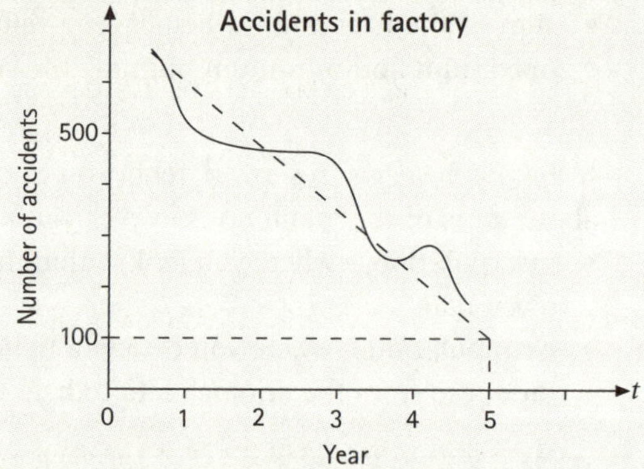

a What is the trend of accidents experienced over the years?

b Extrapolate from the graph to estimate the number of accidents that can be expected in 5 years, if the trend continues.

a The number of accidents is steadily decreasing, even though there are times when a small rise appears.

b The approximate number of accidents in year 5 would be 100 if the trend continued.

4 The graph on page 209 shows the water consumption of a town for the year 2004.

Month	Jan	Feb	Mar	Apr	May	Jun
Water consumption (in millions of kl)	1 564	1 462	1 420	1 350	1 400	960
Month	**Jul**	**Aug**	**Sep**	**Oct**	**Nov**	**Dec**
Water consumption (in millions of kl)	850	1 320	1 450	1 520	1 580	1 559

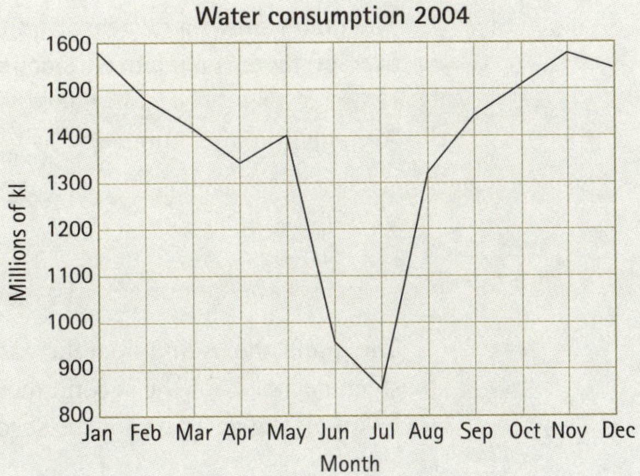

a What is the rate of change of the consumption from April to May?

b What is the average rate of change of the consumption from July to November? Compare your result with the graph.

c What is the average rate of change of the consumption from June to August? Compare your result with the graph.

d What is the average rate of change of the consumption from April to June? Compare your result with the graph.

a Rate of change $= \frac{\text{May usage} - \text{April usage}}{\text{length of interval}}$

$= \frac{1\,400 - 1\,350}{1}$

$=$ 50 million kl per month

b Average rate of change $= \frac{\text{November usage} - \text{July usage}}{\text{length of interval}}$

$= \frac{1\,580 - 850}{4}$

$= \frac{730}{4}$

$=$ 182,5 million kl per month

The graph shows a constant increase over this interval.

c Average rate of change $= \frac{\text{August usage} - \text{June usage}}{\text{length of interval}}$

$= \frac{1\,320 - 960}{2}$

$= \frac{360}{2}$

$=$ 180 million kl per month

The graph shows a decrease and then a steep increase. On average there is a relatively small increase.

d Average rate of change $= \dfrac{\text{June usage} - \text{April usage}}{\text{length of interval}}$

$= \dfrac{960 - 1\,350}{2}$

$= \dfrac{-390}{2}$

$= -195$ million kl per month

The graph shows an initial increase for the first month, but a steep decrease for the second month. The average decrease is less than the decrease for the second month.

Exercise

1 Which of the following graphs might best fit the situations stated? Explain your choice.

a The height of a ball bouncing:

i 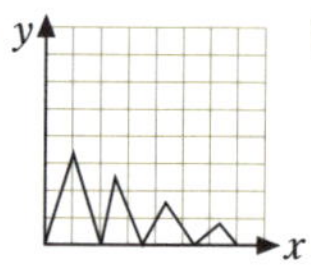
ii 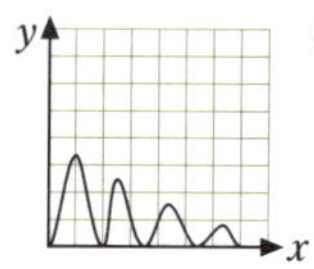
iii 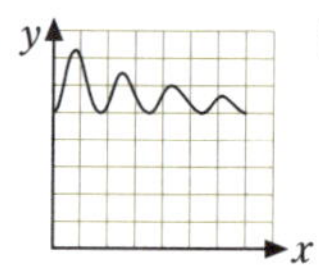
iv 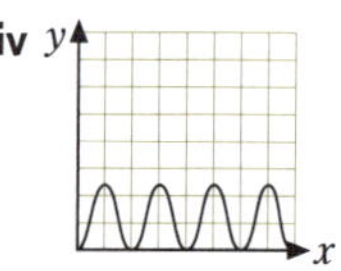

b The distance a car travels in suburban traffic:

i 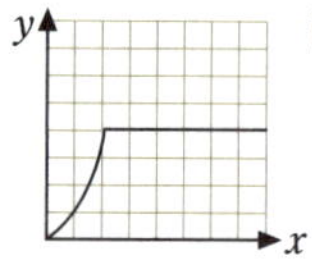
ii 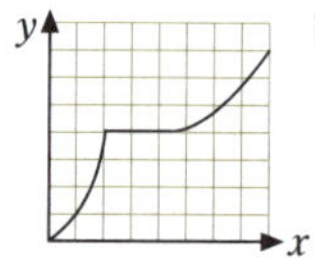
iii 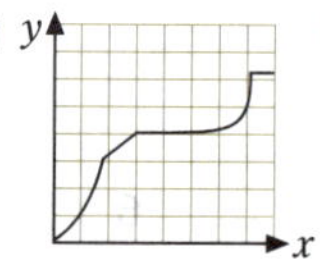
iv 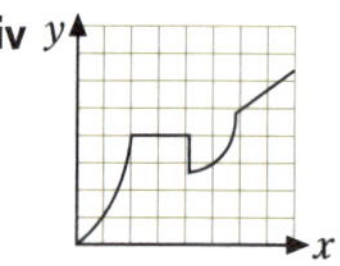

c The distance a car travels on a highway:

i 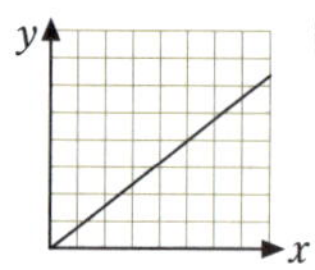
ii 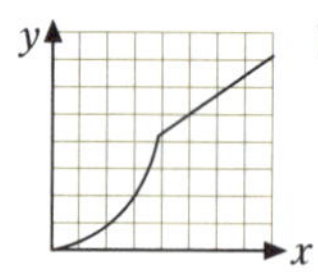
iii 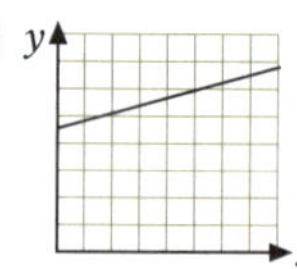
iv 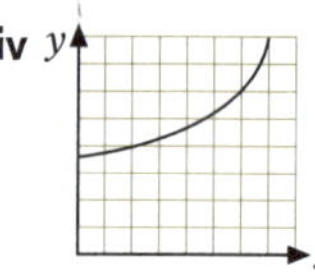

d The height of an arrow shot at a target:

i 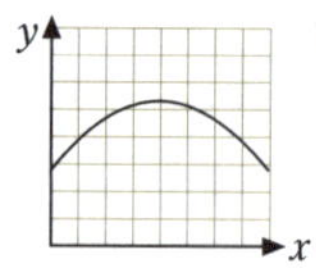
ii 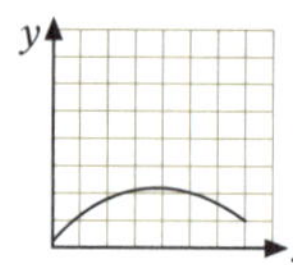
iii 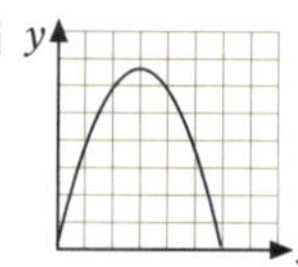
iv 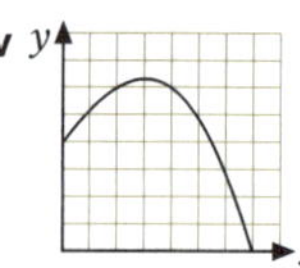

2 A car starts at a set of traffic lights and accelerates to the speed limit, which it maintains for a while before it stops for the next set of lights. Sketch a graph of the car's speed (s) versus time (t). Label and explain the key features of your graph.

3 A lift picks up a passenger on the second floor and travels to the eleventh floor to pick up another. Both passengers get out on the ground floor. Sketch a graph of the lift's position in the building (l) versus time (t). Label and explain the key features of your graph.

4 Sipho drops a ball from a balcony and watches it bounce until it stops. Sketch a graph of the height of the ball above the ground (h) versus time (t). Label and explain the key features of your graph.

5 The graph shows the number of people who own houses and those who own flats in a city.

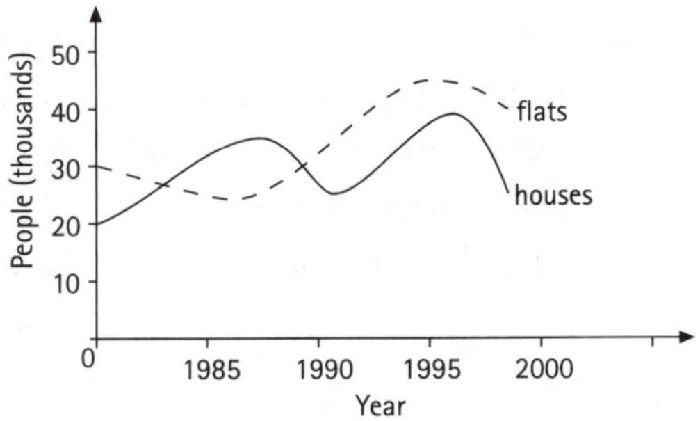

a Explain what each graph shows.

b If you wanted to buy real estate in this city, would you buy a flat or a house? Give a reason for your answer. Use the information given on the graph.

6 This graph shows the number of people killed in road accidents in two provinces of South Africa from 1966 to 1992.

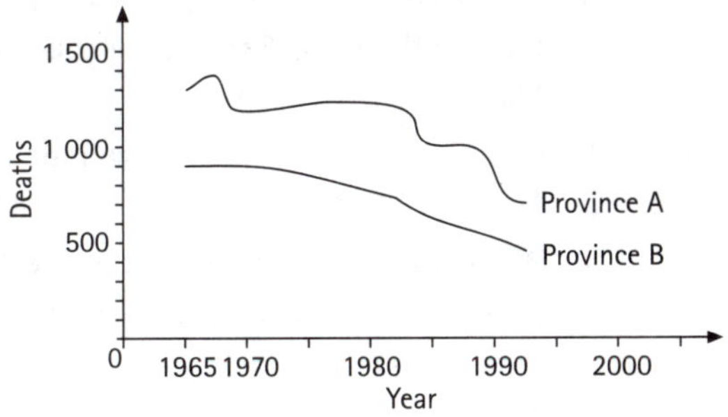

a Discuss the general trend shown on the graph.

b Which province set the best policies to curb road deaths?

c By extrapolating from the graph, estimate the number of fatalities in the year 2000 for each province.

7 On the right is a graph of the numbers of tourists visiting Antarctica.

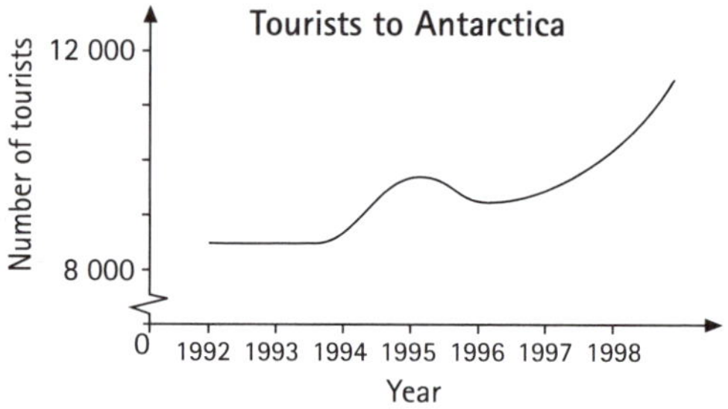

a Discuss the overall trend.

b Over the 7-year period, what is the increase in the number of tourists?

c Estimate the number of tourists to visit Antarctica in 2001.

8 This graph shows the electricity consumption of a household for the first six months of a year.

Month	Jan	Feb	Mar	Apr	May	Jun
Electricity consumption in kWh	405	380	420	410	520	580

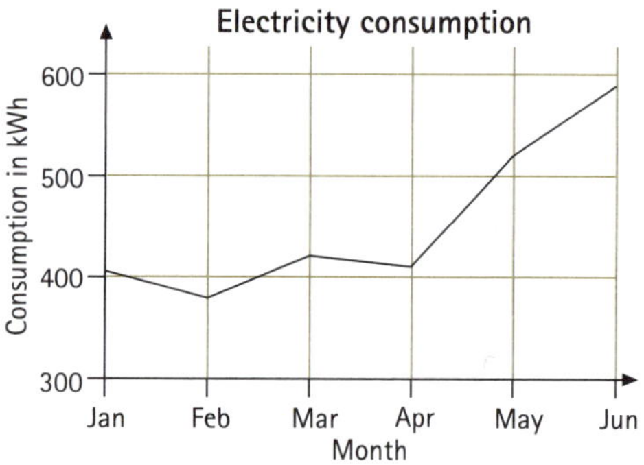

a What is the rate of change of the consumption from April to May?

b What is the average rate of change of the consumption from January to May? Compare your result with the graph.

c What is the average rate of change of the consumption from January to June? Compare your result with the graph.

d What is the average rate of change of the consumption from January to March? Compare your result with the graph.

Modelling relationships

The mathematical process that involves plotting given data and describing the data with lines and equations is called mathematical modelling.

The information in the graph on the right seems to follow a linear trend. So you can draw in a **line of best fit by eye.** When doing this, you need to try and balance out the data by keeping the same number of points above the line and below it.

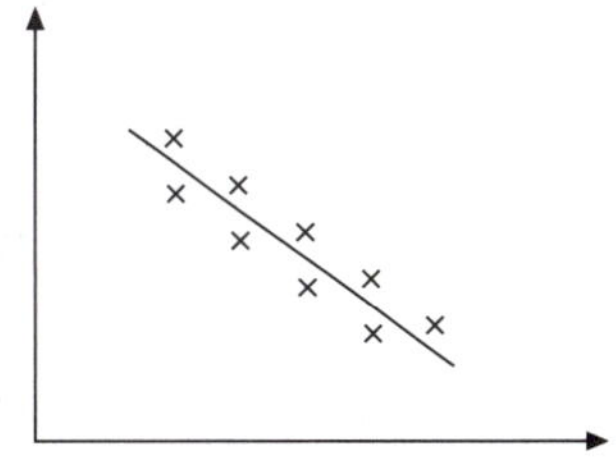

You can determine a linear equation that would model these data. This model will not represent the data perfectly, but will provide you with knowledge of the trend.

In this chapter you can use technology to explore **quadratic and linear models** to see which one best fits the data provided.

EXAMPLE

These data describe the volume of water in a dam during summer.

Time in weeks	1	2	3	4	5
Volume in litres	35 000	27 000	26 000	21 000	14 000

a Draw a scatter plot to show the information in the table.
b Draw in the line of best fit and determine its equation.
c Estimate when there will be no water left in the dam if the trend continues.

a

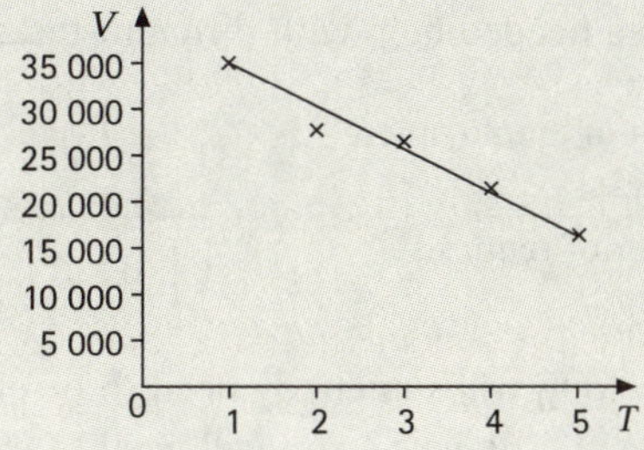

b $m = \frac{35\,000 - 14\,000}{1 - 5}$

$= \frac{21\,000}{-4}$

$= -5\,250$

$\therefore V = -5\,250T + c$

$V = 35\,000$ when $T = 1$:

$35\,000 = 5\,250(1) + c$

$\therefore c = 40\,250$

$\therefore V = -5\,250T + 40\,250$

c $V = 0$: $0 = -5\,250T + 40\,250$

$T = \frac{-40\,250}{-5\,250}$

$\approx 7{,}67$ weeks

Exercise

For the data given in questions 1 to 3:

- **i** Draw a scatter plot.
- **ii** Draw in a straight line of best fit and determine its equation.
- **iii** Answer the questions below each table.

1 In an electrical experiment the voltage (V) and amperage (A) were measured.

Voltage (V)	3,00	4,10	5,60	8,00	10,50
Amperage (A)	15,00	10,86	9,30	3,55	4,60

a What amperage produces a voltage of 9,00 V?

b What voltage produces an amperage of 10,00 A?

2 A chain of stores collected the following data on the amount of money (in rand) spent on an advertising campaign (A) for a certain product and the amount earned per day on the sale of the product in the next week (D).

R(A)	300	150	170	360	320	260
R(D)	4 640	1 740	3 250	5 680	5 570	4 990

a If a store had spent R400 on their advertising, what would you expect the store to earn on the product per day in the next week?

b Will there be a point at which there is no gain in spending less on the advertising, and if so, when?

3 A production manager kept a record of the number of times each employee needed help with different tasks within a given week.

Length of employment (in weeks)	24	11	47	58	3	70	76	44	33	37	26
Number of requests for help	14	20	10	13	25	10	6	15	19	6	10

a How long will somebody need to be employed before they ask for very little help? You will need to explain what you mean by little help.

b From the table, what can you say about the permanency of employment at this workplace? How would you change it? Discuss this with a partner and write down your ideas.

Language skills

axis
curve
equation
function
gradient
graph
hyperbola
inequality
intercept
intersect
linear
maximum
minimum
negative
parabola
parallel
positive
quadratic
reciprocal
simultaneous
substitution
symmetry

Supply the missing words from the list above:

1 A parabola is a curve represented by a ... equation.

2 A reciprocal function graphs as a ... .

3 $y = ax + q$ is the general equation for a ... graph.

4 $y = ax^2 + c$ graphs as a ... If a is ..., the graph is upright. The value of c gives the y-... .

5 When two graphs ..., the coordinates of that point form a ... solution to the two equations.

Write adjectives from:

6 line **7** parabola **8** hyperbola **9** symmetry.

10 What shows that two linear graphs must be parallel?

Explain the difference between:

11 intercept and intersect **12** coefficients and coordinates

13 turning point and minimum value

14 parabola and hyperbola.

Discuss these questions with a partner. Then write down your answers.

15 What is an asymptote?

16 List three methods of solving quadratic equations.

17 Explain the phrase "no real roots".

18 Explain how to sketch a quadratic function of the form $y = ax^2 + bx + c$.

19 How would you distinguish between a cosine graph and a sine graph?

20 Define the amplitude of a trigonometric function.

21 Define the frequency of a trigonometric function.

22 When is a graph continuous on an interval?

Swap your answers with another pair of learners and check each other's answers.

SELF-ASSESSMENT

Also see page 293 (rubric 9).

Check yourself

Make a mindmap of the key facts and formulas of this module under the headings listed here. It will help if you look at the key skills given at the beginning of this module (see page 141).

Solving algebraically:

- linear equations
- linear inequalities
- non-linear equations
- simultaneous equations
- exponential equations.

Graphing and sketching:

- linear equations
- linear inequalities
- non-linear equations
- simultaneous equations
- exponential equations
- trigonometric functions.

Then complete the Review that follows. The exercises in the Review will help you consolidate your knowledge and understanding of the topics in this module.

MODULE 3

REVIEW

1 Solve these equations:

a $3(2x + 4) = 24$ b $\frac{x}{2} + \frac{2x}{5} = 9$

c $3(2x + 4) - 4(x - 3) = 22$

2 Determine the gradient of the line that passes through (3; 5) and (–2; 15).

3 Determine the equation of the line that passes through the points (1; 1) and (3; –3).

4 Determine the equation of a line that has a y-intercept of 2 and an x-intercept of 4.

5 Factorise:

a $3x - 12$ b $2x^2y^2 + 4xy$

c $x^2 - 9$ d $27x^2 - 243s^2$

6 Simplify $(3x - 2)^2 - 49$.

7 Factorise:

a $18x^2 + 9x$ b $x^2 + 7x + 10$

c $x^2 - x - 20$ d $-5x^2 - 20x - 15$

8 Plot the graph of $y = x^2 + 5$

9 Sketch the graphs of:

a $y = 4x^2$ b $y = -5x^2$ c $y = x^2 + 4$

10 Determine the equations of these graphs:

a

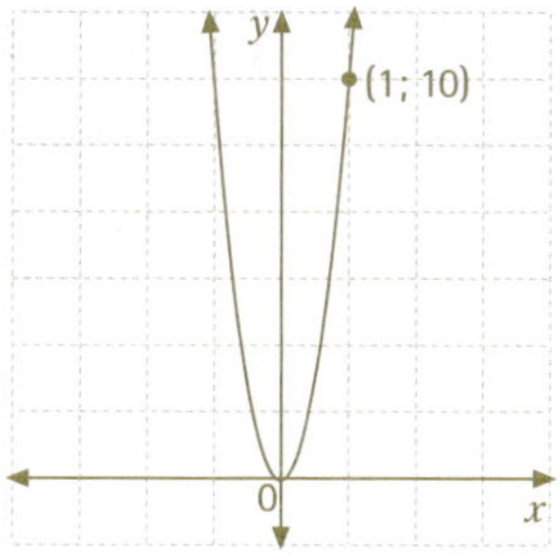

b

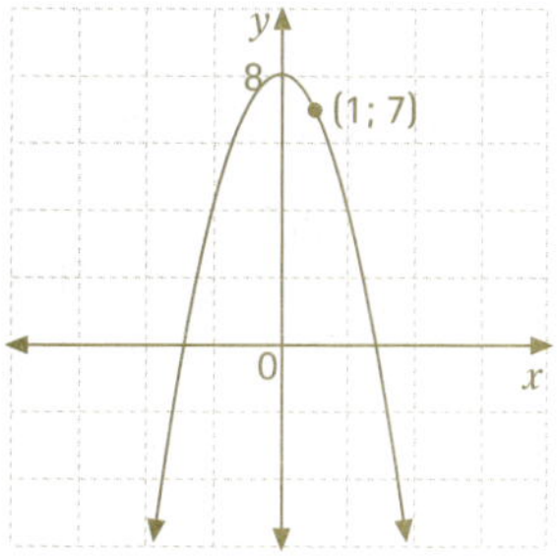

11 Write an equation that results from the transformation of $y = x^2$ being reflected in the x-axis and translated 4 units down.

REVIEW

12 Solve the equations and check your solutions:

a $x(x - 3) = 0$ b $(x - 4)^2 = 0$

c $x^2 - 4 = 0$ d $x^2 + 7x + 12 = 0$

13 Write the equation for each graph:

a A b B

c C d D

Write the coordinates of the intersection of:

e A and B f C and D

g Find the simultaneous solutions of $y = -x$ and $y = x^2 - 2$.

h What coordinates satisfy B and D at the same time?

i Which graph has an x-intercept of $-\frac{1}{2}$?

j Which graph has a minimum value for y of -2?

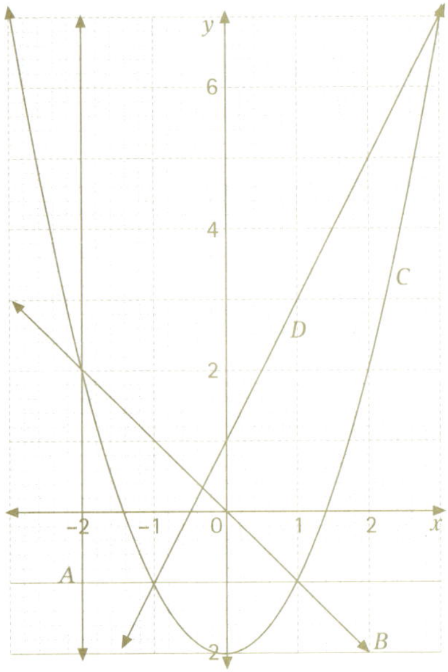

k Which graph is symmetrical about the y-axis?

l Which graph has neither a negative nor a positive gradient?

m Which graph would match the shape of $y = x^2$?

14 Sketch the graph of $3x - 2y + 4 = 0$ using the gradient and y-intercept.

15 Sketch the graph of $x + y = 3$ using intercepts with the axes.

16 Solve $y = -x + 1$ and $x + 2 = 2y$ simultaneously by graphing.

17 Graph $y = \frac{3}{x}$ using this table:

x	-3	-2	-1,5	-1	-0,6	0,6	1	1,5	2	3
y										

18 Graph $y = -2x^2 - 1$ for $-3 \leq x \leq 3$.

19 Solve $y = 2 - x$ and $y = 2 + x^2$ simultaneously by graphing ($-3 \leq x \leq 3$). Check by substitution.

20 Classify each of these equations as linear, quadratic or reciprocal:

a $xy = 9$ **b** $3x^2 = y$ **c** $\frac{5}{x} = y$ **d** $y = \frac{x}{3}$

21 Match each graph with the correct equation:

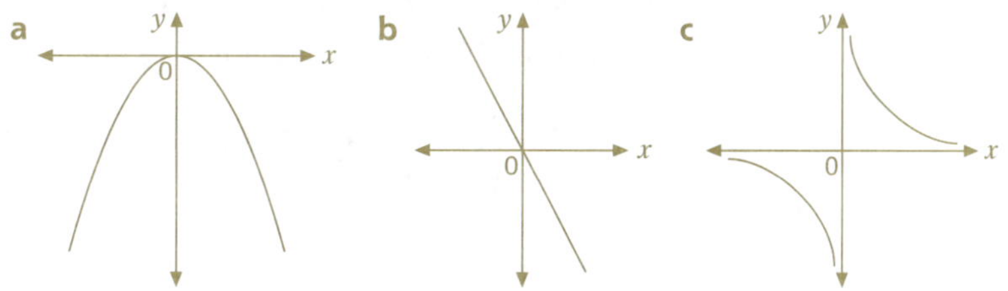

A $y = \frac{3}{x}$

B $y = -3x^2$

C $y = -3x$

22 Plot $y = 4^x$ for $-3 \leq x \leq 3$ and determine the approximate value of:

a $4^{1,5}$ b $4^{-1,5}$ c $4^{0,7}$ d $4^{2,6}$

23 Use your calculator to evaluate the following to 3 decimal places.

a $2^{2,3}$ **b** $6^{0,4}$ **c** $0,27^{7,3}$

24 Use trial and error to solve $7 = 2,3^x$ to 2 decimal places.

25 A car holds 72 litres of petrol. If the fuel in the tank is leaking at a rate of 15% per hour:

a determine the equation for the volume of fuel in the tank (V) after t hours

b calculate the amount of fuel after 3 hours

c determine when the fuel tank will contain less than 2 litres.

26 Determine the point of intersection for $y = 3x + 6$ and $y = 2x - 1$, by plotting the two linear graphs.

27 Solve this pair of simultaneous equations using the method of substitution:

$y = 5x - 13$ $3x + 2y = 13$

MODULE 3

REVIEW

28 Solve this pair of simultaneous equations using the method of elimination:

$3x + 2y = 8$ $\quad\quad$ $x - 2y = 0$

29 A money box contains 20-cent and 50-cent coins. If the amount in the money box is R50 and there are 160 coins, determine the number of 20-cent and 50-cent coins.

30 Solve the inequalities and graph their solutions on a number line.

a $x + 3 \geq 2$ $\quad\quad$ **b** $4x + 7 \leq 15$

c $-2(3x + 4) \leq 4$ $\quad\quad$ **d** $\frac{3x + 4}{2} > -1$

31 Make neat sketches of the following graphs:

a $y = 5 \cos \theta$ $\quad\quad$ **b** $y = 6 \sin \theta + 4$ $\quad\quad$ **c** $y = \tan \theta + 4$

32 This table and its graph show the water consumption of a household for a particular year.

Month	Jan	Feb	Mar	Apr	May	Jun	Jul	Aug	Sep	Oct	Nov	Dec
Water consumption (in kl)	156	146	142	100	140	96	92	132	145	152	158	155

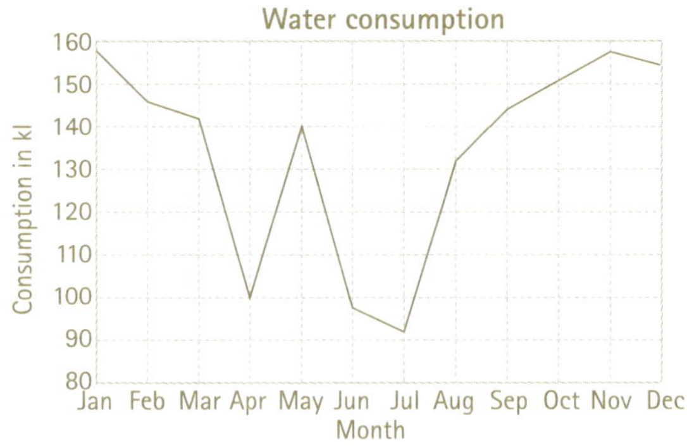

a What is the rate of change of the consumption from April to May?

b What is the average rate of change of the consumption from July to November? Compare your result with the graph.

c What is the average rate of change of the consumption from June to August? Compare your result with the graph.

d What is the average rate of change of the consumption from April to June? Compare your result with the graph.

PROJECTS

ASSESSMENT PORTFOLIO

See page 292 (rubric 8).

1 Divide the towns in your province (and the towns in the neighbouring province, if necessary) among the learners in your class. For your town, collect the daily maximum and minimum temperature readings from newspapers or television for 3 weeks.

Then draw graphs of maximum and minimum temperatures, showing both graphs on the same set of axes.

- **a** On which day was the highest maximum temperature recorded?
- **b** On which day was the lowest minimum temperature recorded?
- **c** On which day was the difference between the minimum and maximum temperature the greatest?
- **d** Calculate the average minimum and average maximum temperature for the 3-month period.
- **e** Compare your results with those of the rest of the class and write a paragraph about your conclusions.

ASSESSMENT PORTFOLIO

See page 292 (rubric 8).

2 **Expansion patterns**

You learned to expand brackets involving linear and quadratic expressions in this module. Now you will take this concept one step further.

$$(x+y)^1 = x + y$$

$$(x+y)^2 = (x+y)(x+y) = x^2 + 2xy + y^2$$

$$(x+y)^3 = (x+y)(x+y)^2 = (x+y)(x^2+2xy+y^2)$$

$$= x^3 + 2x^2y + xy^2 + yx^2 + 2xy^2 + y^3$$

$$= x^3 + 3x^2y + 3xy^2 + y^3$$

Expand $(x+y)^4$, $(x+y)^5$ and $(x+y)^6$.

Finish placing these expansions in the triangular pattern started for you.

$(x+y)^0$	1
$(x+y)^1$	$1x \quad + \quad 1y$
$(x+y)^2$	$1x^2 \quad + \quad 2xy \quad + \quad 1y^2$
$(x+y)^3$	$1x^3 \quad + \quad 3x^2y \quad + \quad 3xy^2 \quad + \quad 1y^3$
$(x+y)^4$	
$(x+y)^5$	
$(x+y)^6$	

MODULE 3

PROJECTS

Look for a pattern in the expansions and try to extend the table to determine the expansions of $(x + y)^7$, $(x + y)^8$ and $(x + y)^9$.

Draw a triangle of the coefficients of the expansions only. Again, the exercise has been started for you.

1
1 1
1 2 1
1 3 3 1
1 4 6 4 1
....
....
....

Comment on any patterns you see in this triangle. There are many.

SELF-ASSESSMENT

Copy the rubric into your workbook and put a tick (✓) in the column that describes how you feel about the work in this module.

Self-assessment rubric

How well do you understand these concepts from Module 3?	I understand this concept very well	I understand this concept	I need more help to understand this concept
Linear equations			
Graphs of linear equations			
Substitution			
Elimination			
Linear inequalities			
Quadratic graphs			
Parabolas			
Hyperbolas			
Factorising quadratic trinomials			
Exponential graphs			
Graphs of trigonometric functions			

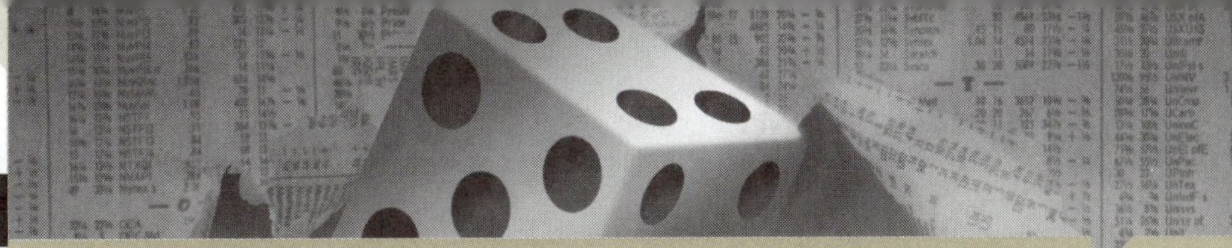

MODULE 4

DATA HANDLING

MODULE CHAPTERS

1
Conducting a statistical investigation
2
Organising and displaying data
3
Measures of central tendency
4
Probability and chance

KEY SKILLS

collect data and organise it • determine the range of a set of data • determine the mean, median and mode • draw appropriate graphs to display data • compare the relative frequency of an outcome with the probability of that outcome • draw Venn diagrams to solve probability problems • understand the sample space of a random experiment • calculate the probability of mutually exclusive events occurring • calculate the probability of complementary events occurring • identify biased data • analyse and explain how and why statistics and charts can misrepresent data • interpret a given set of data • explain clearly the conclusions and predictions that can be made from a set of data • do an investigation involving the collection and interpretation of data

In this module, you will work through activities, projects and exercises that enable you to collect and use data to establish statistical and probability models to solve related problems (Learning outcome 4).

By the end of this module you will:

- collect, organise and interpret numerical data to determine
  - measures of central tendency (mean, median, mode)
  - measures of dispersion (range, percentiles, quartiles, interquartile and semi-interquartile range)
- represent data effectively, choosing an appropriate format: bar graphs, compound bar graphs, histograms, frequency polygons, pie-charts, line and broken-line graphs
- use probability models for comparing relative frequency of an outcome
- use Venn diagrams to help solve probability problems, and identify:
  - the sample space of a random experiment
  - an event of a random experiment as a subset of the sample space
  - the union and intersection of two or more subsets of the sample space
  - P(S) = 1, where S is the sample space
  - P(A or B) = P(A) + P(B) – P(A and B), where A and B are events within a sample space
  - disjoint (mutually exclusive) events; calculate the probability of mutually exclusive events occurring by applying the addition rule for disjoint events: P(A or B) = P(A) + P(B)
  - complementary events; calculate the probability of an event not occurring: P(not A) = 1 – P(A)
- identify potential sources of bias, errors in measurement, and potential uses and misuses of statistics and charts and their effects

MODULE 4 IN REAL LIFE

data collection is important because it helps groups of people (for example, the government of a country) to find answers to questions that affect people's lives • real-life questions that can be answered by means of statistics include: How fast is a disease such as HIV/Aids spreading? How does South Africa's inflation rate compare to that of other countries? How are weather patterns changing? What kind of weather can we expect in 30 years' time? • data collection helps people answer these questions so that appropriate plans can be made for the future

- effectively communicate conclusions and predictions that can be made from the analysis of data
- use theory learned in this grade in an integrated form of assessment, e.g. an investigative project.

Integration

In this module, you will also work with numbers and number relationships (Learning outcome 1) and graphs (Learning outcome 2).

KEY VOCABULARY

statistics • investigation • data • census • sample • interview • questionnaire • observation • survey • opinion poll • table • graph • representative

What are statistics?

Statistics are a collection of information, called data, on a particular subject. Data are usually collected to help people find answers to questions about issues that affect your life, such as social and economic issues. You have worked with graphs and tables before, for example, in Module 1 you used data about inflation rates and the consumer price index (CPI). You can find different kinds of graphs in newspapers, for example, pie-charts, line graphs, bar graphs and pictographs. These represent data in different ways so that the information is easy to read and understand. Researchers use graphs all the time to represent a wide variety of statistical information such as government expenditure, changes in inflation rates, literacy rates, population figures, HIV infection rates, employment, and so on.

DID YOU KNOW?

5% of the South African population is involved in car accidents every year over the Easter weekend.
An estimated 5,5 million people in South Africa live with HIV/Aids.
Each of these is called a statistic.

Stages in a statistical investigation

Statistical data have been collected for centuries. Today, all governments have a statistics department that regularly collects and analyses a huge number of different data about industrial, agricultural, political, social and other issues. For example, it is important for government to know how many young people there are in a country to help it decide how many schools and teachers are needed in which areas. Data collection is a very time-consuming process, but it is a branch of mathematics that helps people deal with real questions for which answers need to be found.

Here is a step-by-step explanation of the stages in a statistical investigation:

1. You will first identify the problem you want to investigate. Problems can be simple ("How much chocolate does a shop owner need to order each week?") or complicated ("How is the weather changing and how will this affect agriculture?").
2. The next step is to choose a sample from the population to work with. You do this because it is time-consuming or difficult to research the whole population.
3. Then you collect the data you need for solving the problem. You do this by choosing one of various methods: interviews, questionnaires, observations, archives, pictures, surveys.

4 Once you have collected your data, you need to organise it. Data can be organised and displayed in a variety of tables and graphs.

5 Finally, you analyse your data to draw a conclusion and make a decision.

Note: This investigation will help you find out how much you already know about statistical concepts and which ones you can identify in real-life contexts.

Investigation: Statistics in real life

Work in groups.

You will need: different newspapers.

Collect different newspapers and bring them to class. Read through the newspapers and identify some statistics used in them. How are these statistics used? Is there bias in how they are used and interpreted? Discuss how they could be biased. Present your group's findings to the class.

Note: Do you believe that all statistics are true? It is important to remember that statistics can be distorted or misrepresented. This is called "bias". Always ask critical questions about statistics you find in newspapers, for example, "Who researched these statistics?"; "Where was the research done?" and "Was the research representative of the population?"

Note: The singular of "strata" is "stratum".

Sampling

Sampling is a process whereby you choose a sample to use in your research. It is important that the group you select for your sample is representative of the whole population. A sample that does not reflect the whole population is called a **biased sample**. Selecting as large a sample as possible can minimise bias. There are different ways of sampling. These include:

- **Random sampling:** individuals in the sample are chosen randomly. All the members of the population have equal chances of being selected.
- **Stratified sampling:** the population is divided into groups (strata) according to their common characteristics. For example, Grade 10 learners in the school can be divided according to learning area. Once strata have been identified, then the members belonging to those strata are chosen randomly. This is done so that the sample will be a true representation of the group chosen, e.g. a country's population. Stratified sampling is also known as quota sampling.
- **Cluster sampling:** random groups are chosen that have similar characteristics. It is used in cases where random sampling would be too expensive and time-consuming, for example, if you wished to conduct a survey to find out what South African people thought about improvements in their lives after 10 years of democracy. The disadvantage of this method is that the chance of selecting sample groups that are not representative of the whole population is greater than for other methods.

Definitions

Before conducting a survey you need to decide whether you will use the whole **population** (census) or a **sample**.

- In a **census** the whole population is surveyed.
- In a **sample** a smaller group that is representative of the whole population is surveyed.

In these definitions, the word "population" is used to mean the entire group we are interested in, not necessarily the population of the whole country. The population may not be people; it could be cars passing a road intersection, packets of sweets, or winning times for a race.

What is a survey?

A survey is an investigation of public opinion on different things or issues. **Opinion polls** are often carried out to determine the views of the population on many issues, for example, who people will vote for in the next election.

In October and November 2001 the South African government conducted a census known as "count us in", which is conducted every five years. Different statistical information was acquired from the whole population. This information included the number of people living in a household, the number of people working in that particular household, whether they have a television, radios, computers, etc.

Another type of survey is **market research**. Market research companies collect information to help companies decide what products to sell, how to advertise their products, what price people are prepared to pay for which products, which brands are most popular, and other issues.

PEER ASSESSMENT

See page 288 (rubric 3).

Investigation: Census or sample?

Work in groups.

Decide whether a census or sample would be appropriate for each investigation:

1 determining the favourite breakfast food of South African teachers

2 finding out the most popular drink at the school tuckshop

3 determining the number of pedestrians hit by cars in your area in a year

4 estimating the number of people who will attend a protest rally
5 deciding which CD was the most popular last month
6 determining which sport Grade 10 learners at your school prefer to play
7 finding out the average lifespan of a specific type of battery
8 predicting the results of an election
9 finding out how many children under the age of 18 there are in South Africa
10 determining the number of students sitting for the Senior Certificate examination this year.

Now think up three questions of your own that could best be answered using a sample, and three for which a census would be more appropriate. Give your questions to your partner and check each other's questions.

PEER ASSESSMENT

See page 288 (rubric 3).

Investigation: Bias in questionnaires and interviews

Work with a partner.

Find out more about how questionnaires and interviews are used in statistical investigations. You can do some research on the Internet or at your local library, or ask your teacher to give you some guidelines on questionnaires and interviews.

Your investigation should answer these questions:

1 When would you use a questionnaire, and when would interviews be more suitable? Give an example of one issue you could investigate with a questionnaire, and one issue you could investigate through interviews.
2 Now draw up a set of questions for your questionnaire and for your interview. Is there a difference between the type of questions you would ask in a questionnaire and in an interview? If so, how do they differ? Your answer should be based on your sets of questions.

Now discuss questions 3 to 6 with your partner and write down your answers.

3 What are the advantages and disadvantages of using a questionnaire?
4 How could your questionnaire be biased? (Give examples, or change some of the questions to make them biased.)

5 What are the advantages and disadvantages of doing an interview?

6 How could your interview be biased? (Give some examples, or change some of the questions to make them biased.)

SELF-ASSESSMENT

See page 286 (rubric 1).

Swap your investigation with another pair of learners and assess each other's work.

CHAPTER 2 ORGANISING AND DISPLAYING DATA

KEY VOCABULARY

raw data • class interval • tally • line graph • broken-line graph • scale • axis • frequency distribution table • bar graph • trend • histogram • pie-chart • percentage • box plot • frequency polygon • stem-and-leaf plot

Data that have been collected but not yet organised are called raw data. Raw data is usually difficult to interpret. This is why it is necessary to organise information. Organised data can then be displayed in various ways. Examples are a frequency distribution table, bar graph, histogram, pie-chart, frequency polygon, stem-and-leaf plot and box plot. In this chapter, you will work with each of these.

Frequency distribution tables

A very useful way of organising raw data is in a frequency distribution table. You organise the raw data by counting the number of times an observation occurs.

EXAMPLES

1 Jonas interviewed people from 20 different households in his township to find out the number of flush toilets in each. The number of flush toilets in each household in his sample was:

Number of flush toilets per household									
1	1	1	2	1	3	1	0	0	1
0	0	3	1	1	1	0	1	2	1

Note: If the number of observations is small, each number can be written and counted.

Here is the frequency distribution table for Jonas's research:

Number of flush toilets per household (x)	Tally	Frequency (f)
0	𝍸	5
1	𝍸 𝍸 |	11
2	||	2
3	||	2

2 All Grade 9 learners write Common Task Assessments (CTAs) for mathematics at the end of the school year. CTAs are set by the Department of Education. One Grade 9 class obtained these marks out of 50 at the end of 2002:

22	25	10	43	47	37	20
33	9	28	39	42	45	39
34	27	35	19	18	29	28
31	33	27	33	18	33	26
33	23	33				

You can now arrange the marks in ascending order:

Grade 9 learners' marks out of 50									
9	10	18	18	19	20	22	23	25	26
27	27	28	28	29	31	33	33	33	33
33	33	34	35	37	39	39	43	47	42
45									

The scores can now be counted to draw up a frequency distribution table.

Score	Tally	Frequency
0–9	|	1
10–19	||||	4
20–29	~~||||~~ ~~||||~~	10
30–39	~~||||~~ ~~||||~~ ||	12
40–49	||||	4
		Total: 31

Note: In this example, the scores have been arranged into classes of specified length, called class intervals.

Note: The total frequency column should be the same as the total number of scores collected (or the total number of people surveyed).

Remember: A survey is a study in which a researcher asks a question or questions about a particular topic to find some statistical data about the topic.

Exercise

1 Josephine is 16 years old. She uses a wheelchair because she lost the use of both her legs after a car accident when she was 9. She is studying the Constitution in her Social Studies class, and decides to do some research to find out about opportunities for disabled employees. She chooses companies that each have around 100 employees, and surveys these companies to find out how many of their employees are disabled. Here are her findings:

1	1	2	0	2	1	1	3
1	0	3	2	4	1	9	0
0	1	1	0	2	5	3	2
1	0	3	2	0	1	1	3
0	1	4	6	2	3	1	0

a How many companies did she survey?

b Copy and complete the frequency distribution table on the next page for Josephine's data:

Score (x)	Tally	Frequency (f)
0		
1		
2		
3		
4		
5		
6		
7		
8		
9		

c What is the most common number of disabled employees in these companies?

d How many companies have more than 2 disabled employees?

e The South African government has an Employment Equity Act that states that in companies of 100 employees, at least 2% should be people with disabilities. Why do you think this policy is important? Discuss this question in your groups.

f How many of the companies surveyed are not complying with the Employment Equity Act? Why do you think they aren't?

g Make some suggestions on how to make companies comply with the Employment Equity Act.

2 Josephine and her friends also decide to monitor a parking bay at a shopping centre that is reserved for disabled people. Each person monitors the bay for an hour, from 9 a.m. until 6 p.m. on a weekday. Here are their results:

Time of day	Number of able-bodied drivers using the bay	Number of disabled drivers using the bay
9–10 a.m.	0	2
10–11 a.m.	2	2
11–12 p.m.	3	2
12–1 p.m.	4	0
1–2 p.m.	1	3
2–3 p.m.	2	3
3–4 p.m.	1	4
4–5 p.m.	4	3
5–6 p.m.	2	3

a Draw up a combined frequency distribution table that shows the frequency of people using the parking bay each hour.

b How many people used the parking bay between 9 a.m. and 6 p.m. altogether?

c What fraction of the total number of people who parked in the bay were disabled people?

d What was the maximum number of people who used the bay in a 1-hour period?

e Is it possible to determine, from this frequency table, for how much time the bay was occupied by people from each group? Give reasons for your answer.

Bar graphs

Bar graphs represent data in a series of bars that are equally wide. The bars can be vertical or horizontal. You need to leave a space between the bars. (If the bars touch each other, you are drawing a histogram.) The height of the bars represents the magnitude or frequency of the observations. You can use a single bar graph when showing one type of information, for example, the amount of rainfall in a particular year, or a double bar graph when comparing the monthly rainfall of two years.

Constructing a bar graph

Follow these steps to construct a bar graph:

- Organise the data in a table.
- Decide whether the bars will be vertical or horizontal.
- Draw axes that have a convenient length.
- Choose an appropriate scale for the axes (horizontal and vertical).
- Divide the axes equally according to the scale you have chosen and label the divisions.
- Decide how wide the bars will be and how much space you will leave between them.
- Construct the bars using the correct numerical values on the scale.
- Label the axes and give the graph a proper title.

EXAMPLES

1 The marks 50 learners obtained in a class test are given below. Draw up a frequency table for the given marks. Draw a bar graph to represent the data.

10	3	6	4
7	7	4	5
6	9	4	6
8	6	7	5
5	6	7	5
4	6	5	7
6	9	1	8
2	3	4	1
7	5	4	7
6	7	6	8
7	5	6	1
6	5	4	5
6	6		

Marks (10)	Tally	Frequency (f)
1	III	3
2	I	1
3	II	2
4	~~IIII~~ II	7
5	~~IIII~~ ~~IIII~~	10
6	~~IIII~~ ~~IIII~~ II	12
7	~~IIII~~ IIII	9
8	III	3
9	II	2
10	I	1
		Total: 50

From the frequency distribution table, you can draw a bar graph. On the horizontal axis, you can write the marks and on the vertical axis you can write the frequency.

Note: This is a single bar graph.

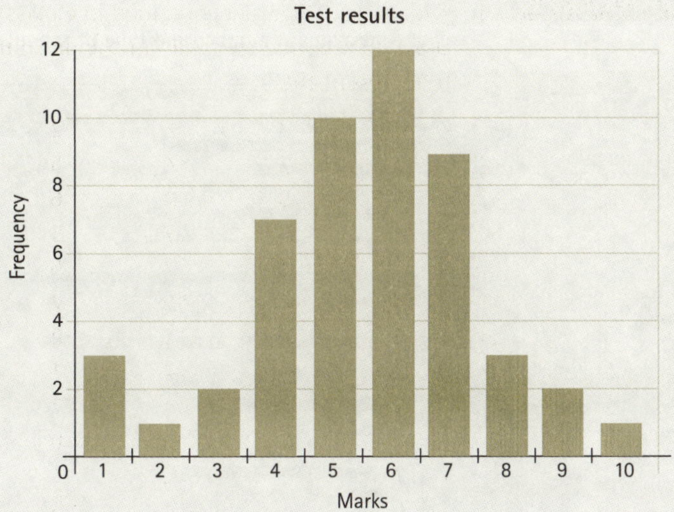

2 You can use double bar graphs and multiple bar graphs to show information that has been grouped into categories. This table shows the approximate number of schools in South Africa, per province, in 2003.

Province	State schools	Independent schools	Farm schools
Eastern Cape	6 1000	900	4 600
Free State	2 450	300	1 500
Gauteng	1 800	400	300
KwaZulu-Natal	5 600	300	600
Limpopo	4 200	250	300
Mpumalanga	1 700	200	500
North West	2 200	100	500
Northern Cape	450	100	150
Western Cape	1 300	100	450

(Source: *South Africa at a Glance, 2003-2004*)

You can use the information in the table to draw up a multiple bar graph.

The bar graph is far easier to interpret than the table.

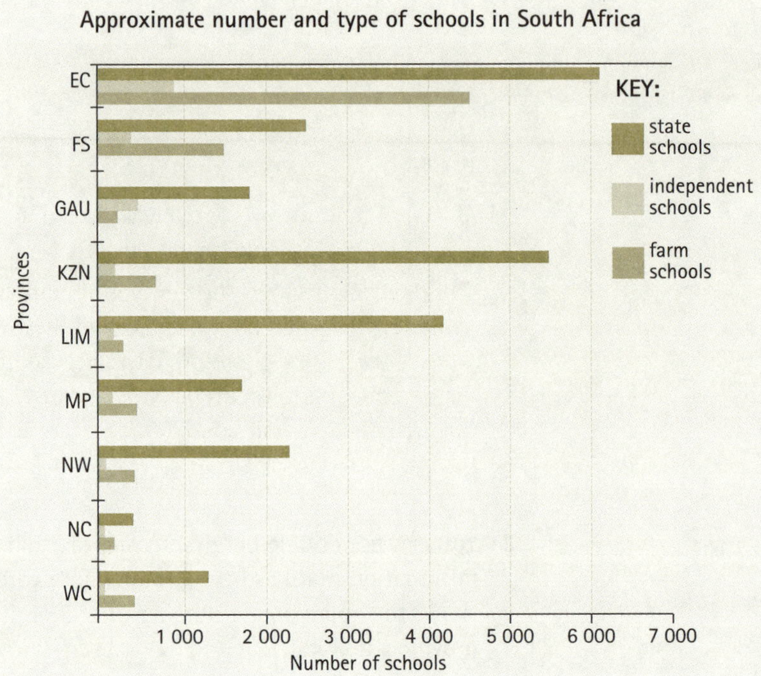

Exercise

1 The following graph shows the analysis of revenue collection:

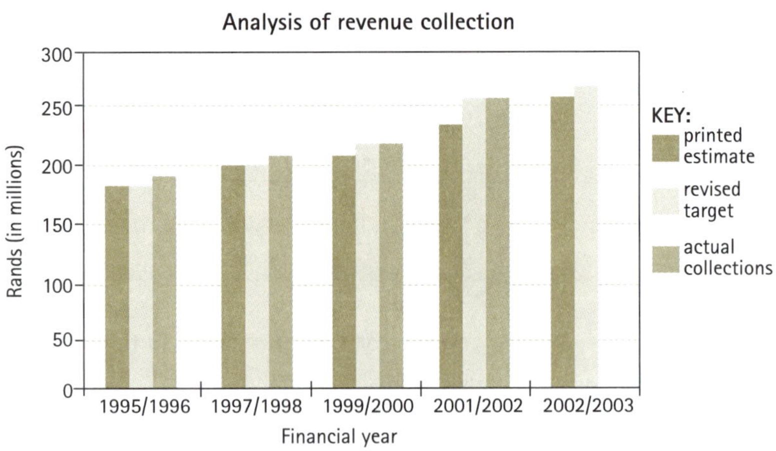

(Source: *The Mercury*, 27 February 2003)

a In which year was the amount of revenue collected R200 000 000?

b In which year was there a considerable difference in the printed estimate and actual collection of revenue?

c What amount was actually collected in the financial year 2001/2002?

d What amount was actually collected in the financial year 2002/2003? Explain your answer.

DISCUSS

Do you think smaller families are a good thing? Do you believe that if people have fewer children, education and unemployment levels will necessarily improve? Give a reason for your answer.

2 This bar graph shows the number of married women using contraceptives in a particular country between 1960 and the late1990s. Family planning has increased dramatically, because it has become more accessible, and because more people believe in the benefits of smaller families.

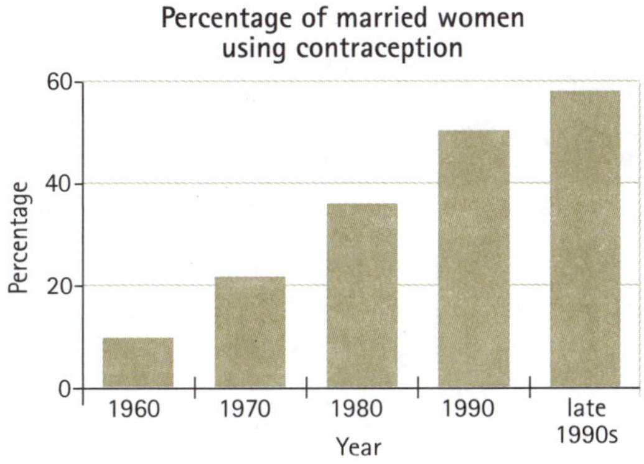

a In which period are the highest number of married women using contraception?

b In which period are the smallest number of married women using contraception?

c Study the bar graph and give some reasons to explain why the graph looks the way it does.

3 The Grade 12 results in 2002 improved tremendously in South Africa. The national percentage pass rate in 1999 was 48,9%. In 2000 it was 57,9%, in 2001 it was 61,7% and in 2002 it was 68,9%. The table shows the Grade 12 results in the 9 provinces in 2001 and 2002.

Province	Pass rate in 2001 (%)	Pass rate in 2002 (%)
Eastern Cape	45,6	51,8
Free State	59,0	70,7
Gauteng	73,6	78,1
KwaZulu-Natal	62,8	70,8
Limpopo	59,5	69,5
Mpumalanga	46,9	55,8
North West	62,5	67,8
Northern Cape	84,2	89,9
Western Cape	82,7	86,5

a Draw a bar graph showing the national percentage pass rate for the year 2002.

b Draw a double bar graph showing the percentage pass rate in the 9 provinces in 2001 and 2002.

c Of the 9 provinces, which one improved the most?

d Which province had the best Grade 12 results?

PEER ASSESSMENT

See page 289 and 290 (rubrics 4 and 6).

Histograms

A histogram looks almost like a bar graph. It describes a frequency distribution in terms of a series of bars (drawn next to each other without any spaces between them). Each bar represents the number of class frequencies in a particular class.

Remember: When you draw or interpret a histogram, it is important to consider the **height and width** of the bars. Bars that are equally wide represent the same class limit. Bars that differ in width represent different class limits. Remember that bars whose areas differ (that is, they may be equally high but not equally wide) do not represent the same number.

Construction of a histogram

- Class intervals, where possible, should be the same size. (Sometimes unequal class size is used to prevent empty classes.)
- If the data is raw, then work out the class interval:

$$\text{suggested class interval} = \frac{\text{highest value} - \text{lowest value}}{\text{number of classes}}$$

- Scale the class frequencies on the vertical axis and the class limits on the horizontal axis.
- When drawing the bars, do not leave spaces between them.

EXAMPLES

1 The ages of the learners in a Grade 10 class are shown in this table. Draw a histogram to represent the ages of the learners. Between which ages are the learners in this class?

12	15	17	18	15	19	16	20	17
16	14	17	15	14	18	19	16	14
21	20	19	17	15	14	13	18	19
16	14	13						

Work out the range first:

range = highest value – lowest value = 21 – 12 = 9

Since the ages do not vary much, you could choose four classes. Now work out the class interval.

$$\text{class interval} = \frac{\text{highest value} - \text{lowest value}}{\text{number of classes}} = \frac{9}{4} = 2{,}25 \approx 3$$

There will be four classes with a class interval of 3. To create the class limits, you start off with the lowest class limit of 11,5 to accommodate the lowest value. Then add 3 to find the upper class limit of 14,5 for the first class.

DISCUSS

What is the age of most learners in your class? Think of reasons why some learners in a class may be older than others. Do you think this can create problems? Why? How could you resolve these problems?

Class limits	Tally scores	Frequencies
11,5–14,5	𝍸 III	8
14,5–17,5	𝍸 𝍸 II	12
17,5–20,5	𝍸 IIII	9
20,5–23,5	I	1

Most learners in this class are between 14,5 and 17,5 years old.

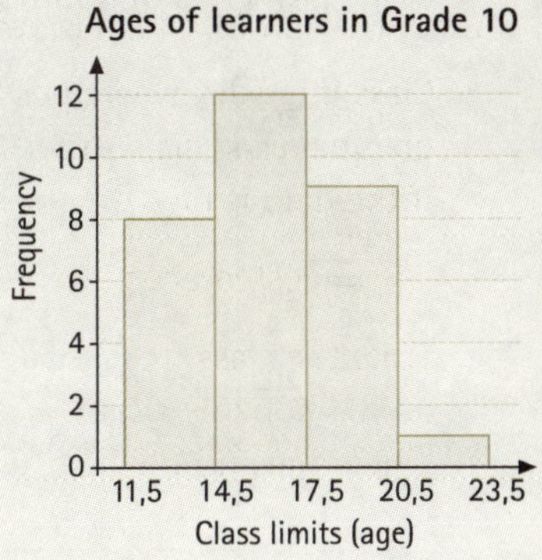

2 *The World Factbook* (http://www.odci.gov/cia/publications/factbook/geos/sf.html) states that the South African population was 43 647 658 in 2002. This table shows three age groups according to gender. Note the different widths of the bars in the histogram.

0–14 years		15–64 years		over 65	
Male	Female	Male	Female	Male	Female
6 943 761	6 849 745	13 377 011	14 300 850	816 222	1 360 069
13 796 506		27 677 861		2 176 291	

DISCUSS

For the purposes of this histogram, the average life expectancy in the 65+ age group was taken as around 80 years. The last bar of this histogram is therefore not strictly accurate. Would you consider this histogram biased?

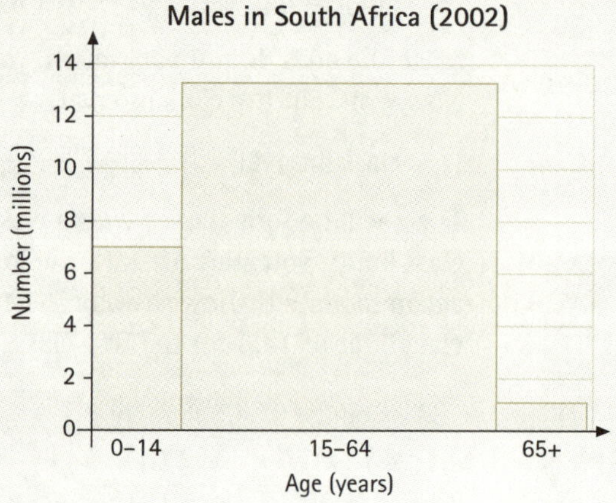

Exercise

1 This histogram shows the rainfall during the summer of 2003 in some provinces of South Africa.

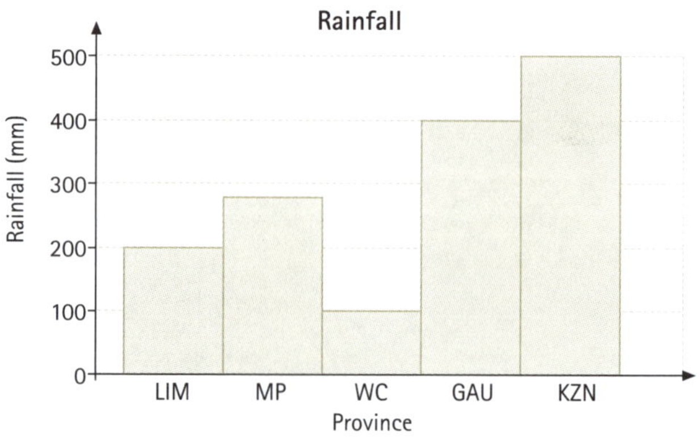

a Which area has the largest rainfall?

b What is the rainfall measured in Limpopo province in 2003?

c If no more rain falls until the end of summer in Limpopo province, what would be the effect on crops that are cultivated there? Can you think of other sectors of the economy that would be affected by too little rain? Explain your answers.

2 This histogram illustrates the financial performance of a company in a particular year.

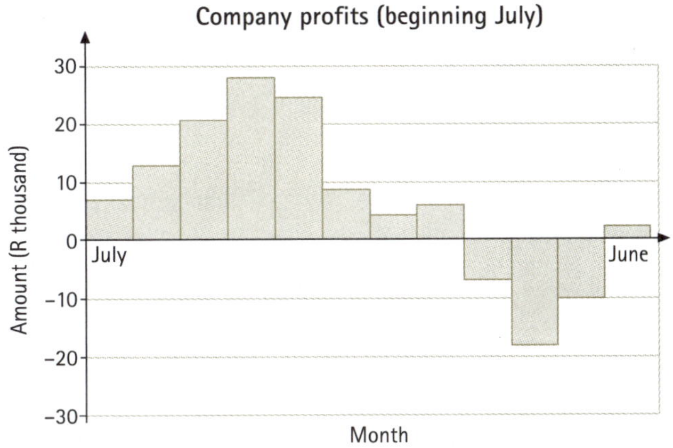

a In which months did the company record these profits:

i more than R20 000?

ii more than R10 000?

b In which months did the company record these losses:

i more than R10 000?

ii more than R0?

c In what fraction of the year did the company record a loss?

d Do you think that the company's profit depends on the time of year? Give a reason for your answer.

PEER ASSESSMENT

See page 290 (rubric 6).

Frequency polygons

Note: Class mid-points represent the middle value that represents each bar of the histogram.

A frequency polygon is a special type of line graph. The class mid-points of a histogram are joined to create a frequency polygon. The first and the last points are then joined to the horizontal axis as shown. This is the frequency polygon for the histogram on page 239.

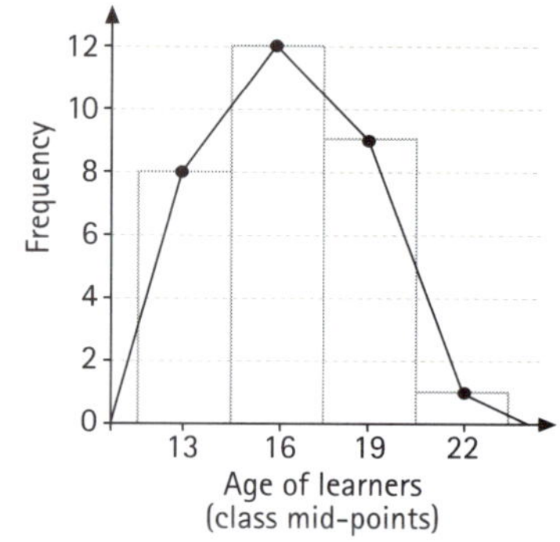

Broken-line graphs

In Chapter 3 of Module 3 you already investigated how line graphs can be used to represent various situations. When you draw a line graph for data that do not indicate a tendency, it must be drawn as a broken-line graph. The broken line means that you cannot use the graph to make any conclusions about information between the indicators on the system of axes.

Look at this graph showing the ages of learners in a Grade 10 class. You cannot use it to read off the number of learners aged, for example, $13\frac{1}{2}$ years or $16\frac{1}{2}$ years. That's why this graph is a broken-line graph.

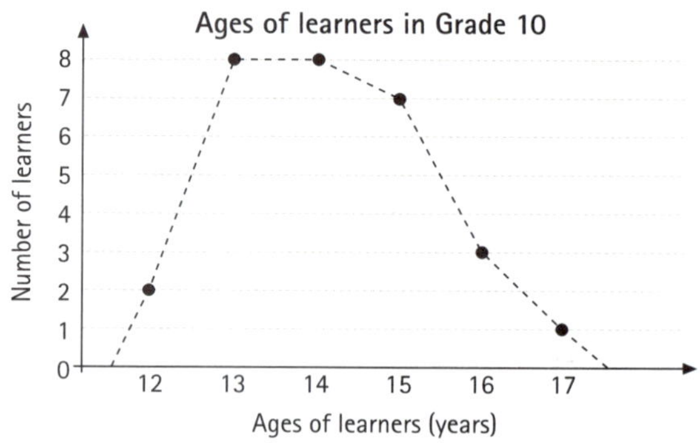

Exercise

1 A company's ability to make a profit often depends upon the number and quality of staff. The graph illustrates staffing and profit details for six years for a particular company.

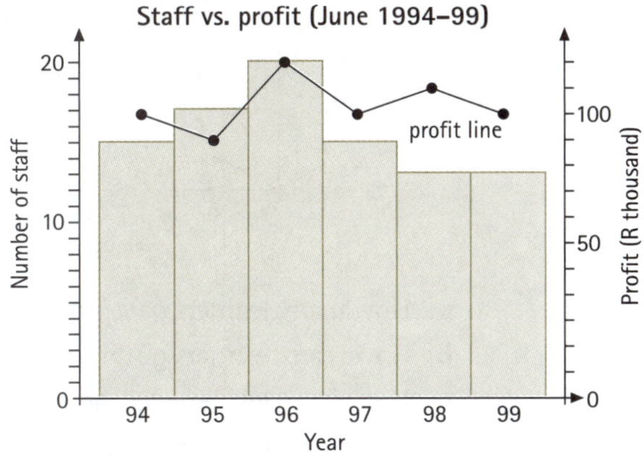

In which year did the company have:

a the highest number of staff?

b the highest profit?

c the lowest number of staff?

d the lowest profit?

e The company decided to cut staff after 1996 to reduce costs and increase profit. Did the company succeed in achieving this? Why do you say so?

2 Draw a frequency polygon to show the number of times 50 families ate take-aways last week. The results are shown in the table:

Remember: Always join the ends of the line to the axis.

Number of times families had take-away food	Number of families
0	13
1	20
2	12
3	4
4	1

3 Mrs Molope was concerned that her Grade 10 learners had too little awareness of HIV and Aids. She put together a quiz to test their knowledge. Her multiple-choice quiz had 6 questions. This frequency polygon shows how the learners scored in the quiz.

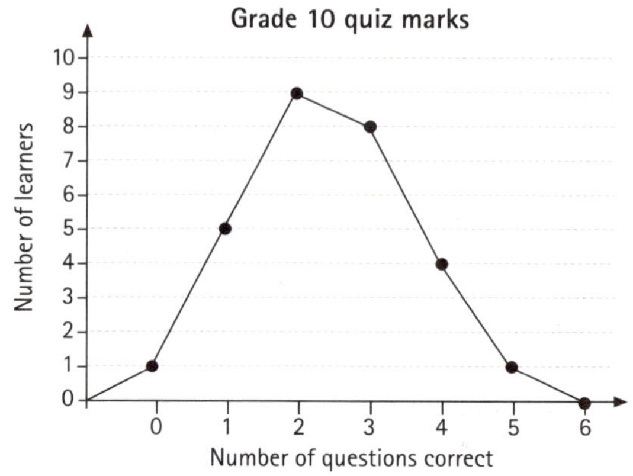

a How many learners gave no correct answers?

b How many learners got more than half the questions correct?

c How many learners did the quiz?

d Do you think the quiz was too easy, too difficult or at the right level?

e Devise your own quiz to test your classmates' awareness of HIV and Aids.

f Your teacher will select which quizzes to use to test your class's awareness. Complete the quiz.

g Capture the scores and represent them in a frequency polygon.

DISCUSS

What is the level of HIV and Aids awareness in your school? Do you think learners in your school need more information? Who do you think is responsible for creating HIV and Aids awareness in schools?

Pie-charts

A pie-chart (or pie graph) is circular in shape. The sectors in the circle represent the frequencies. Sectors can be represented as percentages or degrees. The whole circle will be out of 100% or 360°. Pie-charts are an ideal way of representing data that tells us percentages of a total amount.

Constructing a pie-chart

Follow these steps to draw a pie-chart:

- Add all the frequencies and write each frequency as a fraction of all the frequencies.
- Change each fraction into a number representing percentages or degrees by multiplying by 100% or 360° respectively. The percentages or degrees tell you how big the angles of each sector in the pie-chart will be.

- Tabulate the angles in ascending or descending order.
- Draw a circle of convenient size. Then draw a radius as a starting point.
- Use a protractor to construct the angles at the centre corresponding to each sector.
- Label each sector and give the pie-chart an appropriate title.

EXAMPLE

This table shows how Mrs Nadir spends her day:

Teaching at school	Preparation for school	Housework (cooking, cleaning, washing)	Spending time with her own children	Eating	Sleeping	Other
6	2	3	4	1	7	1

Show the information on a pie-chart.

Start by working out the fractions. The total number of hours in a day = 24.

$\therefore$ Teaching at school: $\frac{6}{24}$
Preparation for school: $\frac{2}{24}$
Housework (cooking, cleaning, washing): $\frac{3}{24}$
Spending time with her own children: $\frac{4}{24}$
Eating: $\frac{1}{24}$
Sleeping: $\frac{7}{24}$
Other: $\frac{1}{24}$

Then change each fraction to degrees.

$\therefore$ Teaching at school: $\frac{6}{24}$ of 360° = 90°
Preparation for school: $\frac{2}{24}$ of 360° = 30°
Housework (cooking, cleaning, washing): $\frac{3}{24}$ of 360° = 45°
Spending time with her own children: $\frac{4}{24}$ of 360° = 60°
Eating: $\frac{1}{24}$ of 360° = 15°
Sleeping: $\frac{7}{24}$ of 360° = 105°
Other: $\frac{1}{24}$ of 360° = 15°

Now you can draw a circle. Arrange the percentages in descending order. Then divide the circle into segments according to the angles you worked out.

Mrs Nadir's day

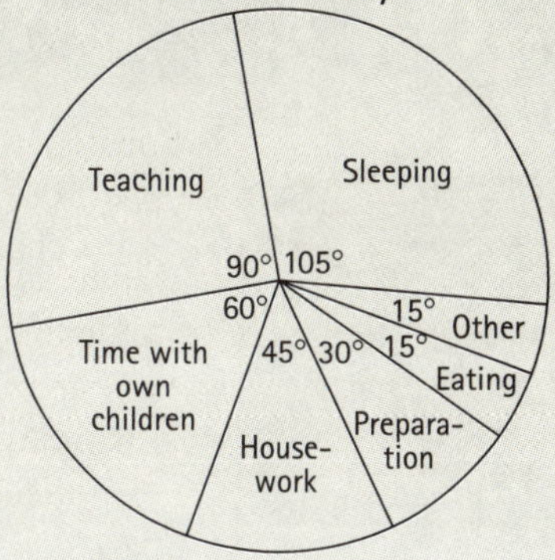

PEER ASSESSMENT

See page 289 (rubric 4).

Investigation: Schools in South Africa

Work in groups.

Look at the multiple bar graph in example 2 on page 236 again. Then discuss the questions that follow in your groups. Present your views on question 6 to the class. You could have a class discussion about this topic too.

1. Which province has the highest number of schools? Which province has the lowest number of schools?
2. How many schools are there in your province?
3. Draw a pie-chart to show the proportions of state, independent and farm schools in your province.
4. What do you think are the factors that influence how many of each type of school there are in each province?
5. What is the ratio of independent schools to state schools in your province? Is it higher or lower than that of other provinces?
6. Try to find out more about how state schools differ from independent schools (e.g. funding, ratio of learners to teachers). What do you think are the advantages of state schools? What are the advantages of independent schools? Do you think it is a good thing that South Africa has independent schools? Should the South African government encourage or discourage independent schools?

Exercise

1. This table shows how Mr Groenewald, a teacher at the same school as Mrs Nadir in the example on page 245, spends his day:

School	Preparation and marking	Gym or squash	Eating	Watching TV	Reading	Sleeping	Other
6	2	2	1	2	1	8	2

DISCUSS

Have a class discussion about how gender roles affect the way men and women spend their time, even if they have the same job. If you can, find some statistics on the Internet to prepare for this discussion.

a. Use the same method as in the example to represent in a pie-chart of how Mr Groenewald uses his time.
b. Now describe the differences and similarities between the way the two teachers spend their day.
c. Have a class discussion about how gender affects the way people spend their time (see the "Discuss" box on the left).

DID YOU KNOW?

Cholera is a water-borne virus that causes diarrhoea (a "runny tummy"). People become dehydrated (lose too much water) and therefore they need to be rehydrated. Here is a recipe for a rehydration solution: Mix 1 litre of boiled water, 2 pinches of salt and 8 teaspoons of sugar. Let the patient sip this mixture at regular intervals. If people are not treated for dehydration, they can die.

2 In KwaZulu-Natal there was an outbreak of cholera in 2001. Many people who use water from the rivers, were affected. The table shows the number of people who were affected by the disease and were treated in rehydration centres around KwaZulu-Natal on a particular day.

Areas affected by cholera	Number of people treated in rehydration centres
Eshowe	219
Mtubatuba	96
Gingindlovu	105
Ndwendwe	16
Malukazi	15

a Draw the pie-chart to represent the information in the table.

b In which area were most people affected by cholera?

c Which area had the lowest number of people affected by cholera?

3 According to *South Africa at a Glance, 2003–2004*, South Africans' religious affiliations are as follows: Christian 75%; No religion/unspecified 21%; Muslim 1,4%; Hindu 1,4%; Jewish 0,2%; African traditional belief or other faith 1%. Draw a pie-chart to represent the information.

Stem-and-leaf plots

When data is organised into a frequency distribution table, the actual observations are lost. But in a stem-and-leaf plot the actual observations are kept once the data is organised. That's why stem-and-leaf plots are quite a popular choice for organising data.

Constructing a stem-and-leaf plot

- The final digit of each score is shown as a leaf.
- The previous digit of each score is shown as a stem.
- The stem is written to the left of a vertical line and the leaf to the right of the same vertical line.
- The "leaf" digits are then rearranged in numerical order.

EXAMPLE

(34)	23	40
35	25	28
18	32	37
29	19	17
32	55	36
42	33	20
25	34	

These data show the ages of women attending a gymnasium to keep fit. Display the data on a stem-and-leaf plot and comment on the distribution of the ages of the women attending the gym.

stem	leaves							
1	8	9	7					
2	3	5	8	9	0	5		
(3)	(4)	5	2	7	6	3	4	2
4	0	2						
5	5							

The tens digit is shown as the stem, and the units are shown as leaves. The first score (34) has been circled to show how it fits onto the plot.

Now redraw the plot, rearranging the leaves in numerical order.

stem	leaves							
1	7	8	9					
2	0	3	5	5	8	9		
3	2	2	3	4	4	5	6	7
4	0	2						
5	5							

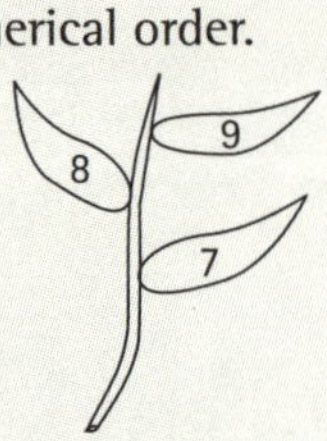

The 1-stem and its leaves

From the plot you can see that the youngest woman attending the gymnasium is 17 and the oldest is 55 (this person is 13 years older than the second oldest person). The gymnasium is most popular in the age group 30 to 39, followed by the age group 20 to 29.

Exercise

1 These data represent points scored by a basketball player:

33	11	26	3	42	49	35	11	49	42
10	3	2	47	32	29	6	18	10	29
6	18	10	49	1	26	49	38	25	9
1	34	39	45						

a Organise the data into a stem-and-leaf plot.

b Transfer the data to a frequency table. Use the table shown here.

Class interval	Frequency
0–9	
10–19	
20–29	
30–39	
40–50	

c Draw a histogram for the data. Show the class intervals on the horizontal axis and the frequencies on the vertical axis.

d Add a frequency curve to your histogram.

e Which class interval(s) of 10 units is (are) the most common for the data?

2 As part of a scientific study, a group of students catch a particular type of fish, measure their lengths, and then release them back into the river. The stem-and-leaf plot shows the lengths of the fish, in centimetres, measured at one location.

stem	leaves
1	2 4 4 6
(2)	0 1 3 3 (4) 5 8 9
3	3 5 6 6 7 8 8
4	0 2 5

a What length was the fish whose score is circled?

b What was the length of the shortest fish?

c What was the length of the longest fish?

d How many fish were measured at this location?

DID YOU KNOW?

Many of South Africa's rivers are polluted. This has a severely detrimental effect on living organisms in and around the rivers. Find out more about river pollution in your area and what your local authorities and communities are doing about it.

SELF-ASSESSMENT

See page 286 (rubric 1).

Box plots

Note: Box plots will be discussed in more detail on page 260 in Chapter 3.

Box plots (or box-and-whisker plots) are a special type of graph that is used to summarise single data sets. They clearly display the minimum and maximum value of the data set, and give a clear picture of how the data is spread. Box plots are usually used to compare two sets of data visually.

KEY VOCABULARY

average • mean • mode • median • grouped data • ungrouped data • dispersion • range • bimodal • score • tendency • interquartile range • semi-interquartile range • box plot

Mean, median and mode

In this chapter you will find out how to analyse numerical data by finding different averages, also known as measures of central tendency. A **measure of central tendency** is a single value that represents a set of data. Measures of central tendency include the mean, mode and median.

Mean

The mean is the **most commonly used** measure of central tendency. It is calculated by adding all the observations. The sum is then divided by the number of observations.

$$\text{mean} = \frac{\text{sum of all the observations}}{\text{number of observations}}$$

EXAMPLE

Calculate the mean of the following observations:

19 28 46 25 26 23 29

$$\text{mean} = \frac{19 + 28 + 46 + 25 + 26 + 23 + 29}{7} = \frac{196}{7} = 28$$

The mean can also be represented as a formula:

$$\bar{x} = \frac{X_1 + X_2 + X_3 + \ldots + X_n}{N}$$

$\bar{x}$ is read "bar x", which is the mean.
$X_1, X_2, X_3 \ldots, X_n$ are the observations.
N represents the number of observations.

The mean from a frequency distribution table

In a frequency distribution table, scores occur more than once, so that the total of the scores column does not tell us the total of all the scores.

To calculate the mean, you need to know the total of all the scores, and so another column is added to the table to show **frequency** × **score** (fx). These values show the subtotals for each score.

Note: The usual abbreviation for score is x and for frequency is f.

EXAMPLE

These data show the number of people living in each unit in a particular block of flats. Calculate the mean number of people per unit.

Score (x)	Frequency (f)	Frequency × score (fx)
1	4	4
2	5	10
3	4	12
4	2	8
5	1	5
Total	16	39

Add on the fx column to the frequency distribution table as shown. The four scores of 1 contribute 4 to the total (1 + 1 + 1 + 1). The four scores of 3 contribute 12 to the total, and so on.

Before you do the calculation, first do an estimate:

Most of the scores are in the range 1–3, so the mean is probably around 2.

$$\text{mean} = \frac{\text{total of all scores}}{\text{total number of scores}}$$

$$= \frac{39}{16} \qquad \frac{\text{Total of } fx \text{ column}}{\text{Total of } f \text{ column}}$$

$$= 2{,}4375$$

The mean number of people per unit is 2,4 (to 1 decimal place).

Note:
- Nominal-level data are data that can be classified into categories.
- Ordinal-level data are data that are represented as a rating, for example, from "excellent" to "poor".
- Ratio-level data are represented in the form of a ratio.
- Interval-level data are represented with intervals of constant size.

Median

The median is the middle score (for an odd number of scores) and the average of the two middle scores (for an even number of scores). The scores should first be arranged in either ascending or descending order before the median can be calculated.

The median has these properties:
- There is only one median for a set of data.
- It is not affected by extremely high and low values, whereas the mean is. It is therefore a good measure of central tendency.
- The median can be calculated for ratio-level, ordinal-level and interval-level data.

EXAMPLE

The masses of a group of Grade 10 learners were recorded in kilograms as follows:

45	56	55	68	53	55	48	49	53	54
56	59	60	63	67	55	57	55		

Calculate the median.

Arrange the data in ascending order:

45; 48; 49; 53; 53; 54; 55; 55; 55; 55; 56; 56; 57; 59; 60; 63; 67; 68

$$\text{median} = \frac{55 + 55}{2} = 55$$

The median lies between these two values, as the set of data contains an even number of observations.

Mode

The mode of a set of data is the value that occurs the most.

- The mode is useful when describing nominal-level and ordinal-level data, such as data in bar graphs or histograms.
- The mode is not affected by extremely high or low values.

EXAMPLE

Note: If no observation value appears more than once, there is no mode. If two values occur the same number of times, then the data is said to be **bimodal**. If more than two values appear most frequently, there is also no mode.

Look again at the masses of the group of Grade 10 learners you considered above, for the median:

45	56	55	68	53	55	48	49	53	54
56	59	60	63	67	55	57	55		

What is the modal weight?

If you work through all the data, you find that the number 55 appears most often. Therefore, the mode is 55.

The mode for grouped data

The mode can be approximated for data that is arranged in a frequency distribution table. The mode lies in the mid-point of the class with the highest frequency. In the example on page 239, the class with the highest frequency was 14,5–17,5. This means that most learners are between the ages of 14 and 17. The mid-point of this class is 16, which is the **estimated** mode. If you look closely at the original values before they were arranged in the frequency distribution table, you will see that the mode is 14, since it occurs 5 times. The value 16 is among three other values that occur 4 times. The mode is therefore an estimate if data is already arranged in a frequency distribution table.

Exercise

1 Calculate the mean for the data in this set:

3	4	5	1	2	8	9	6	3	2	1	6	4	7
8	1	1	5	5	2	3	4	5	7	8	3	4	2
5	1	9	4	5	6	7	8	9	2	1	5	4	3
4	5	6	1	4	4	8							

2 Draw up the frequency distribution table for the same set of data. Then calculate the mean. Is the mean in question 1 the same as the mean you have just calculated?

3 The Department of Education calculated the number of teaching hours per week of the teachers in a certain school. The hours were:

16	22	24	23	17	19	22	21
18	20	21	15	9	18	22	23
20	24	21	18	19	17	19	20
23	24	22					

a Calculate the mean, mode and median.

b Draw up the frequency distribution table and then calculate the mean and mode.

4 This table is part of a table given on the joint report by UNAIDS and UNICEF (*Children on the Brink 2002*, page 22). It shows the estimated number of children in Africa who are orphans as a result of Aids-related diseases in 2001.

2001 Africa: Orphan estimates by year, country, type and cause

Country	Total number of orphans because of Aids
Angola	104 000
Botswana	69 000
Burkina Faso	268 000
Cameroon	210 000
Democratic Republic of Congo	78 000
Namibia	47 000
South Africa	662 000
Swaziland	35 000
Zimbabwe	782 000

DISCUSS

How reliable do you think are these estimates? Estimates on HIV and Aids are widely discussed and contested in the media. Can you think of reasons why these figures could be problematic?

a Draw a bar graph to represent the information in the table.

b Read the mode off the bar graph.

5 The Telkom statements for households of learners in Grade 12 reveal that the number of calls made by each household last week were:

45	25	28	21	12	16	18	28
28	20	32	43	19	26	28	

DISCUSS

What could be done to decrease the number of orphans as a result of Aids-related diseases?

Calculate the mean and the median of the number of calls made.

Measures of dispersion

The measures of dispersion for a set of data tell you how far apart the observations are. A small value indicates that values are close to one another and therefore around the mean. In such a case, the mean is a true representation of the data. By contrast, a large value indicates that the values are far apart from the mean. In this case the mean is not a true representation of the set of data. Measures of dispersion include the range, interquartile range and semi-interquartile range.

Range

The range is the simplest measure of dispersion. It is the difference between the highest and lowest observation.

For ungrouped data, the range can be calculated as:

range = highest value – lowest value

EXAMPLE

What is the range for this set of shoe sizes?

$7; 6\frac{1}{2}; 9; 6; 7; 6\frac{1}{2}; 6; 5; 8\frac{1}{2}$

range = highest value – lowest value
= 9 – 5 = 4

For grouped data, the range is calculated differently. If data have already been grouped into a frequency distribution table, the range can be calculated by subtracting the lower limit of the smallest class from the upper limit of the largest class.

EXAMPLE

What is the range of distances covered by this group of athletes?

Distance	Number of athletes
5–10 km	200
10–15 km	120
15–20 km	80

range = upper limit (largest class) – lower limit (smallest class)
= 20 km – 5 km
= 15 km

The range of distances covered by the athletes is 15 km.

Interquartile range

The interquartile range is the difference between the third quartile (Q_3) and the first quartile (Q_1) of a set of data.

$\therefore$ interquartile range (IQR) = third quartile (Q_3) – first quartile (Q_1)

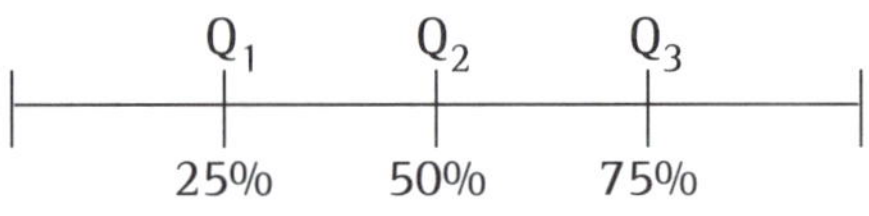

You probably remember that:

- The first quartile (Q_1) is the number above 25% of the ordered data values.
- The second quartile (Q_2) is the median (middle value).
- The third quartile (Q_3) is the number above 75% of the ordered data values.

EXAMPLE

Remember: For a group with an even number of observations, take the average of the middle two.

Calculate the interquartile range for the following set of ungrouped data.

5	9	10	32
25	5	13	4
6	45	39	

First arrange the data in ascending order:

4 5 5 6 9 10 13 25 32 39 45

Q_1 = 5, Q_2 = 10, Q_3 = 32

Interquartile range = $Q_3 - Q_1 = 32 - 5 = 27$

Identifying errors in data

It is quite common for sets of data to contain errors. If you want to draw conclusions from data, it is important to be able to identify errors before you start summarising the data. There could be several reasons for errors:

- If the sample is **biased**, the data can give a wrong impression, as it is not representative of the whole population.
- If the data were collected from a **questionnaire**, the answers people gave may not have been accurate. This can happen if people do not understand the questions, or if they deliberately give wrong information. For example, if a person is asked to give his or her annual income, he or she may give a smaller figure than is really true. (Perhaps people don't want the tax department to find out their real incomes.) Alternatively, a person may give a larger figure than is really true, in an attempt to impress the person asking the question (or anyone else who might be listening).
- If the data were collected by **measurement**, there is a possibility that the measurement was inaccurate.
- A **graph** or **chart** can also be misleading. Here is an example:

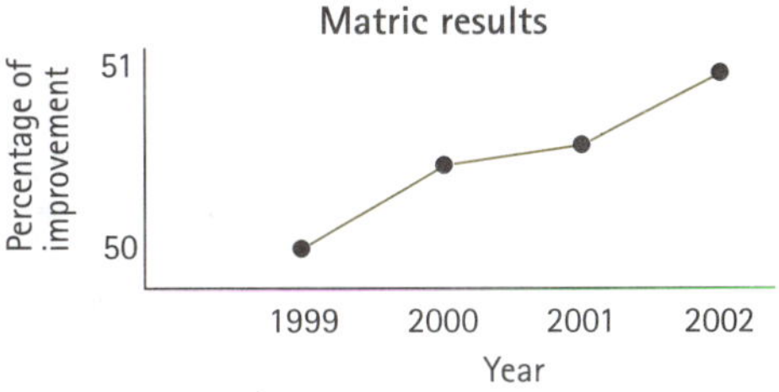

Note: Always be on the lookout for errors. Always ask yourself if the data are reasonable.

The graph on page 257 makes it look as if there was a tremendous improvement in the matric results over the four years shown, when in fact there was only a 1% increase. Therefore, when reading a graph, always ask yourself what it is that the graph is trying to show and whether it is biased or not.

Outliers

When you interpret a set of data, you will sometimes find an odd observation that is either very high or very low. Such a value is called an **outlier**. Sometimes an outlier will be quite easy to spot, but you can also calculate whether a value is or isn't an outlier.

An observation is an outlier if it is either
less than $Q_1 - 1{,}5 \times \text{IQR}$ or greater than $Q_3 + 1{,}5 \times \text{IQR}$

Remember: The median is not affected by any outliers.

It is important to be able to identify outliers, because they can distort the mean and the mode of a set of data significantly.

EXAMPLE

This data set represents the number of lions spotted on each day of a 13-day lion counting operation in the Kruger National Park.

5	1	2	6	3	3	18
4	4	1	7	2	4	

For the given data find:

a the mean of the lions spotted
b the median
c Q1 and Q3
d the minimum and maximum number of lions spotted
e the IQR
f any outlier.
g Can you give a realistic reason why the outlier has occurred?

a mean $= \dfrac{5+1+2+6+3+3+18+4+4+1+7+2+4}{13} \approx 4{,}62$

b When you order the data, you get:

1 1 2 2 3 3 4 4 4 5 6 7 18

(arrows: Q_1 between 2 and 2; Q_2 at 4; Q_3 between 5 and 6)

$\therefore$ median = 4

c lower quartile = $\frac{2 + 2}{2} = 2$

upper quartile = $\frac{5 + 6}{2} = 5,5$

d minimum value = 1 maximum value = 18

e IQR = 5,5 − 2 = 3,5

f The outlier is 18, since $Q_3 + 1,5 \times IQR = 5,5 + 1,5 \times 3,5$
$= 10,75$
and $18 > 10,75$

g Perhaps a particularly large pride of lions was spotted that day.

Exercise

1 Summarise data sets **i** and **ii** by finding:

a the mean

b the median (Q_2)

c Q_1 and Q_3

d the range

e the interquartile range (IQR)

f any outlier.

i	6	8	12	1	17	14	16	23	27	31
ii	3	45	15	20	60	2	37	68	45	32

2 These numbers of cars were counted each day for 15 days on a particular street.

5	12	5	17	13	29	30	35
29	43	68	52	67	14	56	

For the given data find:

a the mean of the cars counted

b the median

c the lower and upper quartiles

d the minimum and maximum number of cars counted

e the range

f the IQR

g any outliers.

Can you think of a realistic reason for the outlier?

3 A group of 10 mountainbikers ride down the mountain at different speeds:

67	80	77	98	54
63	83	87	69	73

Calculate:

a the mean **b** the mode **c** the range

d the IQR **e** any outlier.

Box plots

Remember: Median and quartiles are measures of central tendency.

Box plots are a special type of graph that summarises single data sets. They clearly display the minimum and maximum values, the median, the quartiles and any outliers. Box plots also give a clear indication of how data is spread. They are usually used to compare two sets of data visually.

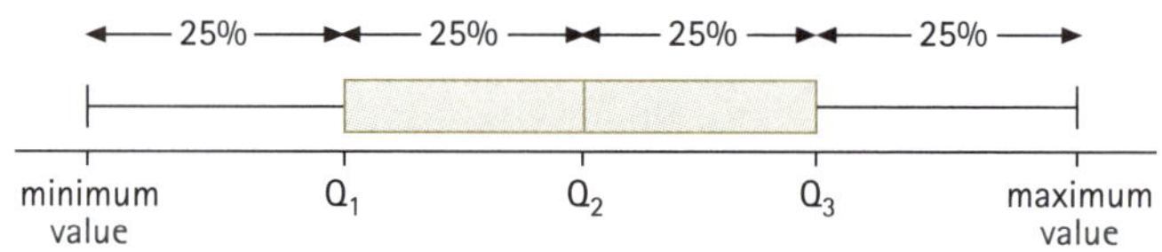

Outliers can be marked with the symbol ×:

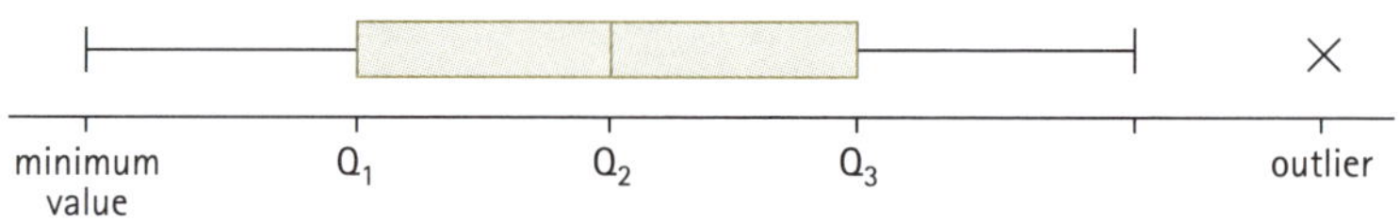

EXAMPLE

Summarise this data set in a box plot. Mark any outliers.

5	9	4	3	5	6	6
5	7	12	2	3	5	

When you order the data, you get:

2 3 3 4 5 5 5 5 6 6 7 9 12

↑ Q_1 (between 3 and 4) ↑ Q_2 (5) ↑ Q_3 (between 6 and 7)

IQR = 6,5 − 3,5 = 3 and 6,5 + 1,5 × 3 = 11

∴ 12 is an outlier.

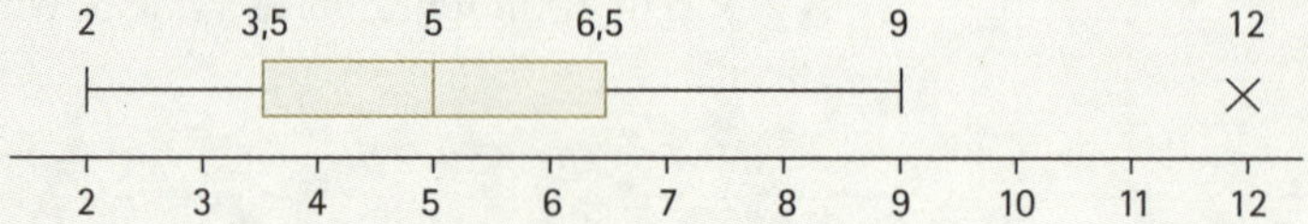

Exercise

1 For the data sets below:

  i determine whether any outliers exist by first finding Q_1, Q_2 and Q_3.

  ii draw a box plot to summarise the data, marking any outliers that exist.

  a 4 6 5 2 3 4 4 13 8 7 6

  b 1,8 1,7 1,8 1,9 1,6 1,8 2,0 1,1 1,4 1,9 2,2

  c 21 23 18 11 16 19 24 21 23 22 20 31 26 22

  d 0,04 0,04 0,03 0,03 0,05 0,06 0,07 0,03 0,05 0,02

2 By first finding Q_1, Q_2 and Q_3, draw box plots for the given data sets. Remember to consider outliers and mark them on your box plot if they exist.

  a 11 15 18 17 12 8 19 15

  b 37 48 52 51 51 42 48 47 39 41 65

  c 0 1 5 4 4 4 2 3 3 1 4 3

  d 124 118 73 119 117 120 120 121 118 122

3 The following weights of 15 lemurs from Madagascar were recorded as part of a conservation project. The measurements are all in kilograms.

  14,4 15,5 17,3 14,6 14,7 15,0 15,8 16,2
  14,9 19,7 15,3 13,8 14,6 15,4 15,7

  a Find the mean lemur weight.

  b Find Q_1, Q_2 and Q_3.

  c Which weights would be considered outliers, if any?

  d Draw a box plot to summarise the lemur weights.

4 Two data sets can be compared using comparative box plots on the same scale, as shown:

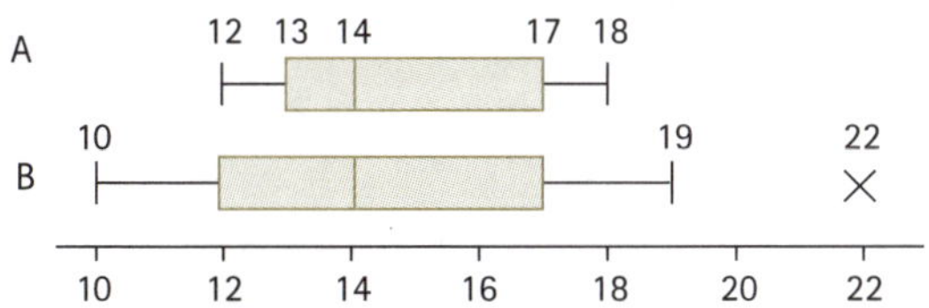

a What statistical measures do the two box plots have in common?

b Which data set (A or B) has a wider range of values?

c Find the IQR for:

i data set A **ii** data set B.

d How would you describe the main difference between the two sets of data for which the comparative box plots have been drawn?

5 Draw comparative box plots for these data (two box plots on one scale) after first finding Q_1, Q_2 and Q_3.

Data set 1:	0,1	0,3	0,3	0,8	0,6	0,4	0,8	0,9
	0,2	0,2	0,5	0,6	0,5	0,3	0,7	0,6
Data set 2:	0,5	0,4	0,4	0,5	0,3	0,4	0,5	0,4
	0,6	0,4	0,4	0,3	0,5	1,1	0,6	0,2

6 An examiner marked 15 essays for spelling errors. She counted these numbers of spelling errors:

3 2 4 6 8 4 6 7 6 10 7 12 7 3 8

Another examiner marked the same 15 essays for spelling errors. He counted these numbers of spelling errors:

12 7 9 11 15 5 14 16 9 11 8 13 14 15 13

a Draw comparative box plots for the data.

b Do you believe that there was a major difference in the ways in which the examiners marked the essays? If yes, describe this difference.

PEER ASSESSMENT

See page 289 (rubric 4).

SELF-ASSESSMENT

See page 286 (rubric 1).

CHAPTER 4 PROBABILITY AND CHANCE

KEY VOCABULARY

probability • sample space • event • lattice diagram • outcome • tree diagram • experiment • set • prediction • Venn diagram • equally likely • fraction • relative frequency • ratio • percentage

DISCUSS

Can you think of any more terms that describe probability?

Remember: Probability is the chance or likelihood of something happening.

You may be familiar with many words and phrases in daily life that describe how likely something is to happen, for example, **probably**, **chance** and **likely**.

In the study of probability, a numerical value on a scale from 0 to 1 is used to describe levels of **chance** or **probability.** This numerical value can be expressed as a proper fraction, a decimal fraction or a percentage.

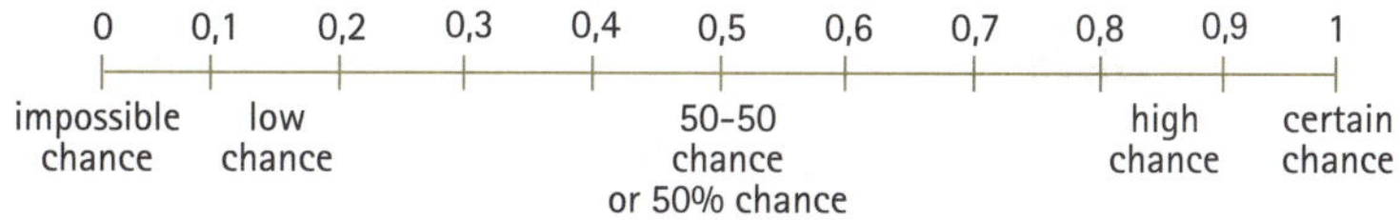

Sometimes the probability of an event happening is difficult to work out, because of many unknown factors. For example, what is the chance that the weather will be fine next Sunday? What is the chance that the South African cricket team wins its next four matches?

Three key terms are used in the study of probability:

- **Experiment:** an activity or the act of measuring something (e.g. throwing a coin twice)
- **Outcome:** a particular result of an experiment (e.g. getting one head [H] and one tail [T] when throwing a coin twice)
- **Event:** a collection of one or more outcomes of an experiment. This experiment has four possible events: HH, TT, HT or TH.

Equally likely events

The probability of an event happening if the outcomes are equally likely is quite easy to describe. For example, if you toss a coin, it is equally likely that the coin will land on heads or tails. The chance of getting a tail is therefore $\frac{1}{2}$. You can also express chance as a percentage. If the weather report predicts a 50% chance of rain, this is the same as a probability of $\frac{1}{2}$.

The probabilities of events with equally likely outcomes can be calculated as:

$$\text{probability of an event} = \frac{\text{number of desired outcomes}}{\text{total number of possible outcomes}}$$

EXAMPLE

A letter is chosen from the word TELEVISION. Find the probability that the letter is:

a a V **b** an E **c** not an E **d** an E or a V.

a $P(V) = \frac{\text{number of desired outcomes}}{\text{total number of possible outcomes}}$

$= \frac{1}{10} = 0{,}1$

b $P(E) = \frac{2}{10} = 0{,}2$

c $P(\text{not an E}) = \frac{8}{10} = \frac{4}{5} = 0{,}8$ Count all the letters except E = 8.

d $P(E \text{ or } V) = \frac{3}{10} = 0{,}3$ Count the number of Es and Vs.

Investigation: Walk the plank

Consider a plank on the side of a ship with five positions as shown in the diagram below. Position 0 is on the deck of the ship where you are considered "safe" and position 6 is in the water where you are considered to be "overboard".

SAFE 0	1	2	3	4	5	6 OVERBOARD

Pretend you are a prisoner standing on the plank at position 3 facing the water. You are instructed to take a step forwards or backwards to position 4 or 2 according to a random toss of a coin. Then, at that position, you take another random step forwards or backwards to the next position. You repeat this until you are either "safe" or "overboard".

1 Use a coin to determine whether you step forwards or backwards. Use heads for forwards and tails for backwards. Repeat tossing the coin until you are either at the "safe" (0) position or at the "overboard" (6) position.

- Repeat the experiment 50 times and count how often you end up "overboard".
- Find the proportion of times that you ended up "overboard".
- Estimate the probability that if you begin at position 3 you end up "overboard".

2 Repeat the investigation by starting at the positions:

a 1 **b** 2 **c** 4 **d** 5.

Exercise

1 A letter is chosen from the word PROBABILITY. Find the probability that the letter is:

a a P **b** a vowel **c** not a vowel

d a T or a vowel **e** not a B.

2 For the following spinners, find the probability that the outcome will be a 4:

a

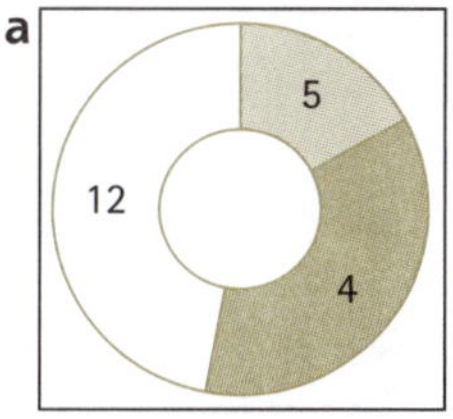

b

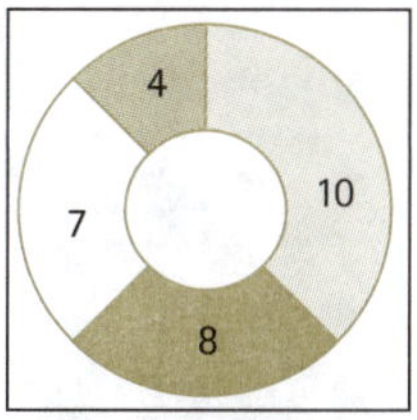

c

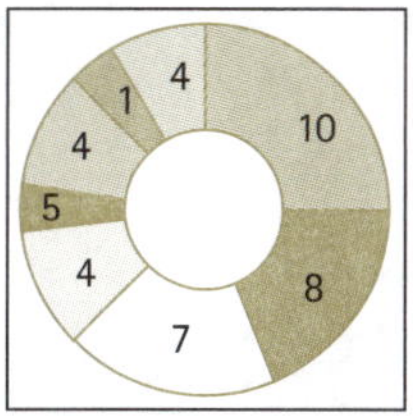

3 A six-sided die is tossed once. Find the probability that the outcome will be:

a a 6 **b** a 2 or a 4

c an even number **d** an even number or a 1

e an odd number **f** a 7.

4 A card is chosen from a pack of 52 playing cards. Find the following:

a P(heart) **b** P(jack of clubs)

c P(an ace) **d** P(king or queen)

e P(neither a heart nor spade) **f** P(11).

DID YOU KNOW?

People were already playing a game similar to today's tennis in the 12th century in France.

5 A tennis player won her match over 3 sets. She had the following number of aces per set. A set consists of at least 6 games.

Set	1	2	3
Number of aces	10	8	11

Use these results to find the probability that the winner hit **an ace**:

a in the first set

b in the second set

c in the third set

d in the match.

e Is it possible from the given information to work out the probability of hitting an ace in the fourth game of the first set?

6 A target board is made up of 3 rings (A, B and C) 10 cm apart, as shown on the right. An experienced archer shoots an arrow at the board and is guaranteed to hit the board, with an equal chance of doing so at any point.

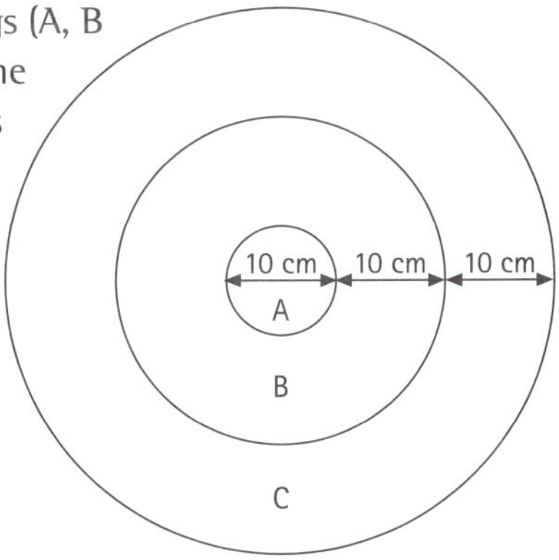

Remember:
Area (circle) = πr^2.

a Calculate the total area of the target.

b Calculate the area of the regions labelled:

i A **ii** B **iii** C.

c Calculate the probability that the region in which the archer's arrow will hit, will be:

i A **ii** B **iii** C

iv A or B **v** B or C **vi** A or C

vii A, B or C **viii** not in B.

Representing probability

In this section you will represent combined events by means of lattice diagrams, tree diagrams, and sets and Venn diagrams.

Lattice diagrams

For experiments that involve two outcomes, a table is useful for organising the list of possible outcomes. Such tables are called **lattice diagrams.**

The list of all possible outcomes is called the **sample space.**

EXAMPLE

A six-sided die is tossed twice.

a List the sample space using a lattice diagram.
b State the total number of outcomes.
c Find the probability of obtaining the outcome (1,5).
d Find the probability of obtaining a double.

a

Toss 2

Toss 1	1	2	3	4	5	6
1	(1,1)	(1,2)	(1,3)	(1,4)	(1,5)	(1,6)
2	(2,1)	(2,2)	(2,3)	(2,4)	(2,5)	(2,6)
3	(3,1)	(3,2)	(3,3)	(3,4)	(3,5)	(3,6)
4	(4,1)	(4,2)	(4,3)	(4,4)	(4,5)	(4,6)
5	(5,1)	(5,2)	(5,3)	(5,4)	(5,5)	(5,6)
6	(6,1)	(6,2)	(6,3)	(6,4)	(6,5)	(6,6)

b There is a total of $6 \times 6 = 36$ outcomes.
c $P(1,5) = \frac{1}{36}$
d $P(\text{double}) = \frac{6}{36}$

Tree diagrams

When you have two or more combined events, it can be very useful to draw up a tree diagram. Tree diagrams list all possible outcomes (the sample space) and help you to determine relevant probabilities.

EXAMPLES

1 A coin is tossed 3 times.

a List the sample space using a tree diagram.

b Find the probability of obtaining exactly 2 tails.

c Find the probability of obtaining no tails.

a

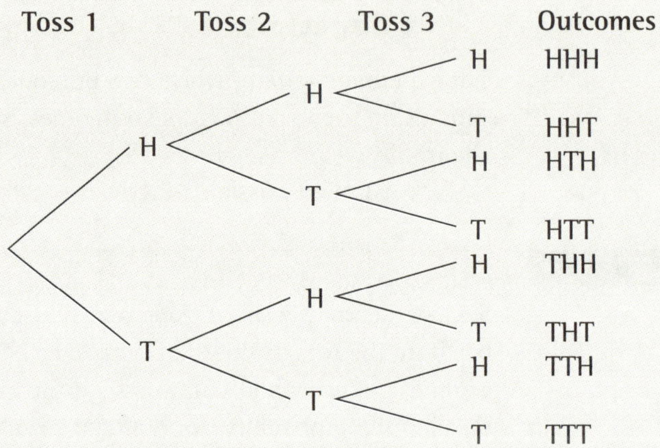

b P(2 tails) = HTT, TTH, THT = $\frac{3}{8}$

c P(no tails) = HHH = $\frac{1}{8}$

2 Two bags (A and B) contain marbles. Bag A has 2 red and 5 green marbles and Bag B has 4 blue and 3 yellow marbles. A bag is chosen randomly and a single marble is chosen.

a From Bag A find the probability of choosing a red marble.

b From Bag B find the probability of choosing a yellow marble.

c Represent the available options using a tree diagram that shows all possible outcomes and related probabilities.

d What is the probability of selecting Bag B and a blue marble?

e What is the probability of selecting a green marble?

a P(red) = $\frac{2}{7}$

b P(yellow) = $\frac{3}{7}$

c

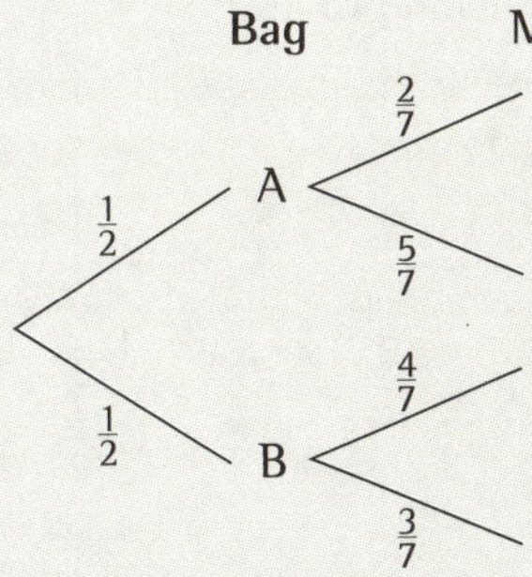

Marble	Outcomes	Probability
red	(A, red)	$\frac{1}{2} \times \frac{2}{7} = \frac{2}{14}$
green	(A, green)	$\frac{1}{2} \times \frac{5}{7} = \frac{5}{14}$
blue	(B, blue)	$\frac{1}{2} \times \frac{4}{7} = \frac{4}{14}$
yellow	(B, yellow)	$\frac{1}{2} \times \frac{3}{7} = \frac{3}{14}$

d P(B, blue) $= \frac{1}{2} \times \frac{4}{7} = \frac{4}{14}$

e P(green) $= \frac{1}{2} \times \frac{5}{7} = \frac{5}{14}$

Sets and Venn diagrams

A **set** is a collection or group of elements that can include numbers, letters or other objects. For example, the set A = {a, b, c, d, e} is a set containing the first five letters of the alphabet.

The **universal set** (denoted by the symbol $\mathscr{E}$) is the set of all possible elements or objects considered in a particular situation. For example, the universal set $\mathscr{E}$ = {a, b, c, d, e, f, . . ., z} is the set of all the letters in the alphabet.

The event A occurs if one or more of the elements in set A is observed. For example, if you select the letter d from $\mathscr{E}$, this means that event A has occurred, because the letter d belongs to set A.

If you want to determine the probabilities for two or more events, you can use a **Venn diagram** to help you.

The Venn diagram for $\mathscr{E}$ = {a, b, c, d, e, f, . . ., z}
A = {a, b, c, d, e} and
B = {a, e, i, o, u} is shown below.

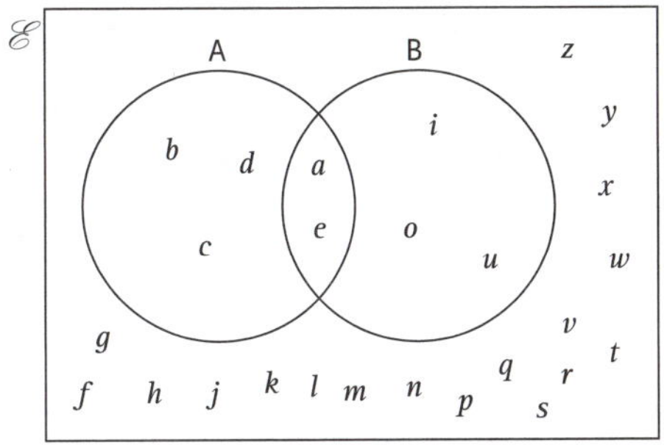

All elements that belong to both sets A and B make up the **intersection:** $A \cap B$.
$\therefore A \cap B = \{a, e\}$

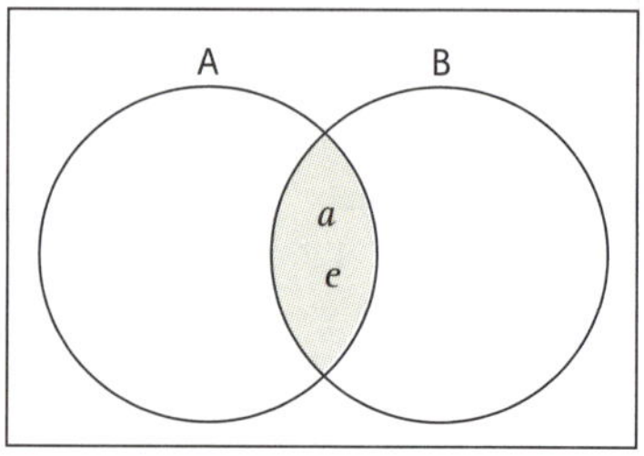

All elements that belong to either events A or B make up the **union** : $A \cup B$.
$\therefore A \cup B = \{a, b, c, d, e, i, o, u\}$

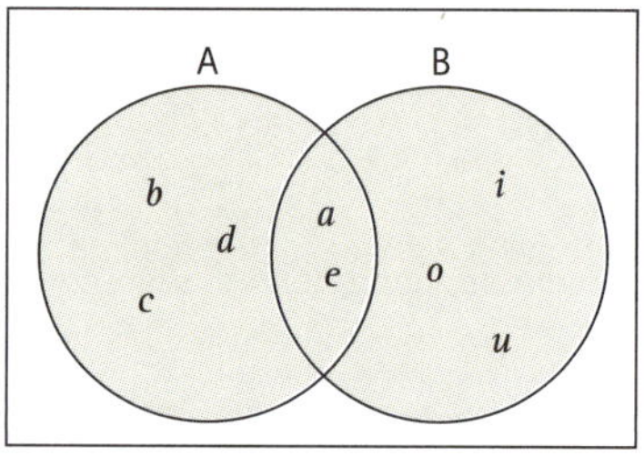

If set $C = \{w, x, y, z\}$ then A and C have no elements in common and therefore $A \cap C = \{\ \}$ or $\varnothing$, the empty or null set. In such a case, the events A and C are said to be **mutually exclusive** and could be represented by this diagram:

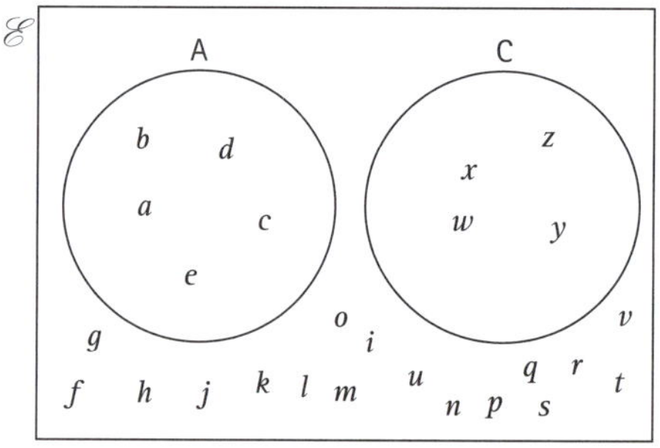

EXAMPLES

Consider the given events A and B, which involve numbers taken from the universal set, including the first 10 positive integers.

$\mathscr{E} = \{1, 2, 3, 4, 5, 6, 7, 8, 9, 10\}$
$A = \{1, 2, 3, 4, 5, 6\}$
$B = \{1, 3, 7, 8\}$

a Represent the two events A and B in a Venn diagram:
  i $A \cap B$ **ii** $A \cup B$.

b If a number from the first 10 positive integers is randomly selected, find the probability that the following events occur:
  i A **ii** $A \cap B$ **iii** $A \cup B$.

a

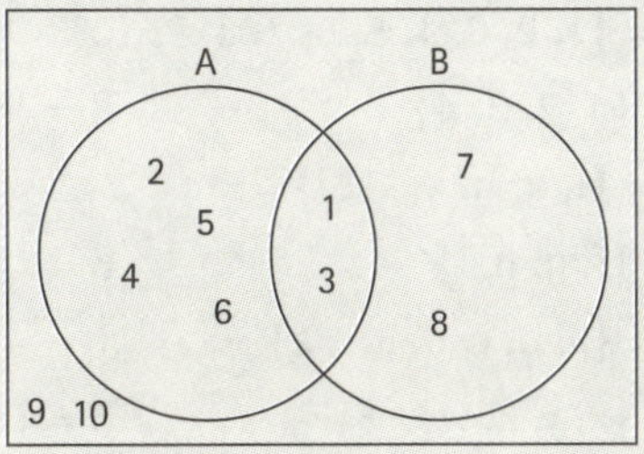

i $A \cap B = \{1, 3\}$ **ii** $A \cup B = \{1, 2, 3, 4, 5, 6, 7, 8\}$

b **i** $P(A) = \frac{6}{10} = \frac{3}{5}$ **ii** $P(A \cap B) = \frac{2}{10} = \frac{1}{5}$

iii $P(A \cup B) = \frac{8}{10} = \frac{4}{5}$

Exercise

1 Copy the Venn diagrams and shade the region described:

a A

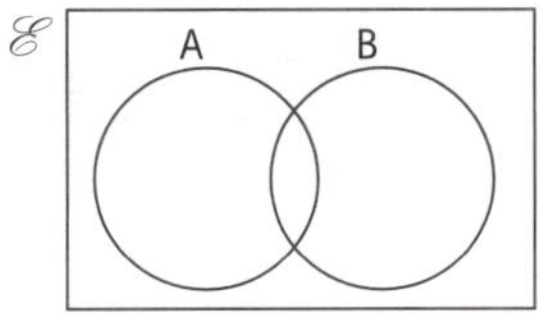

b B

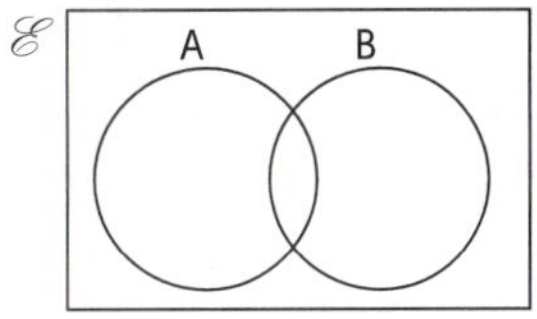

c $A \cap B$

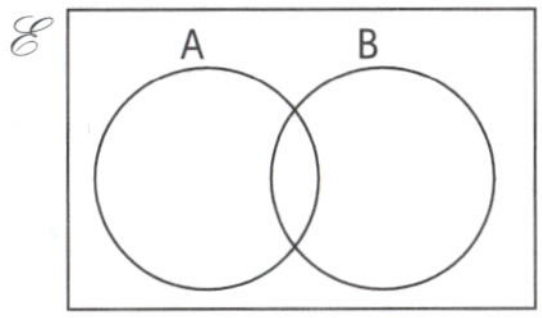

d $A \cup B$

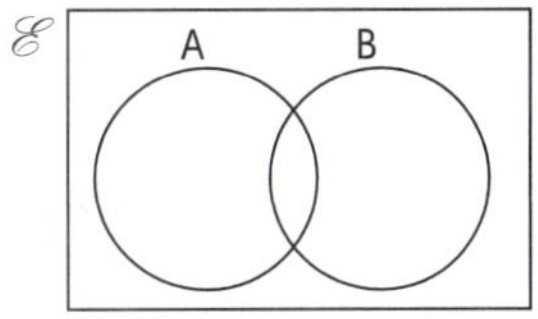

2 For each of the events and universal sets defined below, complete the following:

i Represent the events in a Venn diagram.

ii Decide whether or not the events are mutually exclusive.

iii List the set $A \cap B$.

iv Find $P(A \cap B)$.

a $\mathscr{E} = \{1, 2, 3, 4, 5, 6,\}$

$A = \{1, 2, 3, 4, 5\}$ $B = \{5, 6\}$

b $\mathscr{E} = \{1, 2, 3, 5, 7, 9, 10, 11, 12\}$

$A = \{3, 5, 7, 9\}$ $B = \{5, 6, 10, 12\}$

c $\mathscr{E} = \{1, 2, 3, 4, 5, 6, 7, 8, 9, 10\}$

$A = \{1, 2, 3, 4\}$ $B = \{5, 6, 7, 8\}$

d $\mathscr{E} = \{f, g, h, i, j, k, l\}$

$A = \{f, g, h, i\}$ $B = \{i, j, k\}$

e $\mathscr{E} = \{a, b, c, d, \ldots, z\}$

$A = \{x, y, z, w\}$ $B = \{e, f, g\}$

f $\mathscr{E} = \{1, 4, 8, 9, 16, 18, 32\}$

$A = \{1^2, 2^2, 3^2, 4^2\}$ $B = \{4, 8, 18, 32\}$

Complementary events

Consider a universal set that includes the positive integers from 1 to 10 and event A, consisting of prime numbers:

$\mathscr{E} = \{1, 2, 3, 4, 5, 6, 7, 8, 9, 10\}$ and $A = \{2, 3, 5, 7\}$.

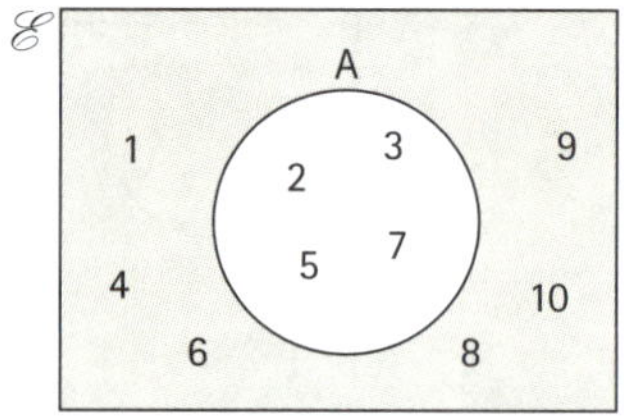

Note: The complement of A is denoted by A' or not A.

The **complement** of A then includes all elements that do not appear in A.

$\therefore A' = \{1, 4, 6, 8, 9, 10\}$.

EXAMPLE

Consider all positive integers from 1 to 16. Let A be the event **choosing an even number** and B be the event **choosing a square number.**

a List the sets **i** A **ii** B.

b Represent the events A and B in a Venn diagram.

c List the sets **i** $A \cap B$ **ii** $A \cup B$.

d List the elements not in A, i.e. list A'.

e If a number from the first 16 positive integers is chosen at random, find the probability that the following events will occur:

i A **ii** $A \cap B$ **iii** A'.

a **i** A = {2, 4, 6, 8, 10, 12, 14, 16}

ii B = {1, 4, 9, 16}

b 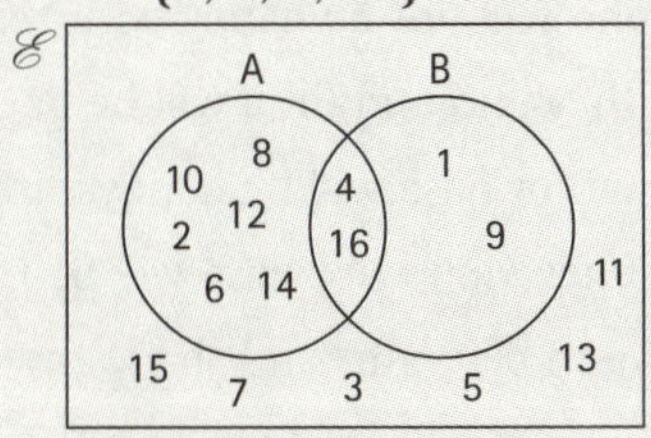

c **i** $A \cap B = \{4, 16\}$

ii $A \cup B = \{1, 2, 4, 6, 8, 9, 10, 12, 14, 16\}$

d A' = {1, 3, 5, 7, 9, 11, 13, 15}

e **i** $P(A) = \frac{8}{16} = \frac{1}{2}$ **ii** $P(A \cap B) = \frac{2}{16} = \frac{1}{8}$

iii $P(A') = \frac{8}{16} = \frac{1}{2}$

Exercise

1 Consider the following universal set and the events A and B:

ℰ = {1, 2, 3, 4, 5, 6, 7, 8, 9, 10}

A = {1, 3, 5, 7, 9}

B = {4, 5, 6, 7, 8}

a Represent the events A and B in a Venn diagram.

b List the sets **i** $A \cap B$ **ii** $A \cup B$.

c List the elements not contained within A, i.e. list A'.

d Find the following probabilities:

i P(A) **ii** P(B) **iii** $P(A \cap B)$

iv $P(A \cup B)$ **v** P(A') **vi** P(B').

2 Mpho and Elisa need to choose a colour to paint the interior walls of their house. There are six colours to choose from, namely white (w), cream (c), navy (n), sky blue (s), maroon (m) and violet (v).

Mpho would be happiest with white or cream, but Elisa would be happiest with cream, navy or sky blue. As they can't decide on a colour, a colour is chosen for them at random.

Let M be the event that **Mpho will be happy with the colour** and E be the event that **Elisa will be happy with the colour.**

a Represent the events M and E in a Venn diagram.

b Find the probability that the following events will occur:

i Mpho will be happy with the colour choice, i.e. find P(M).

ii Mpho will not be happy with the colour choice.

iii Both Mpho and Elisa will be happy with the colour choice.

iv Mpho or Elisa will be happy with the colour choice.

v Neither Mpho nor Elisa will be happy with the colour choice.

3 Copy the Venn diagrams and shade the region described:

a A only

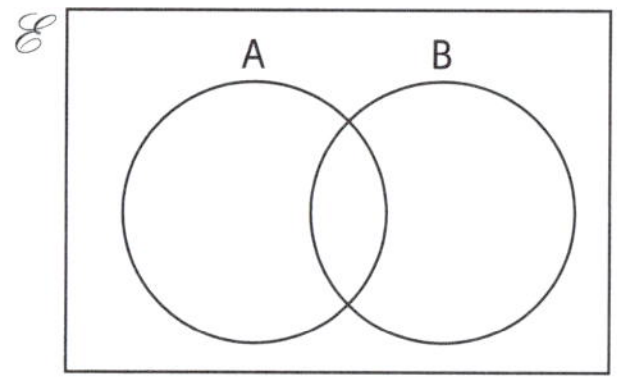

b B only

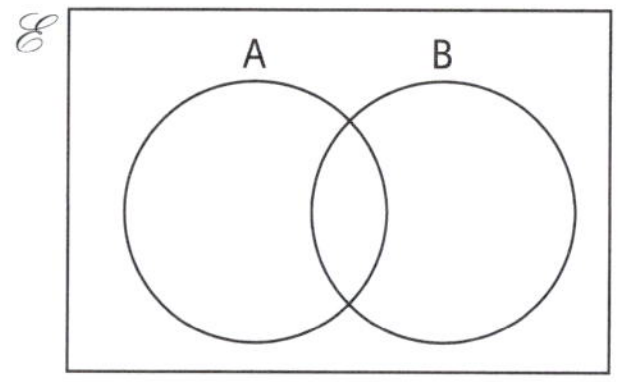

c A′

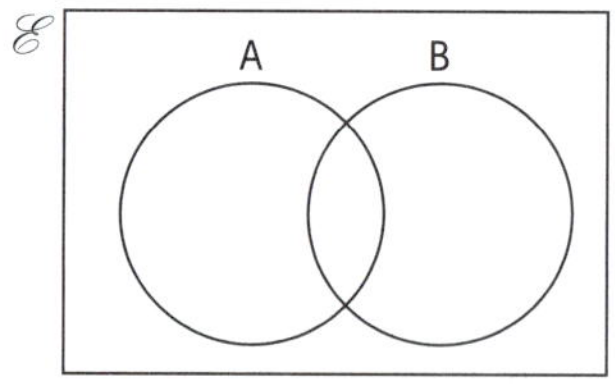

d neither A nor B

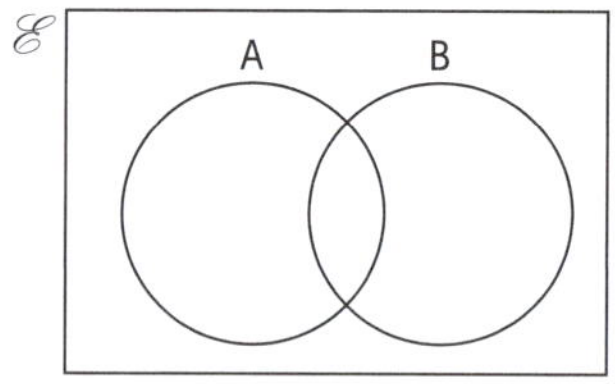

4 From a group of 50 adults, 35 enjoy reading fiction (F), 20 enjoy reading non-fiction (N) and 10 enjoy reading both fiction and non-fiction.

a Illustrate the given information in a Venn diagram.

b State the number of people who enjoy:

i fiction

ii fiction only

iii non-fiction only

iv neither fiction nor non-fiction.

c Find the probability that a person chosen at random will enjoy:

i non-fiction ii both fiction and non-fiction

iii non-fiction only.

Remember: The symbol n stands for the number of elements in a set.

5 For the following information regarding the events A and B, use Venn diagrams in each case to find:

i n(A only) ii n(A').

	n($\mathscr{E}$)	n(A)	n(B)	n(A ∩ B)
a	20	10	15	5
b	50	30	14	9
c	43	6	29	2
d	17	8	7	7

DISCUSS

How does your school encourage learners to recycle materials? Why do you think it is important to recycle?

6 A craft display of items made from recycled goods holds 75 items. Fifty of the craft items are made from recycled cardboard (C), 30 are made from recycled soft drink cans (S) and 18 are made from both cardboard and soft drink cans.

a Illustrate the given information using a Venn diagram.

b For an item chosen at random from the display, find the probabilities of:

i P(C) ii P(S) iii P(C ∩ S) iv P(C ∪ S)

v P(S') vi P(C') vii P(neither C nor S).

7 In a group of 12 chefs, all enjoy preparing fish or red meat. In fact, 7 enjoy preparing fish and 8 enjoy preparing meat. Use a Venn diagram to find out how many chefs enjoy preparing both fish and red meat.

8 In the table there are no elements that belong outside the events A or B. Find n(A ∩ B) in each case.

	n($\mathscr{E}$)	n(A)	n(B)
a	10	7	5
b	15	3	12
c	30	15	15
d	20	15	12

9 Consider a universal set containing 20 elements, i.e. $n(\mathscr{E}) = 20$, and also the events A and B with this information:

- $n(A) = 12$
- $n(B) = 15$
- there are no elements that belong outside A or B.

Use a Venn diagram and perhaps trial and error to find $n(A \cap B)$.

10 Of 10 movies showing at a cinema complex, 5 are rated for adult viewing (A) and 8 are for children (C). Clearly some movies are suitable for both adults and children $(A \cap B)$.

a Represent this information in a Venn diagram.

b Find the number of movies for adults only (A only).

c If you choose a movie at random, find the probability that it is rated for:

i adults only **ii** children only.

Independent events

Two events are independent if they do not affect each other. For two events A and B, the probability that A and B will both occur is found by multiplying the two probabilities. P(A and B) = P(A) × P(B). Each probability of events A or B can be calculated as: $P(A) = P(\frac{A}{B})$ or $P(B) = P(\frac{B}{A})$.

EXAMPLES

A coin is tossed and a die is rolled.

a Show the sample space in a table.

b Find from the table the probability of obtaining both a head and a 4.

c Calculate the probability of event **b**.

a

1	H	T
2	H	T
3	H	T
4	H	T
5	H	T
6	H	T

This table represents the sample space of outcomes. For example, the first row represents 1 in the die and either H or T in the coin. Therefore these are two different events.

DISCUSS

It is obvious that the way the die is rolled does not affect which way the coin lands. Can you think of a case where it is not as obvious that two or more events are independent?

b P(H and 4) = $\frac{1}{12}$ (1 event out of 12 possible outcomes).

c $P(H \text{ and } 4) = P(H) \times P(4)$

$$= \frac{1}{2} \times \frac{1}{6}$$

$$= \frac{1}{12}$$

The way the coin lands does not depend on what the face of the die shows. So the stages are **independent**. Therefore you multiply to find the probability.

Exercise

1 For the events A and B in the given Venn diagrams, find:

i P(A) **ii** $P(\frac{A}{B})$ **iii** P(A and B).

a

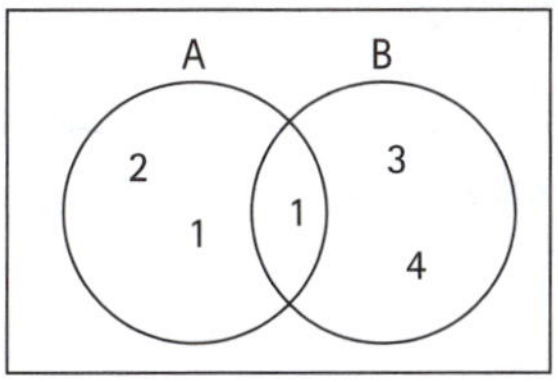

b

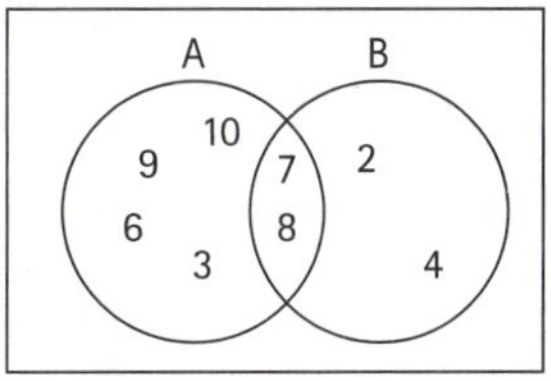

2 A bag contains 8 blue marbles and 2 red marbles. Two marbles are drawn at random. What is the probability of getting:

a 2 red marbles?

b 1 red marble and 1 blue marble?

c 2 blue marbles?

3 A bag contains 12 beads. Five are red and the rest are white. Two beads are drawn at random. Find the probability that:

a both beads are red

b both beads are white.

4 Mahmoud enjoys flying his kite. On any given day, the probability that there is a good wind is $\frac{3}{4}$. If there is a good wind, the probability that the kite will fly is $\frac{5}{8}$. If the wind is not good, the probability that the kite will fly is $\frac{1}{16}$.

a **i**

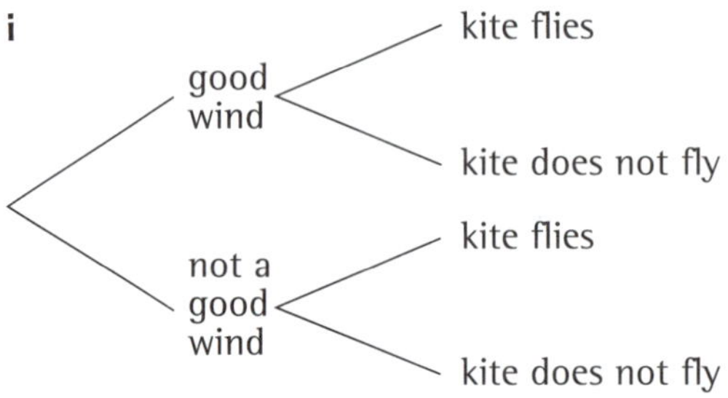

Copy the tree diagram.
Write on your diagram the probability for each branch.

ii What is the probability of a good wind **and** the kite flying?

iii Find the probability that, whatever the wind, the kite does **not fly**.

b If the kite flies, the probability that it gets stuck in a tree is $\frac{1}{2}$. Calculate the probability that, whatever the wind, the kite gets stuck in a tree.

Language skills

interview	*sample*	*questionnaire*	*tally*
survey	*observation*	*census*	*population*
average	*frequency*	*stem-and-leaf plot*	*bias*
histogram	*mean*	*range*	*table*
data	*median*	*tally*	*distribution*
mode	*statistics*	*polygon*	*interquartile range*

1 Work with a partner. Explain to each other how the terms "population" and "sample" are related.

2 Give an example of a biased sample. Discuss your examples with each other.

3 Are there any terms in the list that you do not fully understand? Explain them to each other, or ask your teacher if you are unsure.

4 Select the word from the list that most closely matches the definition:

a The middle score when a set of scores are arranged in order.

b A column graph showing the number of times each score occurred.

c Data collected from a small group that is representative of the whole population.

d A collection of facts, numbers or other information.

5 Use each of these words in a sentence:

a census b bias

c questionnaire d interquartile range.

6 Write two definitions for each word:

a mean b population c sample.

SELF-ASSESSMENT

Also see page 293 (rubric 9).

Check yourself

Make a summary of the key information in this module under the headings listed here. Look at the key skills given at the beginning of this module to help you (see page 223).

Statistical investigations
Ways of displaying data
Measures of central tendency
Probability

Then complete the Review. The questions in the Review will help you consolidate your knowledge and understanding of the topics in this module.

MODULE 4

REVIEW

1 For the first time in 1994, South Africa held democratic elections at which all citizens could vote. Here is a summary of the 1994 election results:

1994 Election Results: South Africa		
Party	Votes	% National Assembly seats
African National Congress	12 237 655	62,65
National Party	3 983 690	20,39
Inkatha Freedom Party	2 058 294	10,54
Freedom Front	424 555	2,17
Democratic Party	338 426	1,73
Pan Africanist Congress	243 478	1,25

Here is a summary of results for the 2004 elections:

2004 Election Results: South Africa		
Party	Votes	% National Assembly seats
African National Congress	10 878 251	69,68
New National Party	257 824	1,65
Inkatha Freedom Party	1 088 664	6,97
Freedom Front Plus	139 465	0,89
Democratic Alliance	1 931 201	12,37
Pan Africanist Congress	113 512	0,73

(Source: http://www.elections.org.za)

DISCUSS

According to the South African Constitution, a two-thirds majority would enable the African National Congress (ANC) to govern by itself. In the 2004 elections, the ANC secured 69,68% of the votes – more than two-thirds of the national vote. Discuss what you think the role of opposition parties should be in a country where one party is by far the most popular.

Note: Some parties have amended their names since 1994. This can present difficulties, as the new parties will want to distance themselves from their old counterparts. How would you present this information in your double bar graph?

a Draw a double bar graph to represent the votes each party secured at the two elections.

b Draw a pie-chart to represent the percentage National Assembly seats for each.

c Is the information sufficient to draw a pie-chart in each case?

d What percentage of National Assembly seats did other parties gain that are not listed in the table?

e Comment on the difference between the figures for 1994 and 2004. (If you have access to the Internet, you can find the full set of figures on http://www.elections.org.za).

2 This bar graph shows the results of a 2002 study of HIV/Aids.

Strategies of sexual behaviour change in the face of the Aids threat by sex (15 years and older), South Africa, 2002

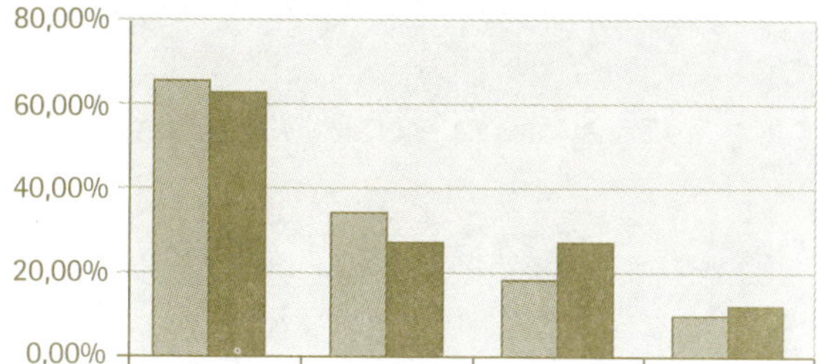

	single partner	condom use	sexual abstinence	reduced no. of sexual partners
male	66,40%	32,80%	17,40%	10,20%
female	63,50%	26,90%	26,90%	12,60%

(Source: http://www.hsrc.ac.za/publish)

a Which sexual behaviour is practised most widely, according to the bar graph?

b Which gender group has a higher percentage in sexual abstinence?

c Is it possible to draw a pie-chart for male or female sexual behaviour change, using the information shown in the table? Explain your answer.

d What can you deduce from the bar graph? Does it mean that the strategies of sexual behaviour change in this age group could help reduce HIV/Aids infections? Discuss this in your groups.

3 A letter is chosen at random from the word HAPPY and a second letter is chosen from the word HEY.

a List the sample space using a lattice diagram.

b State the total number of outcomes.

c Find the probability that the two letters chosen will be:

i H then E **ii** the same **iii** not the same.

MODULE 4

REVIEW

4 A four-sided die is tossed twice and the total is noted.

a Use a tree diagram to list the sample space, including all possible totals.

b Find the following probabilities:

i P(2) ii P(5) iii P(1) iv P(not 1).

5 A universal set and the sets A and B are defined by:

$\mathscr{E} = \{3, 4, 6, 7, 8, 10, 11, 12\}$

$A = \{3, 6, 8, 11\}$

$B = \{3, 4, 7, 10, 11\}$

a Represent the two sets A and B in a Venn diagram.

b List the sets:

i $A \cap B$ ii $A \cup B$ iii A'.

c If a single element is chosen from the universal set, find the probability that the following events will occur:

i A ii $A \cap B$ iii B' iv $A \cup B$.

6 Of 36 people, 18 have an interest in cars and 11 have an interest in motorbikes, while 6 have an interest in both cars and motorbikes.

a Represent the given information in a Venn diagram.

b State the number of people who have no interest in cars or motorbikes.

c If a person is chosen at random from the group, find the probability that the person will:

i have an interest in cars and motorbikes

ii have an interest in motorbikes only

iii not have any interest in cars.

7 Box A contains 2 red and 3 green counters, while Box B contains 1 red and 3 green counters. In an experiment a box is chosen at random and then a counter is selected from the box. From a simulation including 20 selections of a box and then a counter, 6 red counters were observed.

a Estimate the probability of selecting a red counter using the experimental results.

b Draw a tree diagram displaying all possible outcomes and the corresponding theoretical probabilities.

c State: i P(A, red) ii P(B, red).

d Find the theoretical probability of obtaining a red counter.

8 All 26 birds in a bird cage have either clipped wings and/or tags. In total, 18 birds have tags and 14 have clipped wings.

a Find the number of birds that have both tags and clipped wings.

b Find the probability that a bird chosen at random will have tags only.

MODULE 4

PROJECTS

ASSESSMENT PORTFOLIO

See page 292 (rubric 8).

Note: You can also choose your own topic. Check with your teacher that it is suitable.

1 Analyse your own data

Some topics on which you could collect a single data set include:

- the height of students in your class
- the number of hours of TV people watch in a week
- the number of hours learners spend on homework each day
- the number of cars passing your house between 5 p.m. and 6 p.m. spread over a two-week period.

1 Choose one of the above or a different topic and collect a reasonably large data set (at least 14 data elements).

2 Analyse and display your data, applying the following terminology:
   - Q_1, Q_2, Q_3 and the mean
   - stem-and-leaf plot
   - histogram
   - box plot.

3 Report on your data by discussing the important features of your displays in question 2. Discuss any outliers. Comment on the median.

4 Collect a second data set of the same size relating to the topic you chose in question 2. For example, if you measured the heights of students in your class, see if you can collect the heights of students of another Grade 10 class.

   Use comparative box plots to compare and discuss similarities and differences between the two data sets.

ASSESSMENT PORTFOLIO

See page 292 (rubric 8).

2 Life expectancy

These figures relate to life expectancy (in years) at birth in 2001 for 40 countries (Source: *CIA – The World Factbook, 2001*).

Afghanistan	46	Albania	72	Algeria	70	Argentina	76
Australia	80	Bolivia	64	Brazil	63	Cambodia	57
Chile	76	China	72	Denmark	77	Fiji	68
France	79	Ghana	58	Hong Kong	80	Hungary	72
India	63	Kenya	47	Libya	76	Morocco	70
Mexico	72	Nepal	58	New Zealand	78	Norway	79
Pakistan	61	Peru	70	Portugal	76	Romania	70
Sierra Leone	46	Singapore	80	South Africa	48	Sri Lanka	72
Sudan	57	Switzerland	80	Tunisia	74	Turkey	72
Uganda	44	United States	78	Zambia	37	Zimbabwe	37

For the data given, complete the following.

1 a Order the life expectancy data from lowest to highest.

b Find Q_1, Q_2, Q_3 and the mean.

c Draw a box plot for the data, showing any outliers.

2 Summarise and display the data using the following:

a stem-and-leaf plot

b frequency table

c histogram.

3 Consider all your results and write a brief report on the data representing life expectancy for the 40 countries listed. Discuss:

a the spread of the data

b the median

c the outliers

d reasons for outliers.

SELF-ASSESSMENT

Copy the rubric into your workbook and put a tick (✓) in the column that describes how you feel about the work in this module.

Self-assessment rubric

How well do you understand these concepts from Module 4?	I understand this concept very well	I understand this concept	I need more help to understand this concept
Drawing different graphs to represent data			
Interpreting graphs about data			
Measures of central tendency			
Analysing data for bias and errors			
Relative frequency and probability			
Venn diagrams			
Mutually exclusive and complementary events			
Using fractions, ratio and percentages to express probability			

APPENDIX

RUBRICS FOR ASSESSMENT

This appendix provides opportunities for assessment for you to complete and keep in your assessment portfolio. At the end of a school year, your portfolio should contain examples of each type of assessment rubric.

Each opportunity for assessment is identified by an icon in the margin as you work through the chapters and modules in this book. These are the icons:

SELF-ASSESSMENT — This icon indicates opportunities for **assessing your own work**, for example, completing a table or drawing a mind map.

PEER ASSESSMENT — This icon indicates opportunities for **peer assessment**, for example, oral presentations.

GROUP ASSESSMENT — This icon indicates opportunities for **group assessment**, for example, group work or investigations.

ASSESSMENT PORTFOLIO — This icon indicates assessment opportunities that are particularly suitable for your **portfolio**.

SELF-ASSESSMENT

1 Self-assessment rubric

As you complete each chapter of each module, copy and complete this rubric in your workbook. Draw up a list of the most important skills and concepts you learned in the chapter you completed. Then list the concepts under the appropriate headings. Ask someone to help you with the concepts you are not sure about. You can list as many concepts as you wish.

I understand these concepts and skills very well	I understand these skills and concepts	I need more help to understand these skills and concepts

PEER ASSESSMENT

2 Oral presentations rubric

When you give oral presentations to your class, your fellow learners can use this rubric to assess your performance.

Assessment criteria	Level attained			
	1	2	3	Comments
Presentation captures audience interest	Lacked fun or stimulating impact	Interesting at times	Fun! Stimulated interest	
Introduction	No introduction	Muddled, with no clear links to content	Stimulating introduction, with clear links to content	
Body of presentation	Muddled and unclear (incoherent)	Clear at times (sometimes incoherent)	Coherent, clear and to the point	
Summary	No summary	Muddled; did not highlight key points	Highlighted key points effectively	
Support material	None used	Brought but not used in presentation	Brought and linked support materials to content	
Effective use of time	Too fast/too slow	Adequate	Good use of available time	
Responses to questions	No satisfactory answers to questions provided	Answered questions from audience	Gave knowledge-able answers to audience about interpretation of content	
Enthusiasm/effort	Little effort and enthusiasm	Some effort and enthusiasm	Clear evidence of effort and enthusiasm	
Creativity	Little or no creativity	Elements of creativity	Very creative	

RUBRICS FOR ASSESSMENT

GROUP ASSESSMENT

PEER ASSESSMENT

3 Investigations rubric

Assessment criteria	Level attained			
	1	2	3	Comments
Ability to follow instructions	Unable to follow instructions	Instructions followed with occasional supervision	Instructions followed independently, without supervision	
Ability to work with others in a group	Unable to work with others in a group	Sometimes works well within a group	Always works well within a group	
Learner's role in group work	Does not take a leading role in group work	Sometimes took leading role in group work	Almost always took a leading role in group work	
Presentation of ideas	Did not present any ideas for solving the investigation	Presented some ideas for solving the investigation	Clearly presented various ideas for solving the investigation	
Result/final product	No results or answers	Muddled or incomplete results or answers	All results or answers correct and clearly presented	
Ability to work tidily	Cannot work tidily	Work is satisfactorily tidy	Work is very tidy	

PEER ASSESSMENT

4 Presenting data in the form of graphs

Assessment criteria	Level attained			
	1	2	3	Comments
Correct type of graph used (bar/histogram/ line/pie-chart)	Not correct type	Correct type		
Suitable heading	Not present	Incomplete	Complete and correct	
Independent variable on x-axis (horizontal)	Not present	Present, but incorrect	Correct	
Suitable scale on x-axis (horizontal)	Incorrect	Correct		
Labelling x-axis (horizontal)	Incorrect	Correct		
Units for independent variable on y-axis (vertical)	Incorrect	Correct		
Dependent variable on y-axis (vertical)	Not present	Present		
Suitable scale on y-axis (vertical)	Incorrect	Correct		
Labelling y-axis (vertical)	Incorrect	Correct		
Units for independent variable on y-axis (vertical)	Incorrect	Correct		
Plotting points (if any)	All incorrect	1–2 correct	All correct	
Neatness (joining points)	Untidy	Satisfactory	Tidy	
Size of graph	Small, unclear	Acceptable	Very clear	

RUBRICS FOR ASSESSMENT

PEER ASSESSMENT

5 Presenting data in the form of a table

Assessment criteria	Level attained			
	1	2	3	Comments
Suitable heading for tables	Not present	Incomplete	Complete	
Descriptive column headings	Not present	Incomplete	Complete	
Descriptive row headings	Not present	Incomplete	Complete	
Units in heading, not in body of table	Not present	Incomplete	Complete	
Format of table	No horizontal lines or vertical lines in borders	Table format incomplete	Table correctly drawn	
Data entered in table	No data placed correctly	Only some data placed correctly	All data placed correctly	

PEER ASSESSMENT

6 Interpreting and applying data in graph or table form

Assessment criteria	Level attained			
	1	2	3	Comments
Response to questions based on graph or table (reading information off graph or table)	Unable to use graph or table to read off information	Partially correct interpretation of graph or table	Interprets graph or table correctly	
Response to questions related to applying the graph or table	Unable to apply information in the graph or table	Can sometimes apply the information in the graph or table correctly	Alway applies the information in the graph or table correctly	
Doing calculations based on the information in the graph or table	No correct calculations done	Some calculations correct	All calculations correct	

GROUP ASSESSMENT

PEER ASSESSMENT

7 Attitudes and values assessment

Assessment criteria	Level attained			
	1	2	3	Comments
Empathy with other people	Unable to show understanding for other people's feelings	Shows some understanding for other people's feelings	Shows under-standing for other people's feelings	
Respect for other people's viewpoints	Does not consider and listen to other people's viewpoints	Sometimes shows consideration for other people's viewpoints	Considers and listens to other's viewpoints	
Teamwork ability	Does not participate actively; disruptive	Participates effectively sometimes	Participates effectively in teamwork and works towards a good team spirit	
Sensitivity to beliefs of different cultures and environments	Doesn't acknow-ledge other cultures and environments	Sometimes acknowledges other cultures and environments	Is sensitive to other cultures and environments	
Awareness of and recognition of social, political, economic and environmental issues	Little or no awareness of issues	Aware of issues, but does not engage with them	Interested in resolving issues	
Interpersonal skills	Poorly developed communicative skills	Communicates easily at times	Communicates easily and fluently with others	

RUBRICS FOR ASSESSMENT

ASSESSMENT PORTFOLIO

8 Project assessment

Assessment criteria	Level attained			
	1	2	3	Comments
Relevance of material used	No appropriate material used	Appropriate materials used, but not interpreted	Appropriate materials used and interpreted	
Project design	No clear organisation or vision; simplistic	Some parts are clearly arranged; other parts are confusing	Clear vision and organisation; structure is focused	
Media resources accessed	No resources accessed	Limited or insufficient resources accessed	Variety of resources accessed	
Material resources used	No resources used	Insufficient or limited resources used	Variety of resources used	
Standard/quality	Poor quality	Satisfactory, but room for improvement	Of a high standard	
Effective use of time	Did not consider a timetable or time limits	Set up a time-table, but did not adhere to it	Compiled a time-table and handed project in on time	

SELF-
ASSESSMENT

9 Self-assessment checklist

Assessment criteria	Level attained		
	1	2	3
• I find it easy to understand ideas in mathematics	Never	Sometimes	Always
• I find it easy to understand other people's points of view	Never	Sometimes	Always
• I respond well to feedback about my work	Never	Sometimes	Always
• I am interested in learning about mathematics	Never	Sometimes	Always
• I feel safe and relaxed in the mathematics classroom	Never	Sometimes	Always
• I ask questions while I am learning	Never	Sometimes	Always
• I find it easy to participate during group work	Never	Sometimes	Always
• I assume a leadership role during group work	Never	Sometimes	Always
• I enjoy working on my own	Never	Sometimes	Always
• I enjoy taking part in the investigations	Never	Sometimes	Always
• I find it easy to follow instructions when doing an investigation	Never	Sometimes	Always
• I like planning and carrying out investigations on my own	Never	Sometimes	Always
• I find it easy to communicate with people	Never	Sometimes	Always
• I would like to follow a career that involves mathematics	Never	Sometimes	Always
• I appreciate the role of mathematics in my daily life	Never	Sometimes	Always